STUDYING WISCONSIN

STUDYING WISCONSIN

The Life of Increase Lapham

early chronicler of plants, rocks, rivers, mounds
and all things Wisconsin

—|⊢—

Martha Bergland & Paul G. Hayes

WISCONSIN HISTORICAL SOCIETY PRESS

Published by the Wisconsin Historical Society Press
Publishers since 1855
© 2014 by the State Historical Society of Wisconsin

Publication of this book was made possible in part by a gift from the
D. C. Everest fellowship fund.

wisconsinhistory.org

Front cover: WHi Image ID 1944, Increase Lapham inspecting a piece of the
Trenton meteorite.

Printed in Canada
Designed by Nancy Warnecke, Moonlit Ink

18 17 16 15 14   1 2 3 4 5

Library of Congress Cataloging-in-Publication Data
Bergland, Martha, 1945-
Studying wisconsin : the life of Increase Lapham, early chronicler of plants, rocks, rivers,
mounds and all things Wisconsin / Martha Bergland and Paul G. Hayes.
pages cm
ISBN 978-0-87020-648-1 (hardback) — ISBN 978-0-87020-649-8 (ebook) 1. Lapham,
Increase Allen, 1811-1875. 2. Natural history—Wisconsin. 3. Botanists—Wisconsin—
Biography. 4. Naturalists—Wisconsin—Biography. I. Hayes, Paul G. II. Title.
QK31.L3B47 2014
580.92—dc23
[B]
2013036967

To the memory of my husband, Larry Barnett, who called him Decrease
—M. B.

*To my wife Philia, who patiently shared our home with Increase Lapham
for more years than this book was in writing*
—P. H.

Contents

Preface

INCREASE ALLEN LAPHAM, ONE OF THIRTEEN CHILDREN of a New York Quaker family and a self-taught all-around scientist whose curiosity led him into many fields, made his home in Wisconsin from age twenty-five in 1836 until his death in 1875. In that time, he worked as an engineer, surveyor, land agent, cartographer, botanist, geologist, meteorologist, archaeologist, limnologist, and zoologist. A published scientist—his first paper appeared in the *American Journal of Science and Arts* when he was sixteen—he wrote Wisconsin's first scientific paper, a list of plants and shells found in Milwaukee, the year he arrived. In 1844, before Wisconsin became a state, he finished the first book published in Wisconsin, a geography that he hoped would attract immigrants to the territory. He promoted education and culture, being a founder of the Milwaukee Lyceum; the Milwaukee Young Men's Association; the Milwaukee Female Seminary; the Wisconsin Historical Society; and the Wisconsin Academy of Sciences, Arts and Letters; and he was in on the beginnings of the Milwaukee Public Library and the Milwaukee Public Museum.

In his prime, he surveyed, drew, and described the Indian effigy mounds that distinguished Wisconsin's landscape in a book for the Smithsonian Institution in 1855—*The Antiquities of Wisconsin*, his most famous work. In 1869, he warned of the "disastrous effects of the destruction of forest trees now going on so rapidly in the State of Wisconsin" in a prescient report to the state regarding the great cutover of the northern pinery, an example of conservation writing that predated by decades the beginnings of the National Forest Service. His concern over the loss of sailors, vessels, and cargo in Great Lakes storms was a major factor leading to the creation of the US Weather Bureau and as its early employee he wrote the nation's first official weather forecast. Late in life, he was appointed Wisconsin state geologist and he presided over the organization of the Wisconsin Geological Survey.

Today the University of Wisconsin–Milwaukee teaches natural science in Lapham Hall. Lapham Peak dominates the Kettle Moraine formation in Waukesha County. Madison youngsters attend Lapham Elementary School. Lapham Streets carry traffic in Oconomowoc and Milwaukee. The Wisconsin Archeological Society awards its Increase A. Lapham Research Medal to scientists who make important contributions to that science. A fossil trilobite, a plant genus, and distinctive markings in metallic meteorites all have been named for Lapham. Formal portraits of Lapham are prominent at the Wisconsin Historical Society and in a mural in the Wisconsin State Capitol building. A bronze bust of Lapham can be found at the Milwaukee Public Museum.

Increase Lapham may have been Wisconsin's most-respected, most-honored citizen of his time. Yet for almost 140 years after his death, no full biography of Wisconsin's pioneer scientist, educator, and scholar has been published. Graham P. Hawks finished an unpublished biography of Lapham in 1960 as his doctoral thesis in history at the University of Wisconsin–Madison. Partial biographies have appeared as articles, perhaps the best of them by Newton H. Winchell, professor of geology at the University of Minnesota and editor of *The American Geologist*, in which his tribute to Lapham appeared in January 1894. Milo M. Quaife, superintendent of the Wisconsin Historical Society and founder of the *Wisconsin Magazine of History*, ran his "Increase Allen Lapham, Wisconsin's First Scholar" as the first article in the first issue of the magazine in September 1917. Other articles through the years examined specific Lapham contributions: Lapham the pioneer patron of Wisconsin education; Lapham the mapmaker; Lapham the meteorologist and a founder of the US Weather Bureau; Lapham the geologist, the botanist, the agriculturalist, the author, the archaeologist, the surveyor and canal engineer, the promoter of scholarly institutions.

Among the reasons no biography was written may have been the prodigious amount of original Lapham material that was left behind. Shy and modest in person, Lapham was a prolific letter writer, note-taker, datagatherer, compiler of lists of species, and writer of essays, geographies, scientific articles, and histories. He preserved his original, handwritten manuscripts and copies of letters, even scraps of notes. He wrote frequently about Wisconsin for local newspapers and he clipped articles of scientific or geographic interest, pasting them into scrapbooks. Modest as he was he seemed intent on leaving behind all that he had learned about Wisconsin and science, not so much for personal glory but because he believed the information was valuable. After his death, his daughter Julia A. Lapham assumed the curatorial role. She organized his papers and selectively transcribed much of his correspondence, diaries, and journals into hundreds of typed pages, all of which went to the Wisconsin Historical Society before her death in 1921. The Lapham collection at the Society fills twenty-seven boxes, including Julia Lapham's transcriptions.

Whether from youthful courage or naïveté, young Lapham boldly followed his curiosity into botany, meteorology, archaeology, and geology. He did not dabble in these sciences; he delved into them, mastering the basics to a degree that earned the respect of colleagues who had the advantages of having attended universities. He did not measurably advance the sciences but learned from and applied the insights of others. But Lapham had an edge: unlike most of his contemporaries, he had spread before him the vast, diverse, and largely unexplored environment of future Wisconsin. When he stepped ashore at the frontier settlement of Milwaukee at the age of twenty-five, his passion to understand the natural world was at its peak and it would remain undiminished for nearly forty years. He began at once to collect specimens and record daily observations of the unstudied, pristine place. He had a virtual scientific monopoly over the area that was to become the state of Wisconsin, and he was aware that the information he was collecting would be in demand by other scientists of his time, many of whom became congenial, appreciative correspondents. He knew at once that he had found his permanent home and his lifelong occupation.

Remarkably, Lapham on any given day was able to work on several unrelated projects simultaneously, collecting botanical specimens while surveying for a canal, noting geological formations while exploring effigy mounds, monitoring weather while measuring Lake Michigan tides. Moreover, he was also making a living as a civil engineer and surveyor, author, and land speculator, some of them requiring travel, while maintaining correspondence with dozens of scientists as far away as Europe.

Throughout his life, he was deeply involved with his family in an always-changing, busy household that included siblings, nephews, nieces, his wife's mother and later his own mother, as well as his and Ann's five surviving children. This level of activity posed a problem for the authors. Should we tell the Lapham story chronologically as is traditional in most biographies, or would it be easier to sacrifice chronology and write topically, covering Lapham the botanist, Lapham the geologist, Lapham the archaeologist, and so on? While the first half of the book is chronological, each of our chapters 11 through 21 about Lapham in Wisconsin necessarily is a blend of both approaches. However, we have emphasized chronology for ease of storytelling and for clarity.

Both Increase Lapham and his daughter Julia filtered the information that was left behind, especially Julia, who selectively shortened his letters, eliminating portions that she judged to be mundane or irrelevant, as we surmised from the instances when we had both Lapham's original and Julia's truncated versions. We are not certain in all cases whether she omitted material that would be informative but might reflect embarrassingly on her father or reveal him in errors (he was, as we note, a questionable speller in his early correspondence). On the chance that both father and daughter deleted criticism or embarrassments, we were on the lookout for all points of view among his correspondents, his fellow citizens, and other contemporary writers. If Lapham had critics or enemies, we did not find them. We found the opposite: lavish praise, admiration, and affection for the man and his motives, not only from good friends such as Philo R. Hoy, but also from his assistant geologists at the Wisconsin Geological Survey and from correspondents, notably Asa Gray, Benjamin Silliman, and many others. Increase Lapham was regarded as honorable, magnanimous, industrious, and modest. But our heavy dependence upon Lapham's papers and Julia Lapham's transcriptions nonetheless leads us to wonder if we have not written, nearly 140 years after his death, Increase Lapham's "authorized biography."

Martha Bergland began the Lapham biography early in 2009. In writing about Lapham's youth, she filled out the story of the boy worker on New York's Erie Canal along with his father, Seneca, and older brother, Darius; the lonely teenager in Shippingport, Kentucky; the worker and engineer of canals in Kentucky and Ohio; the secretary of the Ohio Canal Commission; the promoter of an Ohio geological survey; and finally, the reasons for his moving to Milwaukee. Paul Hayes joined Martha in writing about Lapham in Wisconsin, covering Lapham's work as Byron Kilbourn's business agent and chief engineer of the Milwaukee and Rock River Canal; as the author of the first scientific paper and first book published in Milwaukee; as mapmaker, weatherman, cataloguer of plant and shell species; and as state geologist. Sometimes we divided the work chronologically, Bergland writing about Lapham in the 1850s and Hayes taking the 1860s. Other times we chose to write about specific Lapham specialties, following our personal interests: Bergland on Lapham the botanist, Hayes on Lapham the geologist. Our collaboration couldn't have run more smoothly.

Lapham's brand of science, the all-encompassing, self-educated generalist of natural history, was in decline even during his lifetime in favor of the college-trained specialist in specific sciences—chemistry, geology, meteorology, and so on. Indeed, if Lapham had succeeded in his youthful ambition to attend Yale College, a quest he could not afford but for which he futilely sought funds, he too might have become a specialist and therefore less fit for the outdoor laboratory of Wisconsin over which he became the master. But first his isolation on the frontier and then the growing respect he earned from other scientists preserved his citizen-scientist status throughout his life. He was fully aware of the emergence of the specialists and their growing importance. His old friend Samuel Sherman related after Lapham's death that when Lapham was asked late in life which field of science he was specializing in, Lapham replied, "I am studying Wisconsin."

Increase Lapham was in every way the man for his time and his place. Not only did he lay the groundwork for sound science based in Wisconsin, but his contributions in weather observation and forecasting, botany, forest conservation, and archaeology spread far beyond the state where he made his home. Not only Wisconsin but the entire nation has benefited immeasurably from his passionate pursuit of scientific knowledge.

CHAPTER I

A Habit of Observation

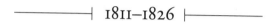

———| 1811–1826 |———

IN 1824, LOCKPORT, NEW YORK, was a great hellish construction site. Here, two of the marvels of the Erie Canal were being built. One, in a bowl-like ravine, was the staircase of the Flight of Five locks, which would raise canal boats sixty feet over the limestone heights of the Niagara Escarpment. The other was the Deep Cut, a seven-mile-long, twenty-five-foot-deep stretch of canal west from Lockport.

The ancient forest lay in ruins along the route of the canal—pulled-out roots of great trees, giant stumps, tangled brush piles infested with rattlesnakes. Rough log shanties were built near mountains of excavated rock and earth. Raw sewage drained from the canal workers' camps. The blasting and "the clinking of a thousand hand drills" were almost constant, filling the air with the smoke of burning powder and with rock dust. Some of the Irish laborers were so hardened to the danger of flying rocks that, instead of running for shelter at the signal for a blast, they held their shovels over their heads to keep off the shower of small stones and were "crushed every now and then by a big one."[1]

In the midst of this din of the "peaceful pursuit of enterprise," a slight, dark-haired boy worked shaping blocks of Niagara limestone to line the canal, bringing forth from the dolomite a slight smell of tar.[2] When his hammer exposed pockets of pink or white crystalline minerals, the boy wondered—in the murk and chaos around him—what these beautiful

The Flight of Five locks over the ridge of the Niagara Escarpment, and the Erie Canal basin where boats arrived from the east. Lockport was in and above the basin. The Deep Cut was above and west of the locks. Lithograph by J. H. Bufford, ca. 1836.

shining things were and how they were formed. It was not from books but here among the rocks of the "Mountain Ridge" that the boy acquired his first lessons in mineralogy and geology. This was Increase Allen Lapham. He was thirteen years old.

Though other boys must have been at work in that Lockport quarry, Increase would have stood out—not for his size, but for his lack of it. Not yet his full adult height of five feet four inches, at age thirteen he must have been a little over five feet, his head looking large for his slight body. (The average height of males at that time was about five feet seven inches.) From later photographs we can see that his fair skin, light blue-gray eyes, and sloping shoulders might have hinted at a delicacy that he didn't have. The boy had stamina; he was capable of working long hours at physical and mental labor. His full and sensual downturned mouth was contradicted by strong, dark, no-nonsense eyebrows and a well-formed nose. His was a handsome face that appeared to move easily between amusement and a frown of concentration. Only his thick black hair seemed unruly.

When it was safe, the boy wandered off to the piles of rock lifted out of the deep canal cut to look for fossils—"petrifactions"—in the softer

Rochester shale that split easily under his hammer, exposing prehistoric sea creatures: cylindrical beads or stems of crinoids, segmented trilobites, and bivalve brachiopods. He began to gather the fossils as well as the pink and white crystals. He carried them home, weighing down the pockets of his jacket, beyond the blasting sounds at the construction site and past the part of town where townspeople had propped stripped saplings against their houses to protect them from the occasional hail of blasted rock.[3]

In the rented house where the large Lapham family lived, Increase showed the fossils and minerals to his father, canal contractor Seneca Lapham, who was building gates and other woodwork for the locks. He showed them to his older brother Darius, who at sixteen was a budding naturalist and an assistant engineer on the canal. And he showed them to his mother, Rachel, who, though busy with her six children, was Increase's earliest teacher, his teacher when there wasn't enough money for the children to go to school. He must have asked a lot of questions. To the Quaker Laphams curiosity was a virtue and education a necessity.

And if his family could not answer his questions, there were canal engineers nearby, friends of his father, who were well educated in the mechanical and natural world. The chief engineer of the canals west of Rochester was David Thomas, a self-taught botanist and contributor to the agricultural journal *The Genesee Farmer*. And there was engineer and architect Alfred Barrett who soon hired Increase away from the quarry to be rodman for his brother Darius. He would earn ten dollars a month and fifty cents a day living allowance.[4] Increase held the leveling staff while Darius made measurements of distance and elevation on the canal. Although this was not a difficult job, Increase had to be disciplined and accurate and he had to use the knowledge of surveying that had been passed on to him from his father and older brother.

When he wasn't working, Increase sometimes wandered off on Indian trails beyond the sounds and smells of the construction into the quiet unspoiled forest of great elms, beech, basswood, and sugar maples. Here he noticed among other beauties the little plants at his feet—the violets, blue phlox, beechdrops, false Solomon's seal, and Jack-in-the-pulpit. "As I found no others of similar [botanical] tastes," he wrote, "these rambles were usually without companions."[5]

In 1824 in Lockport, Increase Lapham was connected to most of the major influences in his life—the great building project of the Erie Canal whose practical-minded canal engineers shared Increase's interests in the natural world; the destruction of the wilderness at the boomtown of Lockport yet also the wonders of its rocks, fossils, and plants; and perhaps most important, his large, close, and supportive family. At Lockport Increase Lapham began his intellectual life as a surveyor, engineer, geologist, cartographer, botanist—an observer, collector, and classifier of nature.

—| |—

Before the 1800s no civil engineering project as ambitious as the Erie Canal had been attempted in this young country. Proposed in 1808, construction began in 1817 on the 363-mile Erie Canal from Albany to Buffalo.

Between 1800 and 1820 a great migration of Americans—more than a million and a half people—moved from the settled areas of the East to the West, beyond the Appalachian Mountains. As historian Daniel Walker Howe writes,

> Never again did so large a portion of the nation live in new settlements. Later generations of Americans would revere the western migrants as "pioneers," a word that originally meant the advance guard of an army, who carried tools to repair roads and bridges or throw up fortifications as needed. It was an apt metaphor insofar as it suggested both occupation of another's territory and construction for later generations.[6]

These million or so people, who for the most part had walked west or had ridden in carts or wagons, needed to be able to cheaply ship their wheat and corn and hogs to markets back east. Before the Erie Canal opened, most produce had to be shipped on flatboats down the Ohio and Mississippi Rivers and their tributaries to be sold for cash in New Orleans. There the flatboats were broken up and the wood sold. The shippers then had to *walk* home along the bandit-infested Natchez Trace. Not only did these farmers need better markets, but the growing eastern cities needed the produce.

The Erie Canal, financed entirely by the state of New York, would connect the ocean port of New York City to the Lake Erie port of Buffalo via the Hudson River, opening travel to Michigan Territory, which at that

WHI IMAGE ID 101357

The Erie Canal excavation of the Deep Cut through the Niagara limestone above Lockport, New York

time included Wisconsin. By the time of the canal's completion in 1825, New York was the largest city in the country. On one day in 1824, 324 ships were counted in New York Harbor; after the canal had been opened for a little more than ten years, more than 1,200 vessels were counted in the same harbor. By 1850, the population of the major cities along the Erie Canal—Albany, Utica, Rochester, Syracuse, and Buffalo—increased up to twentyfold.[7] Within ten years of the completion of the Erie, New York State had paid off the $7,143,789 cost of the canal with the tolls it had collected.[8] As historian Daniel Walker Howe and others assert, the Erie Canal made New York the Empire State.[9]

Building the Erie Canal meant not only digging the channel but also building eighty-three locks, walled with stone and floored with great timbers and layers of planking. These locks raised the canal boats along the forty-foot-wide, four-foot-deep canal. And eighteen arched stone aqueducts carried the canal over rivers.

Besides moving a river of water over another river of water, the canal builders had to solve many other problems: how to take accurate levels of land and water, how to dig and remove massive quantities of earth and rock, how to quickly take down ancient forest trees and cut through their roots, how to safely use blasting powder, how to keep the canal banks from dissolving in water, how to make valves to slowly release water, how to design lock gates, how to make locks and gates that could be operated by one man. The formulation of a cement that would be permanent under water was one of the greatest inventions to come from the building of the canals. Other inventions were the plow and scraper, wheelbarrow, stump puller, and tree feller.[10] The need to solve complex problems made working on the Erie Canal the most important school for engineers and scientists of the time.

Though there were a few schools in Lockport, Increase did not go often if at all while there; he was attending the school of the canal. His father, Seneca, taught him surveying, drafting, and engineering the way he himself and most of the engineers and contractors had learned: on the job. Seneca had been a carpenter by trade, but on his first canal job in Pennsylvania, he learned how to build canal gates. This became his primary work as a contractor on various canals, though he also worked with stone.

Few of the engineers building the Erie Canal were formally trained in canal engineering. Many of the men who became engineers were trained in surveying, the law, medicine, or architecture. Though a few engineers traveled to Holland or England to learn from the European experience of building canals, and though some engineers and builders had worked on shorter canals in the East, most of the engineers on the Erie learned on the job, becoming the first professional engineers in America.[11]

One example of these self-taught engineers was David Stanhope Bates—a designer of the famous five locks in Lockport and an eleven-span masonry aqueduct over the Genesee River in Rochester. Bates, born in New Jersey in 1777, the son of an officer of the Revolutionary War, trained for the ministry but studied surveying and mathematics as he clerked in his brother's store. Employed as an agent and surveyor by a large landowner in Oneida County, New York, he studied law and was admitted to the bar. Bates was elected judge of the Court of Common Pleas in Oneida County; from then on he was called Judge Bates even as an assistant engineer of the Middle Division of the Erie Canal in 1817. Thereafter he learned his engineering working on canals in New York, Kentucky, and Ohio. David Stanhope Bates's son, also named David Bates, became a canal engineer like his father. Seneca, Darius, and Increase Lapham all worked for both of the Bates at one time or another. The two families became trusted friends and colleagues.[12]

The engineers at Lockport noticed that young Increase had become a skilled draftsman. Increase had no formal training as a draftsman, but he must have seen drawings of the canal and locks every day at work and at home. His love of order and observation of detail, combined with an artistic eye and a good hand, made his drawings so good that there was a demand for his plans of the locks. It must have been satisfying for a thirteen-year-old boy to be able to sell his drawings to surveyors like Major David B. Douglass, later the designer of Greenwood Cemetery in Brooklyn.[13] "Quite a lucrative business was carried on at this early day by me," Increase wrote, "in preparing plans of the 'Combined & Double Locks' which were sold to passing strangers, & they are now scattered far & wide over the country."[14] He was proud of making money with his intelligence and talent. And he was proud that his work was scattered far and wide. This was his first publishing.

Four years later, when he was at work on the Ohio canals, Darius remembered in a letter to Increase the drawing of these plans—especially how they got some help drawing the human figure.

> I stopped at the house of Col. John Foble in Lancaster. When he heard the name "Lapham" he enquired if I was ever at Lockport? When I answered, "yes," he said he had a plan of the Lockport locks drawn by me. He showed me the plan, it was yours with your name on it in type, and a plan of Dibble's crane, which I think I drew. Don't you remember, we got a cut [etching] of a runaway [slave] from Turner, with a cane and a pack on his back, we disencumbered him of his pack, and set him to tripping the box and driving horse on the crane in excavating the rock for the canal through the mountain ridge?[15]

The Laphams drew from many sources.

→ | ←

Increase's parents, Seneca Lapham and Rachel Allen Lapham, could trace both of their families to members of the Society of Friends who arrived in this country from England in the 1600s. Quaker George Allen came to the Massachusetts Bay Colony from England in 1635. The weaver John Lapham, one of a family of weavers in Devonshire, came in 1673 to Bristol County, Massachusetts. Weaving at that time was a highly mathematical and skilled trade. Perhaps the Lapham interest and ability in science and engineering goes back to these very early Laphams. In any case the colonial Society of Friends community was a group whose interest in natural science was so common that "almost every Quaker had a scientific hobby."[16]

For the next 140 years or so, the Allens and Laphams lived in Quaker settlements primarily in Bristol County, Massachusetts, and Dutchess County, New York. About 1762, Increase's maternal grandparents, Increase and Mary Allen, moved with a group of Quakers north to the wilderness of the Queensbury Patent, now Queensbury, near Glens Falls, New York, on the Hudson River. But during the Revolutionary War this group suffered from British raids and requisitions and twice their homes were burned, so they moved back to Dutchess County. In 1783 they moved north again to Queensbury and began rebuilding houses, clearing new land, and building grist mills and saw mills. Although Quakers were pacifists, a few of the Laphams and Allens had served in the Revolutionary army, including Pazzi Lapham, Seneca Lapham's father. Seneca Lapham was born to Pazzi and Bethany Lapham in Queensbury in June of 1784. Rachel Allen, daughter of farmer Increase Allen and his second wife, Mary Spencer Allen, was born in Queensbury around that same time.[17]

Seneca and Rachel were married in Queensbury, New York, on August 28, 1803, where they lived until about 1806. In frontier Queensbury, Rachel and Seneca's first two sons, David and Ruel, were born but soon both little boys died. In those days, more than one in six children died in infancy— ten times today's risk; one in four or five children would not live to be twenty-one. The deaths of children were "a commonplace and expectable occurrence" in the "sickly" seasons. In the late winter and early spring, all children were at risk of respiratory diseases; in August and September children were at higher risk of diseases like malaria and intestinal illness.[18]

And on the frontiers the mortality rate was higher for children than in the more settled northeast. As families moved west to wilder places, "a trip westward was almost a demographic journey back in time." Family size was higher, marriage was earlier, the deaths of children more likely. It is cruel that these young parents, moving to places where they could make a better life for their families, taking the work that moved them away from the settled places, were also putting their children and themselves at greater risk of disease and death.[19]

Perhaps to leave the place of such sorrow and to start anew, young Seneca and Rachel Lapham, childless now, traveled for weeks on Indian trails two hundred miles west to frontier land in the Central New York Military Tract in Seneca County, New York—land probably given to Seneca's father for his service in the Revolutionary War. Junius is where Seneca and Rachel's third child, Hiram, another son, was born in December 1806. On the frontier, Seneca's carpentry skills were in great demand. Within two years Rachel, Seneca, and young Hiram had moved about forty miles farther west to Farmington in what is now Ontario County, where Darius Lapham was born December 14, 1808. Family lore says that Seneca then moved his family to nearby Macedon to help a cousin finish his house on "the north side of a hill and about one mile south of the village place known as Freer Farm, near Freer's Bridge."[20]

The next move was to nearby Palmyra where Increase Allen Lapham, the fifth son, was born to Rachel and Seneca March 7, 1811, in Wayne County, New York. Increase was named for his maternal grandfather, Increase Allen. As to the name "Increase"—it could have been worse; some of his male ancestors were named Alce, Ludy, Orpha, Eliaken, Eber, Tohigh, and Pazzi. Brother Pazzi was born in Palmyra in 1812.

After Palmyra and before Lockport—from 1813 to 1822—the Lapham family moved eight times. They lived in Macedon and Galen, New York, at two different times where two more children—Mary and William— were born in 1815 and 1817. Though he doesn't tell us what his memories were, Increase wrote that his first memories were of Galen.[21]

In 1818, when Increase was seven, the family made a longer move, this time back east to Schuylkill County, Pennsylvania, in the mountains near the headwaters of the Schuylkill River, where canals were being built to haul coal down to the cities to the south. For two years Seneca Lapham worked as a contractor for the Schuylkill Navigation Company in Schuylkill County, moving his family among three little towns—Mount Carbon, Pottsville, and Orwigsburg, where for a short time the children attended a common school. In 1819 in Orwigsburg, when Increase was eight, his thirteen-year-old brother, Hiram, died—a loss Increase never wrote about but which he and his close family must have felt deeply.

In 1820 the family moved back to the familiar Galen, New York, where Seneca had a contract to build lock gates on the Erie Canal. For two more years the family lived in Galen, where Margaret was born in March of 1820 and Tasa Ann in September of 1821; both of these little girls also died young.

In only a few years, the Lapham family had lost five children. Life for all young children was precarious, but life on the canals was especially so, with the close quarters of the workers and their filthy living and working conditions making the spread of infectious diseases a constant threat. Though the Laphams would not have lived in the crowded camps on the mosquito-ridden canal, and though Quaker Rachel Lapham most likely kept a clean house, against the fevers and ague she could not keep her children safe. Increase in later life was silent about these hard years of his early boyhood, but these early losses might help explain the exceptional closeness of the Lapham family.

The next move to Rochester, New York, seemed to break the sad spell on the family. In 1822 Seneca was hired by David S. Bates to work on the eight-hundred-foot-long aqueduct over the Genesee River. "One of the arches in the first aqueduct at Rochester was built by my father on a subcontract," Increase proudly remembered.[22] His pride was justified. "The aqueducts were often like the Roman aqueducts of old, combining mass, length, height, strength, and beauty. Piers set in a rushing stream, stone cut for close fitting and bonded with cement, magnificent Romanesque arches supporting a towpath on one side and the wooden trough of the canal on the other, all made for an effect of majesty."[23] The eleven-arch aqueduct in Rochester was particularly impressive.

Though the rest of the family lived in Rochester, young Increase was sent back to Galen to work for a short time for a farmer, "one Elias Langdon," as he called him, communicating his distaste for this work.[24] From Galen, Increase returned the thirty or so miles to Rochester with Darius—the two of them on foot, driving a cow and calf. Increase was eleven and Darius was fourteen—independent and responsible boys. Of all the members of his family, Increase was closest to Darius. Later, when they were separated, working in different places, Darius and Increase wrote long, detailed letters to each other about engineering, minerals, fossils, botany, and everything else of interest to the curious brothers.

For all of 1823 the family lived together in Rochester, where the eleventh little Lapham—Hannah—was born in February. There was a Friends meetinghouse in Rochester at that time, so it's likely the family went to Meeting, confirming in the children the values of the Society of Friends. According to historian J. William Frost, "The Quakers' ideal of living required restraint and composure. The child learned patience, self-control, and moderation from early life."[25] Increase and his brothers and sisters must have become accustomed to sitting still in two-hour meetings in which the silence was broken only by a prayer or sermon.

—| |—

It was after Seneca completed his work on the Rochester aqueduct in 1824 that the family moved to noisy, dirty, fascinating Lockport. They arrived from Rochester with Darius, sixteen; Pazzi, twelve; Mary, nine; William, seven; and the infant Hannah. Young Increase, eleven, had been left behind in Rochester—perhaps to go to school, but also to tend "a petty grocery for one Timothy Miller."[26] This hated work did not last long. Soon

Seneca Lapham helped build the first Rochester aqueduct, drawn by James Eights in 1824.

Increase joined the family in Lockport where he was put to work cutting stone. It must have been clear to Increase's parents that this young, though responsible, boy could earn money for the family.

The evolution of Lockport was typical of the many canal towns where the Laphams had lived and would continue to live. As with later railroad towns, the speculators arrived first. Unlike most canal towns, however, Lockport's speculators were Quakers. When it was clear in 1816 that the Erie Canal would pass through the Lockport area of upstate New York thirteen miles south of the Lake Ontario shore, speculators and farmers began buying land, clearing the forest, building log cabins, planting orchards. Two years later there was a school with a Quaker teacher, a Quaker meetinghouse, and a cemetery. The next year, a road and a tavern. By 1820 Quaker David Thomas, the New York state surveyor and the principal engineer of canal construction west of the Genesee River, had decided that a major set of locks would be built through the sixty-foot drop of the Niagara Escarpment—a dolomite and limestone ledge that arced up between Lakes Erie and Ontario, above Lake Huron and down on the west side of Lake Michigan as far south as Dodge County, Wisconsin. This weathered deposit of a more than four-hundred-million-year-old seabed is what brought Lockport (as well as the cliffs of the northwest side of Door County, Wisconsin) into being.[27]

In 1820, when there was no frame house within five miles of what would become Lockport, a Friend received the contract for rock excavation. In 1821 posters went up in New York City advertising for canal workers. A blacksmith arrived and set up shop. A doctor and his wife built a house of peeled logs painted white. The doctor and a few other prominent members of the Society of Friends met in a tavern and decided to call the place Lockport. In one year the population rose to two thousand people—many of them immigrant Irish canal workers. By the end of 1822, there were two general stores, a baker, a sheriff, a butcher shop, a coffee shop, and three lawyers. Two big hotels, another tavern, and two more churches had also been built by then. In 1824, a traveler wrote, "There are no good roads leading from Lockport; but there will be by-and-by."[28]

In eight years Lockport had gone from being a quiet forested ravine to a boomtown of Quaker entrepreneurs and their families, as well as engineers, contractors, and many hundreds of Irish laborers. West from Lockport was the Deep Cut, two miles of it through solid rock, where piles of dolomite and limestone were being lifted by rows of tall pine-pole cranes each powered by a single horse walking around a capstan—this crane recently improved in Lockport by a man named Orange Dibble.[29] In the center of Lockport, Seneca Lapham was at work on the wood gates of the Flight of Five. From the foot of the unfinished double locks, a completed sixty-four-mile stretch of canal ran east to Rochester "straight and classical as a Roman road," connecting Lockport to Albany and then New York City.[30]

In 1824, Lockport was a lively town. Though the canal was not finished to the west, boats arrived daily from the east carrying goods that were portaged around the unfinished locks and Deep Cut. Settlers came through moving their families and goods farther west. The *Encyclopedia of Albany*, a canal boat set up as a museum and library, was docked in the basin below the locks.[31] Travelers and tourists arrived by packet boat to experience the great canal. In Lapham's words, the locks "were deemed one of the wonders of art and a great triumph of engineering skill. Many strangers visited the place to inspect the work during its progress."[32]

A Lockport man, Mr. Orange Dibble, improved the horse-powered derricks, or cranes, used to lift stone from the Deep Cut. This is Increase Lapham's drawing of the derricks.

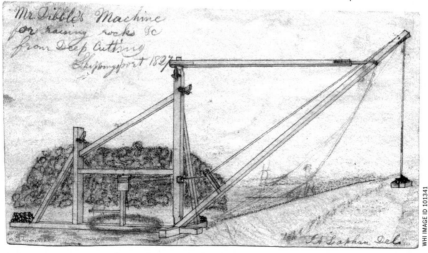

The ingenuity of the work carried on in Lockport while Increase was there, the contrast between the beauty of the surrounding forest and the way it was laid waste by the necessities of development, the stimulation of the company of the engineers, the Irish workmen, the Quaker families— the danger and the comfort and all the interesting manmade and natural stuff—must have made Lockport a wonderful place for a curious and smart thirteen-year-old boy.

On June 24, Increase Lapham was among the four or five thousand people gathered to celebrate the laying of the capstone of the Flight of Five. Young Increase was moved to copy down this inscription carved on the capstone:

> Erie Canal
> Let Posterity be excited to perpetuate our
> Free Institutions.
> And to make still greater efforts than their ancestors to promote
> Public Prosperity:
> By the recollection that these works of Internal Improvement
> Were achieved by the
> Spirit and Perseverance of
> Republican Freemen.[33]

Increase's adult life would be dedicated to the perpetuation of new free institutions and public prosperity, inspired in part by the effort and achievement of the Erie Canal.

The work on the Deep Cut, the last portion of the Erie Canal to be finished, was complete by October. The Laphams celebrated all of this completion with the birth of daughter Lorana (also called Scrana), Rachel's twelfth child, on October 10.

And then they celebrated with the rest of the country when Governor Dewitt Clinton, the chief promoter of the canal, left Buffalo with a flotilla of boats headed for New York City carrying a keg of Lake Erie water that would be poured into the Atlantic Ocean. Taking ten days and stopping to celebrate at towns on the way, one of the flotilla's first stops was October 26 at Lockport. The town fired a cannon that had been used by Commodore Perry's fleet in the War of 1812. And the workers set off thousands of celebratory rock blasts. Increase didn't mention the crowds, the long hyperbolic speeches, the bands, the accidental drowning of a workman in a lock, the first use of the Flight of Five locks. He was fourteen. He mentioned the "telegraphic dispatch sent from Buffalo to New York City; the line consisting of a series of cannon placed at convenient intervals" that announced that the first boat had left Lake Erie on its way to the Atlantic.[34] What impressed him was the planning and cooperation among distant people to arrange such a salute. And probably the noise. And the opening of the amazing Erie Canal.

A great deal had occurred in the United States during the first fourteen years of Increase Lapham's life—1811 to 1825. The population grew from a little more than seven million to more than ten million. The United States fought and defeated the British in the War of 1812. The Missouri Compromise brought in Maine as a free state and Missouri as a slave state. Native Americans were forced by wars and treaties out of most lands east of the Mississippi River. The Great Migration to the West was well under way. Women began to wear corsets. Daniel Boone died. The first steamboat traveled down the Mississippi River to New Orleans. But the most important event during that time for Increase Lapham and his family was the completion of the Erie Canal.

The Erie Canal had placed Increase, along with his father and brother Darius, into a network of canal men who would be employers, colleagues, and friends for years to come. The Erie Canal connected Increase with the discipline and skills required for surveying and drafting. At the Erie Canal in Lockport, Increase Lapham saw the progress that the ideas and the will and the money of men could bring into being against great odds. And this thoughtful boy couldn't help but see, at this time when development was always "improvement," the cost paid for progress in the ravages of the natural world. As a result of his work and associations on the Erie Canal, Increase Lapham was beginning to be recognized as an exceptional, intelligent, talented young man, responsible beyond his years. He was also dutifully contributing to the support of his family who supported him in his interests and abilities, emotionally and intellectually.

But perhaps most important of all, as Increase Lapham himself wrote later in life, "The beautiful specimens I found in the deep rock cut at

[Lockport] gave me my first ideas of mineralogy and initiated a habit of observation which has continued through all my life."[35] This habit of observation laid a strong foundation for Increase's scientific life in engineering, geology, meteorology, and botany, and in an astounding number of other studies.

CHAPTER 2

The Difficulty of Youth

⎯⎯⎯⎯⎯| 1826–1827 |⎯⎯⎯⎯⎯

THE WORK ON THE ERIE CANAL had held the Laphams together in Lockport; its completion flung the older males to places of work and school far from their old home. Pazzi, fourteen, was sent southwest to Cincinnati to study to become a printer. Increase, fifteen, was sent east, probably to a Friends school in Rochester. Darius, then eighteen, went west to Canada to work on the Welland Canal, which would connect Lake Ontario to Lake Erie. Seneca went southwest to Middletown, Ohio, between Dayton and Cincinnati, to work on the Miami and Erie Canal, which would connect Lake Erie to the Ohio River. Rachel stayed in Lockport with Mary, William, Hannah, and the baby, two-month-old Lorana.

In June 1826, Increase left Rochester to join Darius, who was assisting former Erie Canal engineer Alfred Barrett in laying out and estimating a road "to be constructed down the steep rocky banks of the Niagara [River], a short distance below the Falls on the Canadian side."[1]

Unfortunately Increase left no record of his thoughts on Niagara Falls. He and Darius went to southern Canada to work on the first meandering Welland Canal (three more were built) to connect Lake Ontario to the north and Lake Erie to the south via a Deep Cut, again through the Niagara Escarpment and utilizing the Welland and Niagara Rivers. In Canada they worked with Alfred Barrett and another engineer, the tal-

Increase Lapham's early life was shaped by his work on the Erie and Welland Canals.

ented and taciturn David Thomas, who had a strong interest in botany and skill as a mapmaker.[2] Darius and Increase had known Thomas, a fellow Quaker, in Lockport, but on the Welland Canal, Darius must have solidified his friendship with the fifty-year-old Thomas since after that work together they corresponded for years about their common interests in geology and botany. It may be that Increase's serious interest in botany dates from these days in Canada with Thomas, a plant collector particularly interested in apple trees. Thomas had written a detailed and scientific work, *Travels through the Western Country in the Summer of 1816*, which Increase and Darius must have read, whetting their appetite for travel in the Old Northwest.[3] Thomas also wrote about plants and agriculture for such publications as *The Genesee Farmer*. It is possible that it was due to David Thomas that Increase first read Benjamin Silliman's *American Journal of Science and Arts*, which became so important in his life. Increase was learning more than canal work on the Welland.

On August 5, 1826, less than two months after he had arrived in Canada, Increase received a letter from his father saying that he had a position for him in the engineering department of the Miami Canal in Ohio. Being a dutiful son or an unhappy rodman or a boy who loved to travel, Increase left that very day for Lockport, New York, where he spent a few days with his family and made preparations for the trip southwest. Then he left on August 11 in a canal packet boat—floating through the Deep Cut where he had discovered so many interesting minerals and fossils—and arrived shortly in Buffalo where he caught the steamship *Enterprise*, spending two days in travel on Lake Erie. On the fourteenth he took a stage to Columbus, then Dayton, arriving at Middletown, Ohio, at breakfast time on the seventeenth. Increase's arrival at Middletown was a surprise to his father, who expected him weeks later when the job was to begin. So Increase looked around Middletown for a few days and then went down to Cincinnati to spend a week with his printing apprentice brother Pazzi in the big city.[4]

Cincinnati, with a population of more than sixteen thousand, was the largest city and main port in the Old Northwest Territory and the seventh-largest city in the nation, so there was plenty for the boys to see and do. They could walk along the public landing looking at the pleasing confusion of sailing ships and steamships and all manner of homemade boats and flatboats loading and unloading barrels of flour, hams, salted beef, venison hams, beeswax, as well as live cattle and poultry, lime, lumber, and flax—each product with its attendant smell. The slaughterhouse smell of the packing houses on the landing, though they were most likely closed until the cold weather, would have quickened their step. They could have stopped to gaze at the long and shallow "arks"—boats up to a hundred feet long steered at the rear with a sweep or tiller. Each of these arks was a traveling farm or small village with one or several extended families and their livestock, chickens, dogs, tools—all their bundled possessions. These boats usually had a shanty for shelter and a dirt- or stone-lined place for a cooking fire.[5]

The boys could have heard shouts and conversations in German, Irish, Indian languages, southern dialect, and Yankee speech. Along the streets lined with brick buildings they could see, beside the herds of pigs rooting

WHI IMAGE ID 101359

Increase traveled through the Deep Cut on the Erie Canal, enjoying the fruits of his stone-cutting labor on a packet boat like this one on his way to Ohio.

in the garbage, fine carriages and well-dressed men and city women in bright dresses of silk and fine wools and bonnets with feathers and ribbons, so different from the dark clothing of the Quakers and the homespuns of most pioneers. It is impossible to know what the fourteen- and fifteen-year-old brothers did that week, but they had the trust of their strict Quaker parents. Perhaps they had enough money to go into the famous Western Museum near the landing to see the mammoth and arctic elephant bones, "1000 fossils, 3500 minerals, 325 botanical specimens," plus Egyptian and American antiquities and "cosmoramic, optic, and prismorama views of American scenery."[6] All of that is just what Increase would have liked.

On August 30, 1826, Increase "commenced work on the Miami Canal as Rodman under B. Killbourn Assistant & Samuel Forror Resident Engineers, at $12 per mo. & 3 per week for board."[7] This was Lapham's first meeting with both Byron Kilbourn and Samuel Forrer, whose names he would eventually learn to spell. Both of these men would later become very important to Increase Lapham's life in Milwaukee. Kilbourn, twenty-five at the time, was an assistant engineer helping Forrer, the resident engineer, assign contracts and supervise the work on the Middletown-Dayton section of the canal.[8] The fifteen-year-old must have impressed Kilbourn, who ten years later would hire Lapham to work for him in Milwaukee. Yet in December of 1826, when Seneca went to Kentucky to work on the Louisville and Portland Canal, Increase followed him, having worked for Kilbourn and the Ohio canals only a few months. But he would be back. His lifelong connection with Byron Kilbourn had just begun.

—|—

On December 15, 1826, Increase Lapham left Middletown and the Miami and Erie Canal and traveled a hundred miles southwest to Louisville, Kentucky, and the Louisville and Portland Canal. The purpose of this Kentucky canal was to bypass the Falls of the Ohio River at Louisville— the only obstacle to navigation on the Ohio from Pittsburgh to the Mississippi River. Before the canal was built, goods and passengers on the flatboats and steamboats on the Ohio had to be unloaded above the falls in Louisville and portaged two miles downriver around the twenty-two-foot drop of the falls. There, in the little town of Shippingport, goods were reloaded onto other boats that would then pass down the lower Ohio to the Mississippi and then New Orleans. The merchandise that had to be portaged included barrels of flour, pork, whiskey, cider, and wine; bags of onions and ginseng; pounds of tallow, lard, beef, cheese, and soap; coils of tarred rope; bushels of oats, corn, and potatoes; venison hams and bacon; cherry and pine planks; live fowls and horses. And slaves.[9]

An important town even before the canal, Louisville had been founded forty-eight years earlier by George Rogers Clark with soldiers and settlers who built Fort Nelson there for protection against the British and Indian attacks along the Kentucky frontier. Louisville was partly a raw frontier town of drinking and gambling and fights in the streets, where a favorite entertainment was putting horses at stud in the streets.[10] But John James Audubon, who with his aristocratic wife, Lucy, had settled in Louisville twenty years earlier, said that the "generous hospitality of the inhabitants, and the urbanity of their manners" made Louisville one of his favorite places. The beauty of the place, he said, would "please even the eye of a Swiss."[11]

The elder Audubons had moved on by 1826 but there was still "a circle, small 'tis true,... within whose magic round abounds every pleasure that wealth, regulated by taste, can produce, or urbanity bestow."[12] This sounds

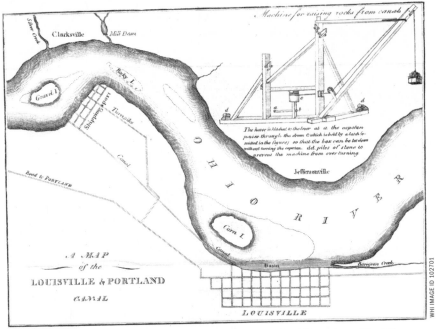

WHI IMAGE ID 102701

Increase Lapham's map of the Ohio Falls, the Louisville and Portland Canal, Louisville, and Shippingport, with his drawing of Orange Dibble's crane in the upper right

like exaggeration, but Mrs. Frances Trollope, a popular British travel writer and severe critic of American life, said of the Louisville of 1828 that it "is a considerable town, prettily situated on the Kentucky, or south side of the Ohio; we spent some hours in seeing all it had to shew; and had I not been told that a bad fever often rages there during the warm season, I should have liked to pass some months there for the purpose of exploring the beautiful country in its vicinity."[13] This charmed circle for the wealthy of Louisville, which was certainly not all of its seven thousand inhabitants, included fine brick houses, a courthouse, a theater, several very good hotels, newspapers, stores, taverns (of course), three churches, and an elite school, the Jefferson Seminary, founded by Mann Butler in 1798.

John James Audubon was in London selling subscriptions to his great *Birds of America* the year the Laphams arrived. Increase's path never crossed that Audubon's, but Victor Audubon, the first son of John James and Lucy, seventeen when Lapham arrived, became a friend of Increase's. Victor Gifford Audubon, a talented artist who later helped his father paint

Mammals of America, worked, when the Laphams were in Kentucky, in his uncle Nicholas Berthoud's store within another less-charmed circle in the area—the French settlers, storekeepers, boat builders, and millers who lived and worked two miles below Louisville at Shippingport. This village below the falls is where goods arrived after having been portaged around the falls, where flatboats were built and provisioned to set out southwest on the Ohio. Steamboats coming up the Mississippi and Ohio Rivers from New Orleans and other points south also stopped here in Shippingport by the falls and unloaded.

Shippingport, downstream from Louisville, was where the "vagabond boatmen & whores" and the brawling canal workers lived.[14] Shippingport's stench seems to have been its most noted feature as several travelers of the time and later Increase himself described it as a "stinking place."[15] Yet this is where Seneca and Increase boarded at Joseph L. Detiste's inn, because Seneca could not afford to rent in Louisville where the canal office was. In spite of the cost, within a week of Increase's arrival, Seneca sent him to school at the Jefferson Seminary rather than putting him to work on the canal. Forty to fifty of the other young men of the area attended, Increase's classmates being the wealthier boys whose parents could afford the twenty dollars for the standard curriculum or the thirty dollars for the classical language training for the six-month term.

Though Darius sent money to his father from Canada, Seneca must have had difficulty supporting the five members of his family in Lockport plus Pazzi, because Increase's full-time attendance at Jefferson Seminary lasted only a few weeks. Increase reports that in January 1827 he was "employed by Judge Bates as Rodman on the Louisville & Portland Canal under John Bates, another of his sons. Went to school when not employed on the canal." At the end of the next month he says he "[l]eft school & commenced divoting [*sic*] my whole time to the canal," making twenty dollars a month as a rodman.[16] It's too bad he couldn't have spent a little more time on spelling. Though he says he quit school at this time—and it isn't clear whether this was his idea or his father's or that of both—he did occasionally attend school after this, but not for more than a few weeks here and there. The few scattered weeks at Jefferson Seminary when he was fifteen were essentially Increase's last formal schooling.

On February 12, 1827, Rachel Lapham arrived in Kentucky with eleven-year-old Mary; William, nine; Hannah, four; and the baby, Lorana, to find that Seneca had rented the family a house in Shippingport.[17] Friend Rachel must have been discouraged. This was as bad if not worse than Lockport. The smell. The mud. The prostitutes and all the drunken men. The fevers and agues. And because of a wharf, the eddies of the Ohio River deposited almost at her door all the floating filth of Louisville. This was a low place in every sense of the word. There was no Quaker Meeting. She was far from Pazzi and Darius and other family and Friends. But Rachel was with her husband again and her good bright boy Increase. She made the best of it; she had no choice. Other women around her were experiencing much the same and making the best of it in much the same way. She hired a woman to help in the house. She bought a cow.

When most of the family were living together in Kentucky, Darius, alone up north, must have written that he was homesick, because his little brother Increase tried to cheer him up: "You must *not* be homesick, if you are I would advise you take a good dose of goodnature [*sic*], mixed with ⅔ weight of content, if this is not sufficient, you are to double the dose, adding a little goodwill towards your friends and your affectionate brother."[18] It's a telling mixture of the languages of science, the Society of Friends, and boyish humor. And a suggestion that Darius rely on what the Laphams always seemed to rely on—their basic "good nature," their ability to "will" themselves happy, and their commandment to think of family before thinking of one's self.

Darius up in the clean northern air of Canada must have also expressed, rightly, concern about the family's health in the reportedly unhealthful Shippingport. He evoked from Increase a long, reassuring, and completely unconvincing letter about the health of the place. His letter to Darius that early August reports the common assumptions of the time—both correct and incorrect—of the causes of disease. "The marsh miasma," which was thought to be the carrier of yellow fever, malaria, and other rampant diseases, was "given out," he said, before it gets to Shippingport. The hot weather, he writes, causes "oppression and lassitude in the muscles with a diminution of appetite all of which disappear on the occurrence of a cool day." But these effects "are truly distinguishable from affections produced by marsh miasma"—that is, the fevers and "agues" or violent chills.[19]

He then tells Darius that "the great depressions of the Ohio in August and September exposes to the sun a quantity of mud with trees and some animals in the state of decay; the exhalations from which are unquestionably prejudicial." Darius must have been plenty worried by this "reassurance" at this point, but Increase goes on: "A wharf which has been built at the river at low water [produces] in high floods an eddy which annually deposits on the beach for nearly the whole length of [Shippingport] a large quantity of filth and mud." If this is not enough, he writes that all the dead domestic animals "are thrown out on the commons between here and Louisville." That's the route Increase walks several times a day. "These and *dissipation*," he laconically concludes, referring to the drinking, brawling, and prostitution, "are the principal causes of disease in this place." But "don't worry, Darius"![20] He tried but convinced no one that Shippingport was a healthful place.

—|—

Though Increase's mother and father would have most likely wanted to send Increase to a good school, they needn't have worried. Increase's subscription to the *American Journal of Science and Arts*, familiarly called *Silliman's Journal of Science* or just *Silliman's Journal*, connected him to the great circle of American and European scientific writing and thought.

Just before he left for Louisville, Increase wrote to the prominent Benjamin Silliman at Yale, sending him three dollars in Ohio bills—it wouldn't be until 1863 that a national currency would be in consistent use—for a subscription to the eight-year-old *American Journal of Science and Arts*. Silliman, a professor of chemistry and natural history, a respected scientist as well as a promoter and organizer of science institutions, published in his journal the work of American scientists and introduced to American readers European advances such as the natural system of classification of plants and animals, the classification of rocks in terms of the fossils they contained, and the chemical approach to mineralogy.

The subscription was Increase Lapham's first independent step to educate himself. After his confidence-building experience in Lockport and the reinforcement of his interests in botany and mineralogy in Canada, Increase was beginning to take himself seriously as one who could study and learn science. Silliman wrote back saying that he did not know what "discount we must pay on Ohio money but believe it is 5 or 6 percent" and he thanked Lapham for the "liberal spirit" in which he encouraged "a work designed to promote the public good," not knowing he was writing to a fifteen-year-old boy who would eventually make that his life's work.[21]

At work on the Welland Canal with David Thomas and his brother Darius in 1826, Increase had read volume II, issues 1 and 2 of *Silliman's Journal*. In issue 1, he would have read a very lively account of volcanoes in Hawaii—written not in a dry or purely scientific way but as a good story studded with scientific fact, speculation, and sound thinking.[22] Another article tells of a trip to Long Lake in Vermont in 1823, to see this lake that had suddenly emptied into Mud Lake in 1810. The writer includes measurements of the lakes and descriptions of types of rocks and soils, as well as the bad meals he ate in a bad inn.[23] Silliman published not only this type of clear, first-person, detailed science and travel writing but also more abstract and theoretical pieces such as "A Review of the Principia of Newton" and "On the Present state of Chemical Science."[24]

Also in these early issues of *Silliman's Journal* were practical pieces on subjects such as where coal had been recently discovered, its types and uses; or the use of shell marl as fertilizer. Some were on engineering— a new type of engine designed by a man in New Hampshire, on the "Problem to determine the Position of a Crank when the tendency of the Power to produce Rotation is a Maximum"—a subject that continued in issue after issue, with writers from around the country sending ideas and corrections and diagrams. Botanists and naturalists sent notices of new species and genii, which were argued and corrected and added to in later issues. A delicately illustrated series of articles by Professor Chester Dewey describes the grasslike carex or sedges. An article on the divining rod discusses the phenomenon and describes, though not a double-blind study, one in which a twelve-year-old diviner is blindfolded and his skills tested in a thoroughly scientific manner by the author, probably Professor Silliman himself. He concludes: "If fountains have been and are

WHI IMAGE ID 44751

Increase Lapham's *carte de visite* of Benjamin Silliman, ca. 1865

discovered according to the predictions of the diviner, (which I allow,) it is because, in this country, men can hardly fail of finding water in from 20 to 50 feet deep, any where: they cannot miss oftener than diviners actually do."[25] This piece and others are accessible—and often still entertaining—accounts of scientific thought and experiment.[26]

The list of subjects is a great range, a mish-mash of science of the day, a clearinghouse, a low-speed web page and blog of scientific thought ranging from "Particulars of the Effects of a Stroke of Lightning" to a clearly written piece by Professor Dewey, added to by Silliman, on the uses of good meteorological data, how to be consistent in collecting such data, and what instruments to use. This later article was of particular interest and use to Lapham, for in Shippingport he began to collect meteorological data. This was one of the tiny beginnings of one of Increase Lapham's—with others'—great accomplishments: the National Weather Service.[27]

In the 1820s only a few disciplines of science were clearly separate from others. Articles on botany, geology, and chemistry in *Silliman's Journal* at that time did not cross into other subjects. More common were articles that included at least two (modern) disciplines. For example, an article on the antiseptic uses of chlorine included chemistry and practical medical applications. At least one of the articles in the journal—"Notes on certain parts of the State of Ohio" by Dr. S. P. Hildreth—includes many later disciplines: medicine, botany, anthropology, meteorology. And Isaac Lea's more general article "On the pleasure and advantage in studying Natural History" is echoed many years later in lectures by the middle-aged Increase Lapham.[28]

Most but not all of the full-length articles in *Silliman's Journal* were written by men with abbreviations before or after their names. They were medical doctors or reverends or professors or lawyers. In volume 11, issue 2 was a writer whom Lapham knew or knew of—James Geddes, an important Erie Canal engineer who wrote "Observations on the Geological

Features of the south side of the Ontario Valley." But the pieces found in the section Intelligence and Miscellanies were often written by men with no titles—smart, observant men, plain Misters, citizen scientists who could read and write and observe well. You would think that as a fifteen-year-old, Increase Lapham would aspire to write an Observation or a Miscellany, a short piece with a narrow focus like "Peach Trees" or "Topaz of Connecticut" or "Heat of July 1825" or "Nourishment of Horses." Amazingly, that's not what he wanted to do.

On February 27, 1827, after he had been in Louisville less than three months, after he had been kept busy, it seems, attending the Jefferson Seminary and working on the canal, he sent a draft of an article in a letter to Darius. This was an ambitious piece that included a description of the falls or rapids of the Ohio at Louisville that made canal navigation necessary, a short history of the charter to build the canal, where the canal was being built, the amounts and methods of excavating earth and stone, the locks, and the difficulty in getting labor on the canal, particularly in the summer, he says—because of the malaria, he doesn't say. The other half of the article is a description of the geology of the area.

> Of the rock strata there are four. 1st. Limerock, probably the common compact lime rock, —as to the depth and extent of this rock I am unable to set any limits. It passes under the slate rock in the bank of the river below Shippingport, but whether it rises again I do not know. It contains a great variety of Petrifactions. The minerals which I have collected from it are few and common, the principal are quartz crystals, calcareous spar and sulpheret of iron. Several springs issue through it, most of which contain a considerable quantity of Oxide of Iron held in a solution by means of Carbonic Acid. This has induced many persons to ascribe valuable medical properties to them. When newly exposed to the air the rock continually gives out an agreeable bituminous odor, occasioned by petroleum or Seneca oil, which is found filling the cavities."[29]

This was the first published mention of oil in this limestone.[30] Increase wasn't yet sixteen. Darius apparently responded favorably and perhaps sent a few suggestions, including corrected spelling.

On June 24, 1827, Increase wrote Silliman acknowledging his receipt of Silliman's December letter and the latest issue of his journal, then offered to send Silliman "any information relating to the canal around the rapids of the Ohio."[31] Silliman responded right away: "I shall be much obliged to you to give me any and every information in your power relative to the canal around the rapids of the Ohio, with permission to insert in the Journal of Science."[32] Increase got this good news near the end of August of 1827, but he didn't respond to Silliman until November "purposely that I might have the time to collect in mass, such matter respecting the Louisville and Portland canal as my humble capacity and circumstances would admit."[33] His circumstances might have been humble, but it's difficult to believe he really thought his capacity humble as well.

When he wrote to Silliman that November 10, he told him that he was sixteen. He said that his "only excuses for not arranging and wording the matter scientifically" were his age and that his "opportunities of intellectual improvement [were] only common"—not inferior, just common. In a slightly unctuous tone he must have thought necessary or appropriate he wrote: "I shall therefore submit the whole to your perusal and correction, and, if by that means I am able by stating facts to amuse any of your intelligent and numerous readers, I shall be wholly compensated for my trouble." He followed this with the draft of his article.[34]

After the manuscript he adds, "Should you think any part of this communication, worthy of a place in the American Journal of Science, I hope you will not let anything incorrect pass." He included three little drawings and told Silliman that he had also drawn "a geological profile extending from Louisville to the Knobs, a profile of the canal and a plan of the locks," but he didn't have the money to mail them so he didn't send them. He also told Silliman that he had found "at last" a new subscriber to the *Journal*—"Mr. Mann Butler, the historian of Louisville," who "teaches a large school and is in high standing in Louisville."[35]

When Silliman received Increase's letter with its manuscript and the drawings in December of 1827, he wrote immediately, telling Lapham that they were "sensible manly productions creditable to your taste and habits of observation." Silliman wanted the geological profile of the country and the profile of the canal as well and told him to mail them cheaply by folding the sheet as a letter. He said, too, that if Increase didn't have the twenty-five cents this would cost, he would pay the postage if Increase sent it collect.[36]

Though Increase wouldn't know this for about a year, when the *American Journal of Science and Arts* volume 14, issue 1 came out in 1828, with one exception, Silliman made only copyediting changes in Lapham's manuscript.

Silliman also wrote in his letter to Increase a statement that was warm, observant, and prophetic. "Time will (perhaps sooner than you realize) remove the *difficulty* of youth and I trust you will improve year by year, in everything that is wise and good and become a blessing to your fellow men."[37]

Increase Lapham had come, if not to a healthy place, to the best schools available—Silliman's circle of men who were American scientists and the circle of men in Louisville and on the canals whom he had the confidence and drive to seek out and learn from. *Silliman's Journal*, the canal, and the towns of Louisville and Shippingport were Increase's best schools. Three years later when Increase left for Portsmouth in Ohio, it was with a man's education and confidence, though not yet a man's social experience. In Kentucky, Increase would be well educated by the men he met, by the books and journals he read, by the work he performed, and by living on his own for the first time.

CHAPTER 3

"A Journal of Science & Arts with Miscellaneous Nonsense"

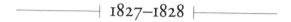

O N A FAIR OCTOBER SATURDAY IN 1827 Increase Lapham worked on the canal in the morning, then walked to Louisville, keeping his eyes open for snails. In Louisville he examined some books that were to be sold at auction the next day. He had his eye on William Nicholson's *Dictionary of Chemistry*, Jedediah Morse's *American Gazetteer*, a pocket dictionary, and a mechanical magazine. He walked back to Shippingport, ate his lunch, and then went to Nicholas Berthoud's store, where he borrowed his friend Victor Audubon's gun, very likely a muzzle-loading fowling piece of the sort John James Audubon used, a gun almost as tall as Increase.[1] Taking the heavy musket out to the forest, carrying the powder horn, wads, and a bag of lead shot, Increase walked on fallen leaves among the great trees festooned with red-leaved vines and the yellow leaves and dark fruit of the wild grape described by John James Audubon in an earlier fall.[2] He might have been hoping to shoot one of the busy squirrels or a wild turkey or a grouse, or a woodpecker to add to Victor's collection of bird specimens, but he was unsuccessful. So he crossed at the falls over to the Indiana side of the Ohio to go look at a cave just east of the village of Clarksville. He found part of it had collapsed, so he couldn't go in and explore.

That evening in his room Increase Lapham opened the little blue notebook where he had notes of zoology, conchology, botany, geology, notes

Basil Hall used his camera lucida to make this drawing of Shippingport in 1828. This view of the village with its frame buildings close to the river was likely very familiar to Increase.

on types of clouds, his meteorological tables, and a formula for a cement for steam engines. Increase turned to page 82, the first blank page, and he wrote, "Journal of Science & Arts with miscellaneous Nonsense By the Auther." On the next page he began his journal with the events of that day, October 6, 1827. "After doing a small job on the Canal this morning I went to Louisville to See about some books that were to be sold at Auction. After dinner I went out with Mr. Victor's Gun but Killed *Nothing.* I went to the Cave on the Indiana Side of the river but found it full of earth; so that it could not be entered."[3]

Many people begin journals and keep them faithfully for a few weeks, but the will and interest flag and the writing slacks off and then stops altogether. Increase Lapham's journal is the opposite. In 1827 he began writing every other day or so, twelve to fourteen times a month, but then in 1828 after a three-month lull the writing picked up, and he wrote in a journal consistently for about five years. These journals of his youth exhibit Increase Lapham's self-discipline and reflection, his passion to educate himself, his desire to write down what he saw and what he knew, to look back and compare the present to the past, to record the books he read, the

(*right*) The first page of Increase Lapham's journal, begun in 1827 when he was sixteen

1827

Journal 1827

Oct 6. Saturday. After doing a small job
on the Canal this morning I went to
Louisville to See about some books that
were to be sold at Auction. After
dinner I went out with mr Victors Gun
but Killed <u>Nothing</u> I went to the Cave
on the Indiania Side of the river but
found it full of earth; so that it could
not be entered.

Monday Oct 8.

This afternoon I went to Louisville to
buy some books <u>a</u>t Aucton. Light
rains All day.

Tuesday Oct 9 This morning I got
the Books that I bought yesterday —
Among them are The Cemical Dictionary —
Morses Gazetteer. — Mechanics
Magazene Va Pocket Dictionary
Fryday Oct 12. Went to Louisville
to day to carry the Engineers reports &c

plants, shells, and minerals he picked up, what he did at work, the letters he sent and received, and his sense of humor. The journal was with him as he grew to be a man in Shippingport and then Portsmouth, Ohio.

The Monday following that first entry, Increase wrote that he went to buy the books he had looked at and the day after that he went to pick them up—"The Cemical Dictionary, Morses Gazeteer, Mechanical Magasene & a Pocket Dictionary." Though spelling was not as standardized then as it is now, the boy by any standards couldn't spell—but he certainly could read. In the front of his little blue notebook, Increase listed his primary reference books, his "field guides," his authorities on animals, birds, insects, minerals, geology, geography, and natural history.[4] Increase might not have owned each of these books; some might have belonged to his father or to Darius, or he knew someone who owned them and he borrowed them occasionally. But at sixteen he knew who the authorities were in each of these areas of science.

On birds he read Alexander Wilson's *American Ornithology*, the first volume in 1825, possibly borrowed from Victor Audubon or Nicholas Berthoud. Wilson not only drew the birds with great accuracy, but his commentary is lively and detailed. Though John James Audubon had not yet published his *Birds of America* and Increase would not have been able to afford it if he had, his local authority on birds was an Audubon, John James's son, Victor, who though a clerk in his uncle's store, was an artist and a collector of birds.

Increase read and coveted the beautifully illustrated and expensive 1824 volume of Thomas Say's *American Entomology*, the most comprehensive work on American insects to that date. Thomas Say, like Lapham and many other naturalists of the time, was interested in almost everything. A self-educated man, Say corresponded with European scientists, took part in extensive collecting trips, and described for scientific journals insects, fossil shells, birds, and mollusks. In a few years Lapham himself was corresponding and exchanging mollusk specimens with Say, though he still couldn't afford to buy the book on insects.

For mammals, Lapham's resource was Richard Harlan's 1825 *Fauna Americana*, which was a useful compilation of the descriptions of North American mammals by Europeans like Carl Linnaeas and Georges Cuvier, but Harlan included new information from scientific expeditions such as Major Stephen Long's up the Missouri and Platte Rivers to the Rocky Mountains in 1820—which Thomas Say participated in. Harlan's inclusion of fossil species, some of which were found at nearby Big Bone Lick in Kentucky, was useful to a young man who was developing an interest in fossils or "petrefactions." This "guidebook" to the mammals, though not illustrated, listed the mammal's genus and species according to various systems, as well as common English names. Harlan included each mammal's essential character, dimensions, description, habit, and sometimes habitat. Using such a reference would require from Increase—or anyone— close and careful reading.

For mineralogy, Lapham lists two sources. First, Silliman's *American Journal of Science and Arts,* in which about a third of the articles were on mineralogy and geology. And, second, Bowdoin College professor Parker Cleaveland's 1816 or 1822 *Elementary Treatise on Mineralogy and Geology.* Reading Cleaveland would provide Increase Lapham not only with careful reasoning on the necessity for systematic classification in all the natural sciences, but also with a detailed presentation of the controversies surrounding the three current systems of classifying and describing minerals: according to the German Abraham Werner's "external characters," to crystal structure, or to chemical composition. Cleaveland pointed out the strengths and weaknesses of each system and used a hybrid system that combined useful aspects of each. Cleaveland's *Elementary Treatise* was at the time not just Mineralogy 101; it took the reader up to graduate school.

Increase's geology authorities were *Silliman's Journal* again and two important authorities to anyone who studied geology at the time. Philadelphian William Maclure, born in Scotland, was called by some the father of American geology. His crude color-coded map of part of the eastern United States using the Wernerian system of rock classification was the first widely available geologic map in the United States. Increase probably read an 1817 revision and expansion of his 1796 map with its "Observations on the Geology of the United States." Though in his list of geologic resources Increase just lists "Eaton," the Lapham family almost certainly had owned since Erie Canal days a copy of Amos Eaton's 1824 *A Geological and Agricultural Survey of the District Adjoining the Erie Canal in the*

State of New York. Educator Amos Eaton was a founder of what became the Rensselaer Polytechnic Institute and the author of several textbooks on geology, which Increase probably also read.

For natural sciences Increase lists the *Journal of the Academy of Natural Sciences of Philadelphia*—an authority for the whole country. Important members and contributors in the Philadelphia area included the botanist William Bartram as well as Richard Harlan, William Maclure, Thomas Say, and Alexander Wilson. Members who sent articles from other parts of the United States and Europe included Thomas Jefferson, Thomas Nuttall of England, Georges Cuvier of France, and Alexander von Humboldt of Prussia.

How did this sixteen-year-old know to read *these* works—in most cases the very ones he ought to have read? First, he sought them out. His mind *needed* this scientific knowledge the way his body needed meat and bread. Second, he knew educated men who had read and recommended some of them, though probably only David Thomas in Canada had the breadth of knowledge and interest to have read all of these works. Seneca Lapham, Increase's father, and his brother Darius were readers of wide-ranging scientific interest, though neither of them was as systematic, voracious, or driven as Increase. Increase worked with educated men on the canals who had secondary interests in the natural world, men like David Thomas and John Bates who were readers and companions in conversation, who shared with the Laphams stories of natural oddities—mineral springs, mastodon bones—that went up and down the canal lines like scientific gossip. And so books or mention of them would be passed along.

Though in Kentucky at that time a wide variety of books was sold by general merchants—dictionaries, grammars, medical treatises, handbooks of surveying, translations of the Greeks and Romans, and novels like Samuel Richardson's *Pamela*[5]—Increase politely ransacked the libraries of the men he knew in Louisville and Shippingport, looking for works to satisfy his curiosity, reading what was handed to or loaned to him. In December 1827 and January 1828 Increase borrowed from storekeeper Nicholas Berthoud James Lee's *Introduction to the Science of Botany*, which included translations of some of the works of Linnaeus; an unnamed book on natural history; and George Gregory's 1821 *Dictionary of Arts and Sciences* from which he copied drawings of plants. Over the next few years it became clear to those around him that Increase was exceptional. Gifts of books and scientific instruments began to come his way; men of means or knowledge who could see his potential wanted to participate in his education.

And with reading, as we know, one thing leads to another. This was particularly true as Increase read *Silliman's Journal* and the *Journal of the Academy of Natural Sciences of Philadelphia*, which had sifted and winnowed the most important science writing in this country and Europe. Without a doubt these were his best sources for finding out who were the scientists he wanted to read. His standard was high-quality scientific thought. He read to learn how to think.

Increase Lapham was reading at a time when all scientists' understanding of geologic and climatic forces on the earth was in transition. A Scot named James Hutton had in 1788 proposed that all geological phenomena had been produced by the "uniform operations" of the slow processes that anyone could see around him—wind, waves, sediment, the pressure of earthquakes and heat of volcanoes. Hutton's theory of earth, uniformitarianism, meant that the earth must be millions rather than thousands of years old.[6] But most scientists active at this time professed to be Christians and did not openly accept a non-biblical theory of an ancient earth. For example, Silliman and Morse saw that some of Hutton's evidence was correct, but they couldn't reject the biblical idea of the flood. More than twenty theories of the earth were floating around in the scientific journals, most trying to reconcile the evidence in rocks, minerals, fossils, and land formations with the idea of a great flood or successive inundations. As a boy in Shippingport, Increase too was trying to reconcile what he saw in the rocks and the rivers around him with the idea of the biblical flood. But with time and the weight of evidence his ideas changed. Not until the 1860s did most scientists accept the ideas of glaciers, that the earth was millions of years old, and that the forces that acted on it then were acting on it now.

⊣ | ⊢

Seneca and Rachel valued Increase's intellectual gifts and his scientific work; they made sure he had the quiet, privacy, and space to carry out his studies in their large and busy household that included four children younger than Increase, the youngest a baby of two. Rachel was helped in the house by a white woman named Mrs. Chapman and a black woman named Louise. For a time Pazzi and Darius were with the family in Shippingport in 1827 and 1828. On February 17, 1828, Increase's journal reads: "Not much occurred. Anew [sic] increase of the family appeared today after the above was writen [sic]."[7] His youngest sister and Rachel's last and thirteenth child, Amelia, was born that day. Rachel Lapham had borne seven sons and six daughters. She had eight living children in 1828.

In his daily journal Increase frequently records that he "stayed in [his] room." He wasn't sulking, which such a phrase might mean today. He was reading. Or he was identifying, drawing, or pressing plants; testing and naming minerals; examining fossils; arranging and recording his collections of shells and rocks; taking notes; copying useful information into a field book for the canal; writing letters. His room was part of his school. Other than one month in late 1827 when his boss, Mr. Henry, moved into his room, it doesn't sound as if he usually had to share a room, except perhaps occasionally with Darius or Pazzi, who were old enough not to disturb his chemicals for identifying minerals, pens and ink and paper, or knives for dissection of plants. Increase's younger siblings either respected his room or were kept out of it.

Few household tasks were required of Increase. Occasionally he did have to go after the cow that hadn't come home, but it's difficult to imagine that Increase minded very much having to go off and wander around the woods looking for it. In those days, cows and pigs were set free to graze and forage in the woods. On the day after "Chrismus," Increase wrote in his journal that he

took a long tour to the South west on the hill or second bank of the river in search of our Cow. I cam [came] across an old rotten log which the hogs had lately been rooting, and found some snail shells marked 4, 5, 6, & 7 in the cabinet. My search for the cow led me to a swamp back in the woods (which now contains a considerable quantity of water). Here I saw a pair of wild ducks and I listened to the song of various others of the feathered creation, such as woodpeckers, crows and a great many that I do not know the names of. I did not find the Cow, at any rate."[9]

Increase was not in his room all the time. He spent much of his day working on the canal. Almost every Friday morning, it was Increase's job to carry the weekly engineer's report containing the progress of the excavation and the workers' hours the two miles from Shippingport to the secretary in the board of directors' office of the canal in Louisville. Increase also worked on the canal taking levels, measuring stone, "tramping about on the canal all day" with his bosses.[10]

On many days when the weather was too nasty for work out on the canal, Increase worked in the Shippingport office of the chief engineer, Judge Bates, or his assistant engineer son, John Bates, or the resident engineer, Mr. Henry. In the office, Increase might make fair copies of letters to be sent and copy reports and statistics into record books. He was also called on to draw large plans of the locks and of a bridge over the canal. On March 24, 1828, he wrote that he "was engaged nearly all this day assisting mr Bates & Henry examining some patterns for the King Stone of the arch bridge over the Canal."[11] The next day's entry gives us a sense of what he would rather be doing than drawing plans. "I had two more plans of the Bridge to draw today: but as I had no paper I was under the necessity of going to Louisville to buy some. In my return I observed three plants in flower neather of which do I know the names of. The birds which ornemented the fields were the Blackbird & the martin the flocks of [passenger] pigeons which flew over head done little more than excite our wonder at their number's."[12]

During this period, Victor Audubon, or "Mr. Victor" as Increase almost always referred to him in his journal, be-

Victor Gifford Audubon was the first son of John James and Lucy Bakewell Audubon and a good friend of Increase Lapham's. A talented artist in his own right, he worked closely with his father.

came a friend and confidant. Victor loaned him books like Basil Hall's *Voyages* and, more telling, several times he loaned Increase his gun. Victor knew how to draw and paint and gave Increase some artistic instruction, starting him off with copying a landscape drawing of the James River in Virginia. Victor showed Increase the geodes he had found near the Ohio. Victor was a punster, telling Increase he hoped his March meteorological table, which was published in the *Louisville Advertiser*, would be more "fair" than the two preceding months.[13] While Increase's parents were in Louisville the relationship between Mr. Victor and Increase was as teacher to taught. Later they become more equal bachelor friends together. Later he became just Victor.

Increase was learning not only from the books he read, from his observations of the natural world around him, and from the educated and experienced men he knew; he learned as well from his somewhat stupid and badly educated boss, the resident engineer, John R. Henry—"Mr. Henry" about whom very little is known except that he came from Rochester. Mr. Henry and Increase spent a lot of time together. In Mr. Henry's room or office, Increase copied into the record book Mr. Henry's letters about the canal, contractors' bills, and lists of workers. They tramped the canal together. They argued about the causes of natural phenomena and these arguments were important enough that Increase wrote them into his journal. The contrast between the way Mr. Henry thought and the way Increase thought must have shown Increase how much he knew, how different scientific thought was from ignorant superstition.

One argument was on October 24, 1827, "a smoky day." Mr. Henry told Increase that the smoke was not the result of prairie or other fires to the west but because of the beginning of the decay of vegetation. Increase told him that "[i]f it is possible for vegetables to be converted into smoke without combustion this will appear very probable." Then Mr. Henry remembered a story about another smoky day at the time of the terrible earthquake at New Madrid, Missouri, in 1811. The smoke, Mr. Henry said, had risen through the ground. Increase said that he supposed "it was owing to peculiar states of the atmosphere which was unfavorable to the decomposition of smoke"—coincidental, in other words, to the earthquake.[14] To this Mr. Henry made no reply. Three weeks later on a gloomy day the argument continued, with Increase getting the last word. Mr. Henry said that this darkness was the forerunner of "our earthquake." But, Increase writes, "the earth did not quake."[15] Mr. Henry is passing on folk stories; Increase is passing on scientific knowledge based in evidence.

Another time Increase wrote with contempt of Mr. Henry's argument that "if an instrument in the form of a funnel, be placed in the air, with the tube downwards, the pressure of the atmosphere downwards upon the funnel is not balanced by the upward pressure through the tube; so that a current of air will rush downwards through the tube of the funnel."[16] Increase at sixteen knew that the air pressure on all points of a funnel is the same, including the narrowing. His boss did not know this.

In Mr. Henry, Increase Lapham encountered the superstitious beliefs common in the popular culture of Kentucky and the southern parts of the Old Northwest Territory. Some common superstitions of the time were that if you tie as many knots in a string as you have warts and bury the string under a stone, your warts will go away. The sight of a white mule when a traveler sets out on a journey causes bad luck. If a horse breathes on a child, the child will have whooping cough. Feeding gunpowder to a dog makes the dog fierce.[17] In contrasting himself, even subconsciously, with Mr. Henry, Increase realized that he was not of the popular and rough culture of the frontier. Not only did he not think like Mr. Henry, but he did not, like many men he knew, drink to excess, get in fights, torture live animals for fun, or brag about his strength. Increase belonged to a very small subculture of Quakers and engineers—educated people. He was not among the mon-eyed elite. For the most part, all his life, Increase Lapham made his own world—a world of science and the mind—from which he was able to be a very keen observer of the natural world and the popular culture around him.

The ignorance of the resident engineer and the scientific acumen of his assistant, only a rodman, caused tension in the relationship between Mr. Henry and Increase. Mr. Henry's propositions irritated Increase enough to report them at some length in his journal—and with a rare tone of contempt. Increase's tone in his journal is almost always the calm objec-tive tone of a reporter of phenomena. Comparing his knowledge to the knowledge of his less-educated "superior" was not without its intellectual benefits and its very human satisfaction.

The relationship between Henry and Lapham was complicated by the occasional forced intimacy of having to share a room. In April of 1828, Increase wrote that "Mr. Henry took an Emetic this morning and I was required to attend to him. After getting through with that Business I went to Louisville to get some Drawing paper."[18] "That Business"!—you can just hear the boy's disgust. And Increase must have quietly seethed at being bossed by this inferior who required him to do much more than a rodman usually did—draw plans, keep and copy reports, everything, he complained at one point, that required that pen be put to paper.[19]

Amazingly, though there was tension between the boy and his boss, the relationship between them was usually amiable. One day Mr. Henry showed Increase "some particulars" in the use of Hadley's Quadrant, which he had borrowed. And another time Mr. Henry gave him "a piece of very fine white clay from Green river Ky."[20] Increase and Mr. Henry spent some very companionable days together, as when Mr. Henry was hired by the US Bank in early March 1828 to survey a piece of land and a road south of nearby Portland. He hired Increase to go with him as rodman. Increase wrote extensively about what he saw that day. In this entry we hear Increase's characteristic reportorial tone tinged with the ecstasy of being in the forest. And we see his close observation of the trees and their state at that time of year, the effects of weather on the trees, his observation of the soils, and his knowledge of the Latin names of the plants and birds. With "translation" into correct English, the distractions of Increase's phonetic spelling and casual punctuation disappear. Without these surface errors, the graceful sentences of an educated person come through clearly. Increase himself comes through more clearly.

> Monday, March 3....I went to Louisville this morning and borrowed a compass of Mr. Joyes.[21] At noon we commenced our survey of a tract of land containing 2000 acres laying south of Shippingport and Portland, and west of Louisville. I came home at night cold and tired with wet feet, and with (having drunk a good mug of beer with Mr. Henry) a head rather above the common order.
>
> Tuesday, March 4. I took breakfast with Mr. Henry in order that I might be ready by the time he was for a trip on the land, which I have been traversing back and forth ever since. It is all covered with forest

trees, of which beech, oak, poplar etc. are the most conspicuous. It contains also a large number of ponds, to get a correct map of which was the principal object of our journey. They are nearly dry in summer and support numerous trees, particularly the red or water maple (*acer rubrum*) which grows very tall and are thickly scattered over the surface. The alder [blank for Latin name] covers those which contain the most water. And we also observed a species of wild rose (*rosa* [blank]) growing with the alder.

The soil is composed everywhere of a yellow clay covered with a layer of vegetable mold to the depth of five or six inches on the arable lands.

On the little ridge which binds one of the largest of these ponds above mentioned I observed several chestnut trees (*Castanea vesca B. Americana*), the existence of which so near us I had not before suspected.

Numerous hogs (*Sus scrofa*) are turned loose in these woods which keep fat all winter on the hickory nuts, beech nuts, etc. which they find in great plenty. They have become very wild.

We saw the crow (*Corvus corone*), ducks (*Anas boschus*), the king fisher (*Alceas alcyon*), two or three species of woodpeckers (*Picies*), [passenger] pigeon (*Columba migratora*), and a great many species of smaller birds which were, as it was a fine day for them, flying about in every direction from bush to bush and enlivening the lovely woods with their sweet and melodious music.

The flowers of the red maple (*Acer rubrum*) and the leaves of the elder (*Sambucus* [blank]) and blackberry (*Rhubus velossus*) have already made their appearance. The heavy frost last night must have been very injurious to them for I saw a great number of the flowers of the red maple laying on the ground.

We chained a road from the west end of the lot to Louisville (3½ miles) where we stopped and took our dinners at a grocery on crackers and cheese.[22]

It was a fine day in the woods for Increase Lapham and Mr. Henry, as well as for the birds.

Increase's prose in the journals—and in all his later writing—is brisk, clear, and usually unadorned—though occasionally the emotion of the moment overcomes him and he resorts to the overwrought writing of the day with a phrase like "sweet and melodious music." His diction is complex, composed of the verb forms of the people around him (*I done* instead of *I did*), as well as Latinate and Latin terms from his studies, and the clear direct language of engineering. He is only sixteen.

The journal entries were not something he wrote and then did not read again. He leaves blanks to fill in later when he finds out the correct, usually Latin, term. He rereads and corrects himself later with notes like "false orthography" and "incorrect." Occasionally he thinks of something he left out in an earlier entry and adds it. On March 15, 1827, he remembers that he also saw mistletoe on the elms in the swamps and ponds he was surveying earlier and cross-references this in both entries. In a few years, perhaps when he was in Portsmouth, Ohio, he began but didn't complete recopying this journal and the following one from Shippingport. Most of this third leather-bound, four-by-six-inch book is blank, but the first few pages are a "neatening up" of the first journal in Increase's hand but smaller. He fixes most of the spelling, but changes little else. It's clear that these journals are an important record to him.

But he also writes down things he thinks are funny. Increase writes on October 23, 1827, "The Steamboat Triton left here with a full load of passengers. She got down as far as Knob Creek where she broke her shaft and the leg of her engineer."

The two journals that Increase wrote in from 1827 to 1830 are not private and confessional diaries. Increase does not write about how he feels—at least not directly. He does not express in the journals the smoldering that was his ambition. He does not complain, except about Mr. Henry. He tells no secrets. He does not say what he likes and doesn't like. He does not write about his family except to say when someone arrives or leaves and when they go on an expedition to the woods or a spring together. The wandering cow gets much more mention than his sisters. He doesn't write about what it was like to be one child in a large household, or the jostling it took to be himself.

Increase wrote down what was important to him, but his whole life was not in the journals. He rarely mentions what he ate or the comfort or discomfort of accommodations. He seldom writes about possessions, though he mentions what he buys and he mentions the books, journals, and tools necessary to his work and his study of the natural world. He doesn't write about the economic or political worlds. Though he lived in a slave state, he doesn't mention the enslaved people he saw around him every day, though, clearly appalled, he told the story of the beating of a slave he witnessed in the street.[23] The Quaker Laphams surely opposed slavery, but this was not an issue that appeared in Increase's early journals or his letters. He didn't write about the incredibly dirty presidential campaign of 1828 when Andrew Jackson, a Westerner from Tennessee, opposed and beat the very Eastern and establishment John Quincy Adams.

Increase also didn't write about the workers who actually dug the canal. Though Increase Lapham's world was, in one sense, wide and extended to scientists and engineers across the country, it was also a world limited by social class. The Laphams' connections, sympathies, and friendships were all with men of a professional class, men—and they were all men—who were canal contractors, engineers, canal investors and administrators, educators and doctors. In Increase's journals there is no mention of a canal worker by name; not one of the Irish or other immigrants or poor whites or slave laborers whom he must have seen day after day is mentioned. Occasionally a man would be killed on the job and this would be noted in the journal, but the name of the dead man was not included—if it was even known. The physical trials, dangers, and filthy conditions of their work are not mentioned.

It may be that part of Increase's drive to make order out of the natural world through knowledge of science and engineering arose from a concentration *against* the disorder and danger he saw in the lives of the ordinary men he saw every day, knee deep in the muck of the canal. His journals were his world of the intellect. His journals were workbooks. His journals were a refuge.

CHAPTER 4

Natural History and the Lapham Brothers

————┤ 1827–1828 ├————

D ARIUS WAS INCREASE'S OLDER BROTHER, his beloved brother, and his closest friend. Increase had lost one older brother when he was eight; he clung—in letters and thought—to this one. The brothers of course lived together for most of their boyhood, but in August of 1826, Increase followed opportunity to Ohio and then Kentucky while Darius stayed on in Canada on the Welland Canal. When they were separated, Increase was fifteen and Darius eighteen.

While the brothers were apart, they wrote to each other, in Increase's words, "a regular, reciprocal correspondence...for mutual improvement."[1] In these letters they told of their work and their worries for each other. They exchanged ideas and news, copied passages from books, sent back and forth books and mineral and plant specimens and at least one draft of a scientific article. Increase's letters to Darius are unlike his letters to any-one else—open, eager, unguarded, informal, funny, and yet full of infor-mation, much of it related to their work. On October 22, 1827, Increase sent a letter to Darius describing a high-pressure steam engine he had studied at the nearby steel mill at New Albany, Indiana. A month later Darius sent Increase a description of an excavating machine in use on the Welland Canal. Increase needed Darius's ideas and he needed to try his ideas against those of Darius. Without his older and more experienced brother to challenge him, Increase might not have become so confident so early. The trust and admiration and love between these two brothers made Increase's relationship with Darius his most important until his marriage years later.

On January 25, 1828, the brothers and the family were together again. Darius had completed his work on the Welland Canal and came south to Louisville hoping to be an engineer on the Louisville and Portland Canal and work with his father and brother. When no job offer came, he worked with Increase and his father informally, without a contract. They must have been a formidable team, the young brothers Lapham. Their combined intelligence could have been intimidating, but there is nothing in the records to suggest that they intended to intimidate. Even Mr. Henry, who must have known he wasn't as smart as the Laphams, wanted to hire Darius though the canal directors said there was no money. Darius Lapham was nineteen, big boned like his father and probably taller than Increase, but with the same black hair and light eyes. Increase was smaller, slighter, more intense.

After Darius arrived in Louisville, Increase's journal is full of Darius. Darius walked with him and Mary and William to school. Darius sent a letter to cousin Phebe Allen. Darius finished a family record that he ornamented with Ionic columns and a Masonic arch. Darius is engaged in drawing perspective. Darius bought a blank book. On many days Darius and Increase worked together either in the room they probably shared at home or in Mr. Henry's office. They calculated the lateral pressure of water against lock gates. They calculated the weight of an iron gate. They figured out specifications for the iron gate designed by their father. The brothers read Stephen Long's *Expeditions* together and commented on it in Increase's journal. They went to Louisville on "arents" and borrowed the *Transactions of the American Antiquarian Society.*[2] Darius began to keep a journal like Increase's. Increase bought a blank book so he could write information relating to canal engineering in imitation of one "lately commenced by Darius."[3]

When the winter began to lift, the brothers worked together on the canal. In late March they set some slope stakes for the contractors and noticed that day that leaves had begun to appear on the *Platanus occidentalis*, the *Populus angulata*, and the *Ulmus Americana*—the sycamore, the eastern cottonwood, the American elm. And the "*Leontodon Taraxacum*" was in bloom—the dandelion. The next day Darius had a "fit of the Ague and fever"—the malaria that plagued almost everyone in Louisville, at least occasionally.[4]

Sometimes the brothers went for walks in the nearby woods. "I took a walk with 'Darius' into the woods south of this place but found nothing to interest the Botenist. The Ornithologist would have been in some degree interested; but verry little."[5] For a time in the journal Increase does an odd little thing. He puts Darius's name in quotation marks. It's a joke that we can't get, a joke between "Darius" and Increase.[6]

Increase might have begun the journal with the idea that he would be the only one to read it, but it evolved into something for Darius to read as well. Here's what he wrote on "St. Paterics Day 1827." First he wrote about the Irish workmen: "Irishmen all drink before night!! An effige appeared on a poll this morning with a string of potatoes about his neck; but he soon disappeared." Then there is a note:

> N.B. Should any person, except perhaps the author himself, be reading this Journal, he is, when he comes to this place requested, and do hereby order him not to read the rest part of the notes for this day. Should this be violated, offender is to be compeled to drag a canal Boat from Lake Erie to Hudson River; and above all he will be under the savere penelty of not being allowed to read any more in the Journal at all. The Auther.[7]

Only Darius, with whom he shared almost all interests and a great deal of writing when they were apart, would want to read this. Increase tells Darius that if he reads this part of the journal, he won't let him read the journal anymore. On June 10, 1828, Darius began his own journal, which Increase said was similar to his, "with quite a humerous preface."[8]

Between April and June of 1828, Darius and Increase went to the woods at least six times on "Botaniseing excurtions." But even the two-mile trip to Louisville from Shippingport was a botanizing excursion. And when Darius and Increase went with their father to an island adjacent to Portland to buy some fish, Increase noted in his journal the "verry tall" cottonwoods and elms, the sycamore, and the willow, using their Latin

The forest near Louisville was drawn by Basil Hall in May 1828 at the time Increase rambled in these woods. Increase would meet Hall that spring.

names. The red and black maples were in bloom, as were the marsh blue violet and the common little-leaved buttercup. The three of them may have bought fish to take home, but what really interested them was a discarded fish—perhaps the shovel-nosed sturgeon.[9]

One of their best days (and one of Increase's longest journal entries) was April 13, 1828, when Seneca, Darius, and Increase went to the "boiling spring" and the "knobs"—the low hills near Louisville. On a fair day after a night of rain, they took a skiff they borrowed from the canal contractors and made their way easily across the Ohio River to the mouth of Silver Creek where they left the skiff and went on foot through the woods toward New Albany. Stepping carefully on last year's brown leaves in the soggy woods, the father and his sons gathered for their herbarium spring flowers, several of which were new to them—the *phlox paniculata*, the azure bluet, and a buttercup. Near a pond they found the spring beauty, rue anemone, and three species of violets. The American beech trees above them were in bloom and the understory dogwood lit the dappled woods with white blossoms that seemed to float in the thick humid air.[10]

Leaving the woods and passing through New Albany, the men met a friend of Seneca's and the four of them set off for the "boiling spring."

Once there the four scientists—like boys—covered the small spring with flat stones and plastered the rocks with mud to seal up the spring and see what would happen. Then they poked a small hole in the spring and lit it. They wanted to see what sort of gas this was, but they were disappointed. Increase wrote that "it did not burn much."[11]

On to the Knobs where Increase saw the woods as "whitened by the involuorm of the *Cornus florida* flowers." Back on the banks of the Ohio, the explorers found the yellow puccoon in bloom and the Jacob's ladder and the cursed crowfoot. On their way home they sheltered from a hailstorm in the

Poa pratensis, or Kentucky blue grass, one of the plants drawn by Increase in 1827 or 1828 while he lived in Shippingport

"nighest shanti." When the storm let up they had some difficulty pulling against the rain-swollen Ohio and had to land on Gravel Island, but they made it home just after the rain began again. It was still raining when Increase was recording the day in his journal in his room at nine o'clock. Not a word from Increase about who was the first to find the yellow puccoon or the cursed crowfoot, not a word about what they said to each other, what they felt for each other, but this was a day Increase didn't want to forget. He couldn't know then how rare such days would be with Darius and with his father.[12]

—|—

Increase continued to grow in his knowledge of the natural world. From example or reading, he learned how to preserve plant specimens. He learned to press them flat between sheets of absorbent paper to dry them, to carefully mount them with glue and small strips of paper on large sheets, to label the sheet with the date and place the plant was found, the Latin name and common name, if they were known. On a Sunday in late March of 1828, Increase "had the pleasure" of finding violets, trillium, bloodroot, and rue anemone in flower. "When I brought them home I brought into my room the same stone to press them with that I used all last summer."[13] The next week he bought chlorate of mercury to dissolve in alcohol to wet the plants he dried for his herbarium to keep them from being attacked by insects. Darius and other members of the family frequently brought home plants for Increase to preserve. This collection of plants Increase began in Kentucky would eventually number in the thousands, some of which still exist at the University of Wisconsin.

Increase also broadened his approach in a new branch of science: meteorology. In December 1827, Mr. Goodwin, secretary of the Louisville and Portland Canal Company, loaned Increase a thermometer so he could record temperatures in his meteorological tables. Up to this time he had been recording the weather in rough charts in his journal but only in general terms: fair, rainy, or cloudy. He also measured the height of the frequent floods on the Ohio. But with a thermometer, he was able to note on January 10, 1828, that the temperature of the river at their door six inches below the surface was 47½ degrees and the air was 52 degrees.[14] From the river in front of their house Increase collected shells off a rotten log that had floated right up to their door. A few days later, Mr. Goodwin had arranged to have Increase's monthly meteorological records published in the *Focus*, a weekly newspaper. His tables were also in the *Louisville Advertiser*. These charts of Increase Lapham's in Louisville are some of the tiny beginnings of the National Weather Service decades later.

One day Seneca Lapham brought home from the canal an aquatic animal caught in a net on the Ohio River, probably a large salamander, the mudpuppy, *necturus masculosus*—a strange but not uncommon creature from eight to seventeen inches long, gray or rusty brown with fuzzy-edged dark blue spots and with feathery maroon gills. "We at first supposed it a species of Siren but we are now confident of its being the *Triton lateralis* of Say Discribed in the first vol. of Longs expedition page 5, or the *Proteus maculate* in Silliman's journal vol. 12 p. [blank]."[15] The sluggish animal was still alive in its tub in the Lapham house almost three weeks later. One evening, gathered around the tub, Darius and Increase and Seneca and other Laphams observed that "the animal we are keeping...Dose [does] not like to be touched on the Tail. When he is thus treated with a stick he jump[s] & springs with great activity."[16] You can see the boy with the stick, poking. You can see the Laphams collectively startling as the animal jumps. Increase writes too that they've noticed it stays on the side of the tub away from the light. And they have heard that it makes a noise like a puppy, hence the name.

A few days after Increase's seventeenth birthday—"My Birth Day!! . . . heigh ho!"—the "water puppy" was found dead. For a few days it had been "more noisy than common," which Increase supposed was due to the increased warmth of the weather. "It was *not certainly* owing to any pain agony or misery: for we got two others of his species today which make full as much noise."[17] The next day they surmised that the water puppy lived mainly on angle worms, because a great number of them were found in the tub, voided by the two new water puppies.

The Laphams collected some little tortoises and observed them, and Increase preserved a lizard in alcohol, later donating it to the Western Museum in Cincinnati.

Natural science in the Lapham house.

On February 13, none of the Laphams went to school because of the rain, but Increase used his time with analyzing "water lime." He pulverized a hundred grams and put it in a saucer, poured muriatic acid over it, whereupon "a strong effervescence took place." Then he let the liquid evaporate, poured hot water over it, and slowly filtered out the water. The next day when the undissolved solids on the filter were dry, he weighed them and noted that he had twenty-four grains of silica and alumina, which he marked precipitate A. Next he dried this precipitate and poured dilute sulfuric acid over it, evaporated it again, and noted that the surface was covered with "a thin black pellicle"—a membrane or scum.[18]

Seventeen-year-old Increase was developing the abilities of observation, collection, identification, classification, and experimentation in four areas of science.

A few days later, Increase worked a little on the canal—the first work in a long time because of rain—and then continued his scientific work with a long letter to Professor Silliman at Yale. He also sent Silliman by way of New Orleans a small keg of mineral specimens and shells. The specimens were numbered and labeled, with the ones he didn't know left blank;

Increase hoped Silliman could supply the correct names. He enclosed also a vial of the petroleum that came from the nearby limestone. "The petroleum occurs in the lime rock from which the specimens 1 to 5 were taken as you will perceive by their peculiar smell."[19] And he offered him a specimen of the mud puppy. In this letter he wrote with an unstudied grace, which easily elicits help.

> You will observe that I have given names to some of the shells, but still I am not certain that they are correct. I have descriptions of nearly all the land and fresh-water shells of the U.S., but I can not determine all the names, until I have become more familiar with the terms which are employed. In return I would be thankful for the nine minerals which constitute the geological alphabet (excepting limestone), and such others,...as you would recommend to a young person acquiring a knowledge of mineralogy.[20]

Not only is Increase's fine mind an asset, but his manner and his tact help him get access to the knowledge he craves—not just in letters but in person as well.

In May of 1828 Increase met and gained the respect of two famous men. Basil Hall, formerly a captain in the British navy, was a world traveler and well-known writer whose account of a voyage to Korea and islands in the Sea of Japan Increase had recently read. Captain Hall was traveling in America and would soon write about the experience in his *Travels in North America in the Years 1827 and 1828*. On May 8, Increase wrote excitedly that "Capt. Basal Hall is said to be at Louisville!"[21] The next day, Nicholas Berthoud, but none of the Laphams, was to have dinner with Hall. On the thirteenth of May, Increase was pleased to hear that Mr. Goodwin had given Captain Hall a copy of the *Silliman's Journal* with Increase's article in it, but the next day Hall left for St. Louis.

That same day Seneca Lapham left for Ohio to look for work. Seneca had designed himself out of a job. Usually he built wood lock gates, but for the Louisville and Portland Canal he had designed an "improved" iron gate—a gate he could not build.

The day after that, on May 15, Hall's visit must have been put out of Increase's mind because the equally famous and well-respected Erie Canal engineer Canvass White arrived from Philadelphia with other stockholders of the Louisville and Portland Canal. For the rest of the month of May, Increase worked closely with Mr. White. This was certainly more exciting and demanding than working with Mr. Henry.

First, Mr. White had serious misgivings about building an iron instead of wood canal gate: "Mr. White is likely to *blow up* the Boiler Iron gate," Increase wrote in his journal.[22] Second, Mr. White was concerned about how much was left to excavate on the canal. For a day and a half,

Canvass White, self-taught engineer and inventor, was an important engineer on the Erie and earlier canals and a mentor to Increase.

WHI IMAGE ID 102702

White, with the help of Increase and Mr. Henry, took levels on the canal to see how much excavation was needed. The levels done, White set Increase to drawing a plan of the locks and bridge. When he finished the plan the next day, Increase took it to Louisville to give to Mr. White, but Mr. Henry took it out of his hands and gave it to White himself. Yet while he was there at the canal office Increase sold a plan to one of the directors for five dollars and got an order for another one. The next day, Mr. Henry brought in an old plan of the wooden gate and a map of the upper end of the canal; White wanted Increase to make copies of them within two hours, which Increase did. Then White left for Ohio, taking the copied plans with him.

On May 24, Mr. White sent orders to Increase to make tables of the number of cubic yards in the canal. Increase began right away. For days he worked on the complicated tables, taking time now and then to look out the window at the steamboats on the river, yet also carrying out his usual work on the canal. On the twenty-sixth he had to work with Mr. Detiste whom he called Mr. *Deteste* in his journal. On the twenty-ninth Darius went on an excursion to Knob Creek; Increase, left out, had to stay and work to get the time lists for the Friday report.

On Friday, May 30, as he usually did on Fridays, he copied the engineer's reports and carried them to the canal office in Louisville, where he found Captain Basil Hall himself! Hall was on his way back up the Ohio. Increase had an "interview" with him; he got to talk to him. Captain Hall showed Increase his *camera lucida* and how it worked. And, Basil Hall told Increase that he was very much pleased with his article in *Silliman's Journal,* particularly the part about the geology.[23]

This was wonderful, but the river was flooding, causing more work on the canal. Yet that Saturday, even though he had to work, Increase noticed that the "*Datura Stramonium* [jimson weed] commenced flowering."[24] He worked on Sunday, too, doing some more leveling that had to be redone because of the flood, and he wrote canal memoranda in his room. Then he wandered about the fields in search once again for the cow. That night, though, he took time to write a long entry about the natural curiosities found as the canal descended into the "blue alluvial earth."[25] He and Mr. Henry and Darius had had long conversations about these curious bricks and chunks of worked iron, human skulls, dog bones, deer bones, a stone ax, and the bone of an animal decorated with beads—probably the skull of an animal worked into an ornament, said Darius.

On the morning of June 3 Increase finally finished the table of the cubic yards of earth in the canal and Mr. Henry sent it off to Canvass White in Ohio. Increase took a little break and went to Louisville with Darius. But the next day Increase was "very unwell all day this day" and he stayed in his room—exhausted after these intense but satisfying weeks.[26]

―｜―

On Friday, June 6, Seneca Lapham returned to Shippingport with the news that he had secured a job on the Ohio Canal halfway between Chillicothe and Circleville as a contractor on a half mile of excavation and a large stone culvert. "He intends to move the family to Chilicothie Ohio Next week!" wrote Increase.[27]

They had a week left in Shippingport and Louisville. In the next few days, Increase completed his plan of the locks for the directors, Darius copied his "Sketches of the Geology of Ohio Falls" into Mr. Henry's book, the moth mullein bloomed, and Increase and Darius went to the woods behind Shippingport where they found some fossils. Mr. Henry paid Increase $43.25 for the past month's work. Increase went to Louisville and bought three shirts for $2.36 and a short coat for $3.75. He bought letter paper, bread, and tape. There was nothing to do on the canal but examine the timber. Increase again went after the wandering cow, which they hadn't seen for a long time.

On Wednesday, June 11, it was raining again, but it didn't matter because all the Laphams were in the house packing. "We expect to be off tomorrow," Increase wrote. He packed his herbarium, his books, and his shells and fossils and minerals into his trunk. Another canal. Another move. Increase gave his father fifteen dollars of his earnings.[28]

But the next day his expectations were cruelly disappointed. He wrote on June 12, 1828: "Today Father with all the family and all the goods belonging to them left this Dirty, nasty, stinking, little place for Chillicothe Ohio. They got off about 12 oclock on the steam Boat Native. I am now left alone boarding with Mr. Henry at Detist's inn."[29]

Increase was to stay behind in Shippingport. No one—not his mother or father or brother Darius—told Increase that he wasn't going until the morning they left. Yet hurt and angry as he was, the only anger Increase could express was at dirty, nasty, stinking Shippingport.

CHAPTER 5

First Year on His Own

⊢ 1828–1829 ⊣

I NCREASE LAPHAM'S FLARE OF ANGER at being left behind by his family in
Shippingport died down quickly, though it didn't burn out. He finished
his journal entry of June 12 on a calmer note, writing that after he helped
the family off, he had his trunk carried over to Detiste's Inn. On his own
in what he called his "new home," there was no place for what must have
been frightening anger at his family.[1] He couldn't let emotion overpower
the working of his mind. Though he was only seventeen, he was prag-
matic and calm by nature.

The next day, a Friday, Increase as usual copied reports and estimates
and then took them to the canal office. While in Louisville, he bought
some paper, a hair comb, and a toothbrush for 68¾ cents. On Saturday
he went back into the woods in search of the wandering cow, but he did
not find her. It would be weeks before he found that cow—then sold her
for seven dollars.

During his first month on his own Increase tried to keep track of his
money. With his salary from Mr. Henry, Increase paid Victor Audubon
for a boat he must have committed to buy before he knew the family was
leaving. By June 27 he had paid for letter paper, bread, soda powder, a short
coat, three shirts, his comb and toothbrush, a new pair of shoes and repair
of his old shoes, "turlled" (toilet) paper, a journal, and postage on three letters. On June 27 when he added up his expenses, they came to about sixty

cents more than his salary. Including the money he gave his father, Increase had spent $43.84¼ from June 7 to June 27.[2] But the next day he received twenty-one dollars from the canal board. He spent two dollars for a book on minerals, five dollars for room and board, and, on the Fourth of July, forty cents for admittance to a museum in Louisville that had on display an Egyptian mummy and "other curiosities and paintings."[3] Though Increase spent money on clothing, books, "ferriage" to cross the river, postage for sending and receiving letters and the journals he subscribed to, he almost always had money at hand. Later in his life, Increase Lapham's financial situation was more mysterious—it's not always clear what he lived on— but in his early life we see him being careful to maintain some savings and careful to limit his expenses to the necessities of his body and of his mind.

After his family had been gone ten days, Increase seemed to have not only adjusted to living on his own, but in a letter to Darius he was celebrating it—though there is an undertone of what-do-I-need-you-for. Living at Detiste's Inn, he was, he said, "as happy as a priest, nothing and no one (with the exception of Mr. H) to trouble me! We have all sorts of things to eat set on the table and we have nothing to do but to choose which we will eat and it is brought to us; I go to bed when I choose and get up when I choose."[4]

Increase had a new entertainment—going to church with Mr. Henry, Victor Audubon, and Thomas H. Taylor, the owner of a store in Shippingport who shared Increase's interests in botany and geology. The four of them didn't just go to one church; they seem to have tried them all out—Methodist, Presbyterian, and Baptist. If they found upon arrival that service was not being held in one church, they headed for another. With his parents gone, Increase was less of a Quaker and more of a bachelor. These four respectable young bachelors went to respectable places where people gathered to hear music and entertaining preaching. And they probably wanted to meet girls. Church was the only place for them to meet nice girls, and it was free, except for the occasional hire of horses and a hack. If they didn't go to church, they went down to the river to look at steamboats or they looked for fossils.[5]

In his journal, Increase paints a clear picture of one of their Sundays together. The four of them walked to Louisville to go to church, but finding there were no services they decided to go to a camp meeting about five miles south of the city. They rented a hack with four horses and drove to the meeting place in the woods at the foot of a gently rising hill. At the end of an oblong enclosure of cabins, a man surrounded by several hundred people stood at a crude pulpit preaching. Increase noted that some of the people there were listening to the preacher, while most were walking about the encampment "attending to their secular concerns"—like those four young men from Shippingport. Though Increase judged that the sermon was "plane," he thought "the solemn chanting of the musitians when heard from a distance was sublime & impressive."[6] The pleasures of being with his close and stimulating family were being replaced with the pleasures of being a young man exploring the world with friends.

Though Increase complained about Mr. Henry's ignorance to Darius and Seneca, he remained on good terms with his boss. Outside of work, they spent a great deal of time together. This relationship is indicative of Increase Lapham's lifelong ability to get along with people no matter what he privately thought of their competence. He could be critical of a person's professional work yet maintain a friendship, generously seeing the man in more than one aspect. Though intense in his interests and activities, Increase seems to have been an easy person to be with—nonjudgmental, funny, easygoing.

When Increase Lapham was on his own, he worked on the Louisville and Portland Canal much more consistently than he had when the family was there. After the family left, there was no mention of going to school during this year and he stayed in his room reading much less often. He was a working man among other working men, not a boy among his family.

Soon after the family left, Judge Bates, a trustworthy friend of the family and easy man to work for, resigned his appointment as chief engineer on the canal because of his differences with the directors of the canal. The directors had wanted to declare the contract with the contractors null and void because of problems with the building of the canal. They had asked Bates, their engineer, to look into this. But after careful study Bates agreed

with the contractors that many of the problems had come from the directors and the direction the contractors had been given—some of them his
own directions![7]

A chronic shortage of workers on the Louisville and Portland Canal was
caused by the increasing competition for canal workers over an increasing
number of projects throughout the Northwest Territory, as well as the
reluctance of workers to live in such an unhealthy place. Other difficulties with work on the canal were the weather and floodwaters. Louisville
summers were hot and humid. Louisville winters were cold and damp. It
rained frequently in hot weather and cold, multiplying mud and flood-
water and mosquitoes. High water slowed the excavation of the rock in
the bed of the canal. Large quantities of driftwood or floating ice filled
the canal excavation during floods. Expensive cofferdams to keep out the
water and driftwood failed.[8]

When Bates resigned, the incompetent Mr. Henry was given Bates's
very big job to finalize canal designs, supervise workers, and make sure
the contractors built what they had contracted to build. For the next year,
Increase worked in Mr. Henry's office in Shippingport and the canal office
in Louisville and more than 140 days outside on the canal.

Increase's work on the canal was both skilled and unskilled. Sometimes
he was sent to Louisville to fetch things—a level, an auger, a measuring
tape, a surveyor's compass. Sometimes he just counted the number of
laborers at work or got lists of the laborers' hours from the contractors or
measured stone. Most often when he was on the canal his work involved
surveying. Because of difficulty purchasing from a Mr. Rowan a parcel
of land on the proposed new route, Mr. Henry put an unplanned curve
in the canal. Increase had to help Mr. Henry take a survey of this new
curve, take levels, set out stakes, and take the height of the rock above the
new angle. He also helped survey the falls to find the best situation for
the bridge. Taking levels, Increase had to wade across the river while Mr.
Henry was in a canoe. Surveying the shore of the river to Goose Island,
Increase waded in the river; Mr. Henry was in a skiff.

Increase's increasing professionalism about his work is illustrated by his
beginning at this time to keep a field book like Darius's—he calls it a book
of "memorandums"—in which he copies information on more than 130

topics that will be useful to him at work on canals. Some of these entries are excerpts from scientific journals, some are from letters from Darius, some are from official canal documents, and some are of his own devising. He carefully indexed the book so he could find things again. Here are the subjects listed under "S" in his index: "Screw, endless, for pulling down trees; Single or Palmer's rail road; Specifications in relation to the Erie & O[hio] canals; Spoil banks; Slope of the banks; Stumps, machine for rising; Steam boats, measurement of."[9] This seventeen-year-old knew how to make an index. To do that he had to think about what information would be useful to him in the future. He had to *imagine* his future. As he did all of his life, Increase Lapham collected information, classified it, and stored it in such a way that he could retrieve it and use it and recombine it with further information—processes basic to the gaining of knowledge and doing science.

During the rest of Increase Lapham's time in Shippingport, besides his work on the canal, he continued collecting shells and minerals. And he shared his interests with prominent men in these and other fields. He continued to correspond and share specimens with Benjamin Silliman at the *American Journal of Science and Arts*. He wrote to and sent shells to Dr. Samuel Prescott Hildreth in Marietta, Ohio, a frequent contributor to *Silliman's Journal*. He dined occasionally with the educator and historian Mann Butler and discussed, no doubt, *Silliman's Journal*, which they both subscribed to. Increase exchanged specimens of freshwater mollusks with the prominent Philadelphia publisher and conchologist Isaac Lea. He exchanged botanical and geological specimens with William Darlington of the West Chester, Pennsylvania, Cabinet of Natural Science.[10]

And Increase read. He read Henry McMurtrie's *Sketches of Louisville and Its Environs* and went across the river to look for the petrified beech log that McMurtrie described. The scope of McMurtrie's *Sketches*—geography, geology, history, flora, fauna, fish, weather, manufacturing, buildings, navigation, commerce—may have influenced the scope of Increase's later book about Wisconsin. He borrowed from a canal director the *Transactions and Collections of the American Antiquarian Society*—the journal of an organization that in the 1850s would be important in backing some of Increase's most influential work in Wisconsin. He bought Major Stephen H. Long's

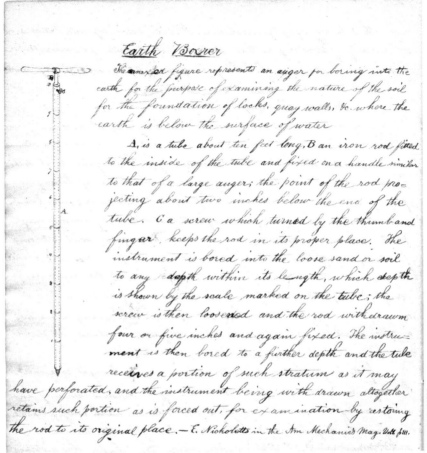

A page from Increase's book of memorandums

An Account of an Expedition from Pittsburgh to the Rocky Mountains, Performed in the Years 1819 and 20 and would soon correspond with Thomas Say, who wrote about the fauna on Long's expedition. He devoured Henry Rowe Schoolcraft's *Travels in the Central Portions of the Mississippi Valley*. Schoolcraft, essentially a geologist, also wrote in detail about the Native Americans, perhaps influencing Lapham's 1855 book on the "antiquities" of the Native Americans of Wisconsin. He borrowed a book of the New York canal laws and copied extracts into his book of "memorandums." He also copied parts of Thomas Tredgold's *Practical Treatise on Railroads and Carriages* into his book of memoranda. Earlier than his older brother

Darius and many other canal engineers, young Increase saw the coming importance of the railroads. When Darius sent him his plan for a book he was writing for civil engineers, Increase suggested that it include railroads: "within ten years you may make yourself acquainted with the practical opportunities of their construction, perhaps Ohio will make a railroad from Dayton to Lake Erie."[11]

The famous Erie Canal engineer, Mr. Canvass White, working on the Louisville and Portland Canal, gave Increase a fine drawing pencil and John Lee Comstock's 1828 *History of the Greek Revolution*, which Increase read right away. The chairman of the board of directors of the canal company loaned to Increase Parker Cleaveland's *Elementary Treatise on Mineralogy and Geology*—the first book on mineralogy and geology to be published in America. Dr. Asahel Clapp, a physician and geologist who lived in nearby New Albany, Indiana, loaned him Frenchman Georges Cuvier's *Essay on the Theory of the Earth*, one of the most influential works on geology of the time, as well as the Scotsman Niel Arnott's *Elements of Physics, or Natural Philosophy*. Dr. Fitch of Louisville loaned him a copy of the *Transylvania Journal of Medicine*, which contained the first of botanist Dr. Charles W. Short's articles on the flora of the Lexington area. Short and Lapham would become correspondents and traders of botanical specimens for many years. Storekeeper and friend Thomas Taylor showed Increase his copy of Isaac Lea's book on the new species of freshwater bivalves, a beautiful book with colored plates. Taylor apparently wouldn't let Increase take the valuable book home.

Though the teenaged Increase did not have the opportunity for a higher education, he had the attention and respect of the more learned and schooled men around him. He was not grudgingly helped or only tolerated; these men went out of their way to be a part of the education of Increase Lapham—a genial young man with talents and abilities obvious to his elders.

—|—

From the time Increase Lapham's family left Shippingport in June of 1828 to October of 1829 when Increase himself left Shippingport, Increase and Darius exchanged at least thirty letters. And most of those letters still exist, at least in part. It's an amazing exchange of ideas between the seventeen- and eighteen-year-old Increase and his older brother, an assistant engineer, twenty to twenty-one years old.

One particularly interesting exchange is a series of letters in which Darius and Increase share observations and speculations about the geology of the Ohio River valley. Were the rocks they saw in the Ohio Valley the same as those Amos Eaton had described along the Erie Canal?

Increase and Darius had both read Eaton's *A Geological and Agricultural Survey of the District Adjoining the Erie Canal, in the State of New York.*[12] Eaton's important work, the first to show transition and secondary strata, or layers of rock, in relation to each other, had probably been well thumbed in the Lapham house since Lockport days on the Erie Canal. In this geological survey Eaton incorporated German and British nomenclature for rocks when he could, but for American rocks new to science, he invented new names. Eaton listed the types of rocks found in four classes of strata, as well as their mineral and sometimes fossil content, and then described them as he found them along the canal route in such a way that it would be useful not only to geologists but also to miners and engineers. The 1820s in America are sometimes called the Eatonian Era of geology.[13]

In a section of his book about graywacke, a fine-grained sandstone containing particles of rock, quartz, and feldspar, which underlay parts of the Erie Canal, Eaton raises several questions for others to study. He asked, "At the northern termination of the greywacke, . . . does the greywacke divide into three great layers? . . . Do the geodiferous [containing geodes] and cornitiferous [Eaton's term for a kind of limestone] lime rocks run in between the middle and upper layers?"[14] The curiosity of the Lapham brothers was piqued by these and similar queries, so that when they had a chance they looked for the graywacke, the geodiferous, and the corniferous[15] along the Ohio River and on their travels.

Four days after the family left Kentucky that June of 1828, Darius wrote to Increase describing the layers of rock he had seen as he sat on top of a stagecoach on the one-hundred-mile trip northeast to Cincinnati. Darius wrote that just past Utica, Ohio, is a sixty- to eighty-foot limestone ledge. One layer of rock he saw there abounded in geodes. "I am suspicious," Darius wrote to "Inc," that "it is the geodiferous lime rock of Eaton."[16]

He suggested that Increase go to the little town of Utica not far above Louisville to trace the stratum downriver back to Jeffersonville. He trusted Increase's eye to look for what he didn't have time to look for—the geodiferous and below that the calciferous.[17]

Increase received this tantalizing letter "replete with useful and interesting information" five days later. That day it rained in Louisville and Increase went back to drawing a plan of a wood gate, complaining in his journal that "neither Mr. Henry or Judge Bates will draw a single line on paper if they can help it; but merely look on me while I am drawing and give directions &c how it must be and how I must draw it. Under these circumstances who could make a good picture?"[18]

The next day he wrote to Darius telling him that he had gone to a nearby quarry to look at the limestone above the sandstone and found a great deal of loose hornstone. He was "forcibly impressed" with the idea that the hornstone, which was reddish and not as gray as that quarried from the canal, was "itself the corniferous lime rock of Eaton." If that were so, "the sandstone would then be merely the sandy variety of the geodiferous limestone of Eaton and the lime rock which forms the rapids would, as you suggested in your letter, be the calciferous slate of Eaton."[19]

Darius wrote to Increase on June 23 telling of his two-day, 120-mile journey from Cincinnati northeast to Chillicothe, Ohio. He had a lot to say because, he tells Increase, "In traveling by stage I always ride with the driver when I can, where I have the pleasure of viewing the country through which I pass."[20] On this midsummer June day, after they came to the valley of the Little Miami River and forded the river into the beautiful bottomland, the stage left the flats and went up the valley of a small clear brook to what was called the High Country. All the while, Darius, bounced around and holding on, was taking notes in his journal. He was looking in vain for a clear view of the rocks, not covered with vegetation or rubble at the bottom of a hill.[21]

When the stage came to the banks of the east fork of the Little Miami, Darius finally saw what he was looking for—"the sight of the rock strata *in situ*."[22] He got down from the stage and saw that the east bank of the creek was "clay slate with an occasional stratum of compact lime rock six or eight inches thick interspersed." Darius was so engrossed in the olive green slate and the lime rock glistening with reddish brown facets that he didn't notice that the stage went off without him. He had to find a horse so he could catch up to the stage.

From the top of the stage again, at the mouth of Paint Creek, Darius saw that the flimsy bridge they were crossing rested on what he thought might be Eaton's corniferous lime rock. Below Bainbridge, he noted two conical ancient mounds and heard from a gentleman he was riding with that there were more, but the stage rumbled on. In a road cut through a hill, before they made it to their destination in Chillicothe, Darius saw "the third Graywacke, the variety which is called *Pyritous Shale*."[23]

Here is the conclusion he comes to from these days of travel and geological observations. He told Increase: "Your rock at Shippingport is the Calciferous of Eaton as all rocks above present themselves in going from Cincinnati to Chillicothe. . . . The juncture of the corniferous and the 3rd Graywacke is on the summit ridge between the little Miami and the Scioto."[24]

What Darius said next *must* have raised a bit of envy in Increase, who was stuck in Kentucky drawing plans for Mr. Henry, though he doesn't say anything to that effect in a letter or journals. Darius wrote excitedly that "[w]e have now got into another country, things are materially changed, different rocks, different plants and people present themselves."[25]

This geological profile of the gradations from Utica and Knob Creek illustrates Increase's ability to systematize and synthesize geological observations and measurements.

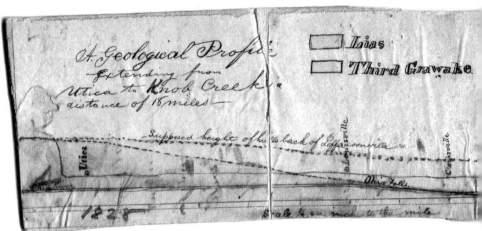

Darius wrote again on July 2 from Chillicothe to affirm Increase's observations that the quarry rock was corniferous and the sandstone sandy geodiferous, and to urge Increase to examine the strata near Utica. He also summarized what they knew so far and suggested that if they should "collect from authentic books and from observation all the facts relating to the geology of the country between Pittsburg and St. Louis...we would soon get facts sufficient to make out a geological profile the whole distance, which would be a continuation of Eatons."[26]

The last letter in this series of geological letters is written by Increase. In his journal on Sunday, July 6, 1829, he wrote that ever since he got Darius's first letter from Cincinnati on June 21, he had been "wishing for an opportunity to go to Utica and to examine the rock which presents itself in the hills a little above that village."[27] So on that Sunday morning he borrowed a horse from one of the contractors and, taking a ferry across the river, set out for Utica on the Indiana side of the Ohio. His horse, he wrote in his journal entry of that day,

is one which is what is commonly called "hard, or hard to ride," that is, he jolts a person very much when on the trot. Not being much used to riding horses, I at first found it necessary to hold on by the saddle to prevent myself from being precipitated entirely off from the horse! However I soon learned to set on without much fear of that but then

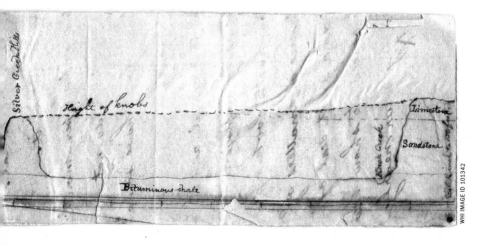

the jolts must of course ware a sore on the parts most exposed to the saddle.[28]

The language in Increase's journal is usually more informal than in his letters to Darius. Darius's more formal and sometimes pretentious language seems to set the tone for their correspondence. However, Increase was not above making fun of Darius's language as he did when he wrote to his brother on July 8: "A few days ago I proceeded up the river to Utica and attentively examined the limestone ledge or 'mural precipice' [quoting a fancy phrase from Darius's recent letter] which presents itself along the river bank."[29]

Increase climbed up the hill on gravelly debris and then climbed from shelf to shelf to the top of the precipice. Walking along the top he came to the quarry Darius had mentioned. He described the rocks he found and concluded that he "can say with the most perfect confidence that the precipice above Utica is composed of the *same general stratum of rock as seen at the Falls of the Ohio.*" And here is his evidence: "The appearance of the rocks when fresh fractured is exactly the same; the same variety of hornstone occurs; the petrifactions are substantially the same kind as those found on Goose Island and the opposite shore in Indiana at the falls."[30] Increase was a little more conservative as a scientist than Darius. He wrote in this letter that he "collected no facts on this excursion which tend either to verify or disprove the idea of calling this general stratum the . . . Calciferous slate of Eaton." And he suggested that Darius test one of the rocks he had labeled. If Darius saved a specimen of the selenite he found there, "I wish you would pour a little acid on it. I presume it will effervesce and thus prove to be *calc.* spar." But he agreed with Darius and supposed that "the geodiferous and corniferous lime rock are wanting in all this region and that the 3rd Graywacke lays immediately on [the calciferous]."[31]

Then, changing the subject of their correspondence for the next few months, Increase closed this letter with expression of his desire that Darius and his father get him a situation on the Ohio Canal. He wanted to get out of Shippingport. He'll tell him why in the next letter.

CHAPTER 6

Until a Better Opportunity Presents

──────────────┤ 1828–1829 ├──────────────

I N SPITE OF THE PLEASURES OF BACHELORHOOD, churchgoing, geology, and "cucumbers for breakfast," Darius's letters from "another country" to Increase, who was doing the same boring work on the canal day in and day out, were having an effect.[1] In the next series of letters in July of 1828—two to Darius and one to their father—Increase asked them to get him a job near them as a rodman on the Ohio and Erie Canal. But he prefaced his four carefully stated reasons for going to Ohio by saying that "you must not think that I am unwilling to remain here any longer; I will remain if I can not do as well or better in Ohio."[2]

Increase's first reason was homesickness, but he exaggerated, saying that there were few people in Shippingport with whom he could associate, so he must spend his leisure hours "in lonesome reflection." He told Darius, "As you know what it is to be homesick I will not pretend to describe it." His second reason was that he wasn't being paid enough money; he wanted thirty dollars a month and was paid only twenty. Third, he said that since Judge Bates had recently given up his responsibility on this canal he must "depend on the president and the directors for his wages." He rightly trusted them less than he trusted their old family friend. The fourth reason was his having to work for Mr. Henry: "I do not like to be under one who *has* to look *up* to me for council and advice and to consult me on matters of importance. A great many little facts might be adduced to show how disgusting such a situation is, for they occur nearly every day."[3] He didn't mention wanting to explore the rocks, plants, fossils, and shells of a "new country."[4] Increase did, however, go on to mention in this letter that he saw another proof of the existence of the third graywacke in Schoolcraft's *Travels*.[5] And in his next letter to Darius, before he got an answer to his Ohio job query, Increase included a long quotation from the latest issue of *Silliman's Journal* on the erratic boulders of Ohio; he then discussed and rejected the idea that they were brought and dropped there by an immense lake.[6] He went on with his life, not pressing his father and brother too hard to get him work in Ohio.

Increase's job did put him in an untenable position. He was right to want to get out of there. Though finally appointed to be Mr. Henry's assistant engineer, he still received a rodman's pay. And because of the ignorance of Mr. Henry, Increase did much of the actual engineering work, though occasionally Canvass White would swoop in from Philadelphia and give technical advice and orders. White had been elected director of the canal company by the stockholders, who hoped that his expertise as a canal engineer could rescue the struggling operation from its construction problems.[7] Increase's position on the Louisville and Portland Canal in Shippingport is the definition of a frustrating job—irregularly receiving the pay of a subordinate for doing the work of an unqualified boss.

Darius responded first to Increase's well-phrased plea to get him out of there. On August 4, he wrote "Increase Allen Laphamicus" that his reasons for wanting to leave his work on the Ohio Falls "are not very weighty except the 4th"—having to work for a man who knows less about the job than he does. "To be sure," Darius wrote, "this is a sore grievance, but if there is no place for you in Ohio you will by a little practice become accustomed to it and it will not appear so 'disgusting.'" Then Darius referred to Increase's own humorous cure for homesickness sent about a year earlier to Darius. "As to the homesick part you know how to prescribe for that, and I think the prescription you sent me, with a little alteration for age, circumstances, habits &c. would probably apply to your case."[8]

That same week Seneca wrote a rare letter to Increase telling his "Dear Son" in typical Quaker fashion that "[t]hy request that I should endeavor to get thee employed in Ohio has not been neglected by me. I have made some enquiries on the subject." And he told him that he had better remain that season at least at the falls. "Thee must recollect," Seneca added, "that perhaps not one person in a hundred thousand, of thy age and size gets $30.00 per month." Seneca was sympathetic; he did not want Increase to stay in Shippingport "against [his] inclination," but he was also stern and the money Increase made was hard to come by. He told his son that "if thee can stay and (with a little exertion) be contented, this will be best until a better opportunity presents." He closed by saying firmly that "[t]hee had better labor under a slight paroxysm of home sickness than to relinquish such a situation as thee occupies for an uncertain one for so small a reason as thee assigns. Thee can not expect to be an engineer without being absent from all thy near relatives."9

Now all three of these male Laphams have expressed two basic tenets of their family (and perhaps Quaker) culture—that personal feelings do not outweigh personal responsibility and responsibility to the family, and that with some "exertion" they can make themselves *feel* differently; they can be contented in difficult circumstances. The heart is subject to the powers of the mind. You feel what you have to feel to get your work done. Increase had articulated his frustra-

A pastel drawing of Seneca Lapham, the father of Increase and Darius

tion, but family culture and family need would not allow him to flee such a difficult situation. After the letter from his father, he seemed, for a little while, to settle in to stay in Kentucky with his work and with excursions on Sunday with Mr. Henry, Mr. Taylor, and Victor Audubon.

But at the end of August he wrote his father again to say that Seneca had misunderstood him; Increase wasn't *unwilling* to stay in Shippingport. Though everything in the tone and intent of those three earlier letters seemed to say *Get me out of here*, it's true that he never actually said that he wouldn't stay in Shippingport working for Mr. Henry. Yet

WHI IMAGE ID 102183

in this letter to his father, he gave another, fifth reason for wanting to leave. It's an interesting letter showing the narrow path he was trying to walk—with filial gratitude on one hand and homesickness backed by professionalism and a subtle new argument for leaving on the other.

Increase very diplomatically wrote to acknowledge to Seneca his "obligation for the favor you have conferred on me by endeavoring to get employment for me on the Ohio canal." Yet "there is very little business done here which occurs on common canals and what little there is, is not by any means done in the usual manner, so that instead of learning anything here I am getting accustomed to an awkward and irregular manner of conducting business." He was right about that. Soon the dissatisfactions with Mr. Henry and some of the work on the Louisville and Portland Canal would erupt into a lawsuit that Increase would be involved in as a witness. Then Increase wrote his father that he did *not* say in his letter that he was "really homesick, although I am 300 miles from home, while at my boarding house." A bit of a guilt trip here on the old man. Finally, in an unconvincing way he dismissed his need for his family, intending to cause a little hurt: "I can by walking one half mile on the canal be as much at home as ever."[10]

He was a homesick boy, still angry at his family for leaving him behind, though he couldn't admit it. He was also becoming a much more responsible canal engineer. And this last reason to leave the Louisville and Portland Canal did the trick. Both Darius and Seneca were convinced. On September 1, Darius wrote to "Dear Ink" to tell him that he and his father agree that his observations on the way business was conducted on the canal were very rational. They "perfectly agree" with his wanting to leave and will make every effort to get Increase a situation that will be "more beneficial both in a pecuniary and intellectual point of view."[11] The intellectual respect of his brother and father must have been very gratifying to "Dear Ink."

Although a new situation did not immediately present itself, Increase got a break from Shippingport in November 1828 to spend a month visiting his family on the Ohio canals.[12] On the morning of November 2, he paid six dollars for a ticket and boarded the steamboat *Talisman* heading northeast up the Ohio for Cincinnati.

From the deck, after a delay caused by fog, Increase was able to observe the height of the hills and the "mural precipices" along the river. In Cincinnati on the evening of November 3, Increase met up with his sixteen-year-old brother Pazzi in a boardinghouse where they "took supper" and then carried Increase's trunk and keg of specimens to the office of the *Western Tiller*, the newspaper where Pazzi worked as a printer—and where Increase saved the cost of a room and slept on the floor.[13] The following day, Increase took some specimens to the curator of the Western Museum, scientist Joseph Dorfeuille, with whom he exchanged several minerals and fossils, just as Darius had done in June when he passed through Cincinnati. After his exchange with Dorfeuille, Increase examined the "curiosities" of the museum.[14] Besides minerals, he must have looked at the fossil bones of the great "Jefferson's ground sloth," found with the bones of other massive extinct mammals at nearby Big Bone Lick in Kentucky.[15] He probably also saw the lurid wax figures that Dorfeuille, desperate for paying visitors, had got up—a "Human Monster and Cannibal" and a representation also in wax of the death of George Washington.[16] Then Increase took a long walk along the bank of the Ohio River up to the reservoir where he found clay slate between layers of lime rock and some new species of shells.[17]

The next day he left Pazzi and boarded a stage for Chillicothe where he traced the route Darius had taken some months earlier. The morning of November 7 Increase met up with Darius in Chillicothe. The brothers had breakfast together and then headed off to the quarry where Darius was working. There Increase saw the third graywacke of Eaton. He stayed at the quarry with Darius all day.[18] The following day Seneca met the brothers with horses, and the three of them rode ten miles north to Yellow Bud Creek on the Ohio Canal, where the family worked and lived.

The three elder males of the Lapham family arrived in the evening and found that several of the younger members of the family were "shaken with the ague and fever."[19] The family members living in Yellow Bud at that time were Seneca and Rachel and their children Mary, thirteen; William, eleven; Hannah, five; Lorana, three; and baby Amelia, ten months old. The next day, a Sunday, Increase wrote that he stayed at home all day; all of his family were there except Pazzi.

Monday morning Darius went back to Chillicothe to work. During the time with his family, Increase continued exploring and collecting along a little creek, taking along his little sisters and brother. He helped his brother William make a grindstone. And at a table he had made, Increase copied information into his book of memoranda and wrote in his journal with his family around him. He also worked a few days with his father on the canal in Yellow Bud and copied a plan for the Yellow Bud aqueduct.[20]

When Darius could get away from work, the brothers went off to look at some pudding stone in a quarry and explored an ancient tumulus at Circleville. And one morning, while helping his mother in the kitchen, Increase cut open a turnip and "discovered a cavity in which was a young sprout of this plant bearing an exact resemblance to this plant when it first makes its appearance above the ground, in the spring."[21]

On a wet and rainy Thursday in late November, Darius took Increase and his trunk in a one-horse wagon to Chillicothe where he got a stage for Cincinnati, then took the steamboat *Shepherdess*—a short and fast trip this time—downstream on the flooded Ohio back to Shippingport.[22] Though Increase Lapham must have thought that any day he would be leaving Shippingport to work near his family on the Ohio Canal, he was to remain in Shippingport for almost a year.

—| |—

That December after his trip to see his family, Increase continued looking for shells and working on the canal. He received a copy of Darius's article on the Welland Canal in the *Ohio State Journal*. On Christmas Day he accompanied some of the most "respectable" citizens of Shippingport, and "six or seven of the most experienced hounds over to Gravel island, for the purpose of hunting a fox, which was said to be on there; but in this we were, as might be expected where foxes are so scarce, disapointed."[23] That afternoon he went to an exhibition of a solar microscope and saw for the first time "animalcula," or microorganisms that, greatly magnified, resembled eels.

The next day Mr. Canvass White "and Lady" arrived. At the beginning of January 1829, Increase worked closely with chief engineer White for several weeks, locating and marking a new line of the canal. Increase ordered and had shipped to his brother and father on the Ohio Canal a leveling instrument and three barrels of water lime to be used on the Yellow Bud aqueduct. He also inquired about buying his own copy of Thomas Say's *American Conchology*. When White left for New Orleans on January 19, he gave Increase "a fine pencil for drawing."[24] Andrew Jackson passed through Louisville on his way to becoming president of the United States, Increase making little note of that.[25]

Increase made only three journal entries in February 1829, mostly about letters he wrote and received and money he was paid and his friendship with ladies. That February, "a lady acquaintance" gave him a tiger cowry shell, which he described and drew. He was introduced to Mrs. Hannah Le Baum Goodwin, the wife of canal engineer Goodwin, who gave him several small seashells. The ladies of Louisville seemed to be taking an interest in this attractive young man, who may have seemed to them lonesome.[26]

He wrote that February to Benjamin Silliman at Yale, renewing his subscription to the *American Journal of Science and Arts* and telling him that on his travels he "only discovered that the lime rock (which I think must be the *Lias* [graywacke] of Prof. Eaton)...continues to be the only rock found *in situ* until we arrive within ten miles of Chillecothe." This is news but he wanted to discover more; he wanted to be able to report more interesting things to Silliman. Instead he apologizes to Silliman for the quality of specimens he had earlier sent him. He confessed that when he took a keg of mineral specimens to Dorfeuille at the Western Museum, Dorfeuille induced him to throw most of them away because they were common, not valuable at all. He was vexed at himself for sending Silliman specimens that were "quite as bad."[27] But he tells Silliman, perhaps hoping to regain what he thinks is lost credibility, that Mr. Canvass White has engaged him to collect shell specimens for himself and Philadelphia

Increase Lapham not only collected, cleaned, and traded fresh-water shells with other collectors, but drew them while he was living in Kentucky.

Cypraea tigris
or Tiger Cowry.

WHI IMAGE ID 99701

scientific institutions. Increase seems depressed in February of 1829. Or bored.

But a few days before his eighteenth birthday, he took a big and independent step—this time without consulting either Darius or Seneca. On March 3, White had arrived back in Louisville from a trip to New Orleans. On March 5, Increase drafted a letter to him revealing a long-standing desire and unspoken dream of Increase's. On March 6, he delivered the letter. On March 7, Increase was eighteen years old. Here is his letter:

> Respected friend,
> Deeply impressed with a sense of the high importance of a thorough course of collegiate education, and knowing that I have now arrived at the age when it must be commenced, if ever; I have for some time past been reflecting on the subject, and endeavoring to form a plan by which I might obtain so desirable an object. But various difficulties present themselves resulting from one source only—the want of sufficient means to accomplish it. My parents are not in such circumstances as to allow my calling on them, and to wait until I could earn it myself would cause too great a delay and perhaps blast my prospects, forever.
>
> The strong desire I have for accomplishing this object is my only apology for addressing you on the subject. I want the means to do it; and if you through your own pecuniary resources or influence in society will obtain for me the amount of money necessary for a collegiate education you will have conferred on me a favour equal to my most sanguine expectations, and which (if life and health are the allotments of heaven) shall be cancelled to the last farthing.
>
> I am Sir, with sentiments of Respect & esteem your most obedient humble Servant, Increase A. Lapham[28]

It's a good piece of writing—just the right amount of passion, confidence, determination, and deference to White. Delay will "blast" his prospects forever. He will pay the money back "to the last farthing." Who wouldn't lend this kid the money? Increase sent a copy of the letter to Benjamin Silliman at Yale—but not to his family—not yet.

On March 9, White visited Increase in Shippingport and "took one suit of shells which [Increase] had collected for him consisting of 30 varieties of the genus Unio." A good follow-up to the letter. Then White told Increase that he wanted him to remain with Mr. Henry on the canal this season while he tried to procure for him "a situation at Yale College: being acquainted with Silliman and others of the Professors." He added that he would be in New Haven before he came back to Kentucky in June—only a few months away—and while in New Haven he would "make such arrangements as are necessary." He would let Increase know about them when he came back in June. The next day White left for Pennsylvania.[29]

Through the rest of that cold and snowy March in Shippingport, Increase worked on the canal setting stakes at the upper end of the basin and working on plans of locks. He was trying new things. He began to play chess. He read the poetry of Alexander Pope. He went to a lecture on colonizing the free black people of the United States on the west coast of Africa. He continued to collect shells. He looked in vain in Louisville for a job for his printer brother Pazzi. By the end of the month the fields began to look green, the elm was in flower, the martins had come back.[30]

On April 1, he wrote to his father telling him of the step he had taken to get a university education. Different in tone from the letter he wrote White, there are tugs on his father's guilt feelings for leaving him behind in Shippingport and a bit of schmaltz.

> Dear Father,—
> Perhaps you are aware that it has long been a favorite project with me to procure a collegiate education.
> Many have been the methods which have occurred to me for ac-complishing it and almost as many have been found deficient and therefore abandoned before they had passed the limits of my own imagination.
> Many of the lonely hours which I have seen since the family left here, in June last, have been spent on this subject, and in the many lonely walks which I have taken between this place and Louisville has my mind been occupied in the same way. And I have at last hit upon a plan which I think is likely to succeed. This plan is to procure the funds from some friend, if I can find one both able and willing, and to return such funds to him after receiving my education....

I am aware that you will lose the profits of my wages while at college, but can not this be repaid? Or can it not be arranged in some way so as not to prevent my going if Mr. White should succeed in getting me in at Yale?

Please let me hear from you soon.

Your affectionate son,

Increase A. Lapham[31]

It would be a month before Increase heard from his father in response to this letter. During April Increase worked on the canal as usual and corresponded with Benjamin Silliman about their exchange of mineral and shell specimens. He dined with his former schoolteacher, Mann Butler. He looked for flowering plants in the woods. Entering a public competition, he and Mr. Goodwin collaborated on a plan to design the Louisville Exchange and Assembly Room, drawing several plans and elevations. He measured a load of cypress timbers for the lock gates. The Ohio flooded after terrific thunderstorms. He wrote in his journal every day of that rainy April.[32]

On May 1, he received a letter dated April 22 from Darius in Yellow Bud, Ohio. "I am authorized by [Father] to say that if you succeed in obtaining funds to defray the expenses of a collegiate education, on the conditions you mentioned, his consent for your going to college will not be wanting. Your letter to Mr. White was excellent."[33] A major obstacle to Increase's college education was out of the way. On those "lonely walks"

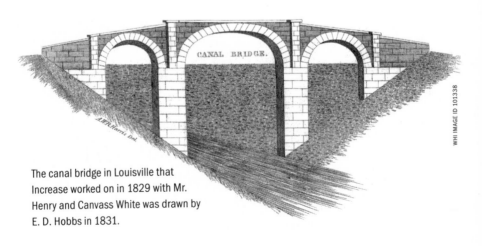

CANAL BRIDGE.

The canal bridge in Louisville that Increase worked on in 1829 with Mr. Henry and Canvass White was drawn by E. D. Hobbs in 1831.

WHI IMAGE ID 101338

between Shippingport and Louisville, he must have begun to think about college, about going to Yale. What would it be like? Could he really do it?

⊣ | �

Increase worked on the canal all through the "boisterous" weather of May 1829. He enjoyed the "curious & delightful" echoes of a flute played within the completed, empty guard lock. He did a side job, leveling a site for a summer house for Joseph Detiste under a big elm called Hall's Tree after Captain Basil Hall.[34] On Sundays Increase continued to attend various churches with his friends and he went for walks with them, continuing to collect plants, shells, fossils, and minerals. Increase presented a plan of the canal locks and bridge and a map of the Falls of the Ohio to Napoleon Buford, a West Point graduate and teacher there, who told Increase that if he wished to go to the West Point Academy, he would give him "any assistance in his power."[35] Increase was clearly college material. The canal board authorized the lock gates and a swing bridge across the guard lock to be built based on plans altered by Mr. Henry, drawn by Increase, and approved by Canvass White.[36]

The last part of May was marked by accidents on the canal and fights in Shippingport. On May 21, Mr. Peter Carney, a masonry contractor on the canal and a friend of Increase's, fell onto rocks from a platform and was badly injured. That same day while Increase was on the canal, another man on a crane jumped down, grabbing onto a rope he thought would lower him down, but it was not attached to anything. He was not seriously injured. Nor was another man who that same afternoon smashed his fingers between large stones. When he got back to Shippingport that day, Increase found that a man who had recently murdered a black man and received a suspended sentence, Jeremiah Phipps, had fought and seriously injured another man. Phipps was released again. The next day there were three fights in the streets of Shippingport but as Increase noted in his journal, "[n]o notice is taken of such occurrences by the Civil authorities."[37] On a night of heavy rain, Increase, Mr. Henry, and Mr. Taylor sat with the very ill Mr. Carney drinking "lemon ade."[38] The next day Increase was sick with a hangover—a rare occurrence for the good Quaker. Though the masonry work on the bridge was proceeding smoothly, the canal work was suffering from a lack of workers. In spite of the hot weather at the end of May, Mr. Carney was improving. Increase still had not heard from White, but Mr. Henry got a letter from him stating that he would probably leave for Louisville on May 26.

But on June 9, Mr. Henry heard that Canvass White wouldn't start for Louisville until June 20. All through the cool, rainy, mosquito-ridden June, Increase worked on the canal, getting news from those around him of canals to be built on the Wabash River in Indiana and in Illinois. On the Louisville and Portland Canal, the contractors continued to have difficulty getting enough workers. The directors had made changes in planned routes and there were complaints of a contractor's "lack of energy."[39] Increase complained in his journal that the "directors of this canal make their engineer also a clerk."[40] He noted on June 12 that his family had been gone for a year. Still he went out to the woods looking for shells and fossils. He found one day an interesting lizard "with feet like climbing birds" and "a tail of a lead blue color."[41] On June 24, he heard Henry Clay give a stump speech. The next day Clay rode across the nearly finished bridge. Increase told the story that "Capt Smyth observed to a man who has the management of the work that he saw Clay on the bridge, at which he was quite miffed not understanding the joke." Increase got the joke and later told and retold the story.[42] At the end of June he began collecting shells for Mr. White's arrival. June came and went and no Canvass White.

Finally White arrived in Louisville on July 18. On that day he came to Shippingport and told Mr. Henry—not Increase—that he had been in New Haven only two hours and therefore he had done nothing toward getting Increase a situation at Yale. Increase was told that White would be going to New Haven again soon and would then have more time to attend to getting Increase a position there. Increase made no comment in his journal but went off to the falls to look for some shells for White. Increase seems to have been good at waiting—but in the end, Canvass White never made good on his promise to Increase.

The following January, a canal engineer and family friend wrote to him saying that he would give him a one-hundred-dollar scholarship toward study at "Geneva college NY" if Mr. White doesn't assist him with a college education. Increase wrote in his journal that as he had no means to pay the other expenses while at the college, he was "compelled to decline this very generous offer."[43] Increase Lapham's chances at a college education had been very tenuous. They had slipped away. This disappointment would last all his life.

CHAPTER 7

A Man's Estate

—| 1829–1830 |—

Though he was only a boy of seventeen when he found that his plans for college were not to come to pass any time soon, Increase Lapham quietly went on with his life that July of 1829. He expressed no bitterness, though his spirits seemed low that summer. With quiet determination he would accomplish a great deal in the following year—a good plan for a family farm, a paper on geology with Darius. He would demonstrate the thinking and writing abilities of a man rather than a boy. And he would demonstrate a good man's honesty and generosity of spirit.

The day after hearing the disappointing news from Canvass White, Sunday, Increase stayed home while Mr. Henry dined with Mr. White. But Increase was back at work Monday copying a map for White and explaining to him a plan sent by Darius for fastening down lock gates in high water. White rejected the plan.[1]

All that month, Increase worked in the canal office as the plans for the basin at the upper end of the canal were made final. He worked on the lock gates in the July heat and between rain showers. He set stakes, copied estimates, and ordered castings. Increase boxed up the shells he had collected for White, visited the geologist Dr. Clapp across the river in New Albany, Indiana, and bought himself a bunch of quills to write with. At the end of this July, White left for Pennsylvania and Increase was paid ninety dollars for several months' or one season's work.[2]

On August 1, he noted that Mr. Samuel Scott, one of the iron contractors, died of "bilious fever"—probably typhus—after being ill for two weeks. The rest of the month was hot and rainy with high water frustrating the canal work. Increase's work was repetitive and his tone in the journal joyless. On two days he says he was "somewhat unwell."[3] He wrote Darius on August 5, sending along a plan of a swing or pivot bridge invented by Mr. Henry. "With this sketch and what I have formerly said concerning the bridge you will be able to discover its principle and the superiority it possesses over other bridges of this kind."[4] And he noted in the journal that Mr. Henry had taken over superintending canal hands for one of the contractors. So Mr. Henry was not totally useless.

In that same August 5 letter, Increase told Darius a little peevishly that he thought "the $25.00 which you would have sent for the back volumes of Silliman's Journal would be better employed in purchasing elementary works, these, by directing our attention exclusively to one subject would give us more perfect and general ideas of things. By reading a little of everything we learn nothing."[5] Increase is betraying his dissatisfaction with the piecemeal way they are learning about science. More often than does his older brother, Increase writes about how to think and how to learn.

On August 7 he wrote to William Darlington, president of the West Chester, Pennsylvania, Cabinet of Natural Science asking if they can exchange mineral specimens from Pennsylvania for shells from the Ohio Falls.[6] On August 11 he counted thirty-seven steamboats on the river and entered that in his journal. Then, for more than two weeks Increase did not write in his journal. At the end of August he received a letter from Darius reporting that he was now assistant engineer on the Ohio Canal (later the Hocking Canal, a feeder to the Ohio and Erie) five miles below Lancaster. Increase wrote back gracefully and probably genuinely, saying, "I am glad that you have received an appointment so near at home, and that your line presents no greater inconveniences than those you mention."[7] But it's understandable if Increase was tired of waiting more than a year for a better job, tired of being away from his family, tired of waiting to go to college, tired of the same work in Shippingport.

In September of 1829, Increase wrote only five times in his journal—brief entries noting that he wrote to Darius and to his father about a level and noting that he received letters from Darius and Dr. Darlington in West Chester, who said he would be happy to exchange specimens, including botanical specimens, with Increase. He made up a collection of shells for a Mr. Fitch. Thomas Taylor received from Isaac Lea his beautiful new book on "the new species of *Unio* and the new genus *Symphynoto*"—a "valuable work with coloured plates"—in exchange for a set of shells Taylor had sent him. Increase quickly began gathering his own collection of more than forty species with twenty-one varieties of univalves to send Isaac Lea; he wanted his own copy of that valuable book. He bought some paper and a new pair of socks. It rained and rained.[8]

Though Increase was not pleased with his own situation, the eighteen-year-old was thinking about that of his parents and his younger brothers and sisters. On August 5 he wrote to Darius that there were "a great many arguments in favor of the family settling on the Scioto river near Yellow Bud Creek, if a farm can be procured there."[9] This opens a discussion between the brothers that goes on for several months. At the end of August, Darius wrote back that he too could see a good many arguments for the family settling on a farm at Yellow Bud Creek near the Ohio Canal where Seneca worked. He wanted to hear Increase's reasons, but in the meantime he was eloquent in presenting his own. There was a neighborhood of Friends, so the family could go to Meeting. Darius would be living near there for a time and would be able to easily get home and work on the farm. It would be a place both brothers could go and be "profitably employed" between jobs on the canals. A permanent home would enable them to preserve natural history specimens and "think of forming a cabinet of minerals or any other branch of Nat. History." And, "a farm when once cleared would pay, over and above the living of the family, for the labor of cultivating it."[10]

Darius's next paragraph is a rare and impassioned expression of the difficulties the family experienced as they lived and worked on the canals:

The only sure way to supply the family in food and clothing is to cultivate the earth, if the seed is planted in its proper season and carefully nursed, in autumn we are sure to reap a rich reward for our labor,—not so with canaling, we there may labor hard for years and be defrauded of half our honest dues, as has invariably been the case with us since my remembrance. Pennsylvania, Rochester, Lockport, and Shippingport are instances, and at Yellow Bud we shall labor two years and make nothing but our living and will probably have to leave here *in a one horse cart.*[11]

Though Darius romanticizes farming, he clearly expresses long-suppressed bitterness at the way he and his father and Increase have been treated on the canals. Among the Laphams, these feelings were almost never written down and probably rarely spoken of. They felt what they had to feel to get the job done; bitterness was not thought to be productive. Darius adds that when he hears Increase's reasons for buying a farm, he will then write a letter to their parents presenting the idea to them that they buy a place near a Friends settlement at Dry Run.[12]

Increase wrote back to Darius right away telling him that Darius misunderstood: Darius thought Increase was proposing that the family *settle on a farm,* but Increase had supposed that the idea of buying a farm was agreed upon; the question was *where.* Increase wanted to make clear that he was suggesting that the farm should be in Yellow Bud, Ohio, on the Scioto River rather than three hundred miles west on the Wabash River, which must have been discussed with his family the previous November. Increase can't figure out where this "Dry Run" Darius writes about is. On the Wabash? On the Scioto? Increase wants the family to settle at Yellow Bud, but primarily he wants them to settle *somewhere* so that they can "accumulate many articles of furniture and other conveniences of life, which their present *migratory habits* do not allow them to retain." Increase chides his older brother: "You have enumerated the advantages that would be gained *by us* chiefly, if the family were on a farm and almost wholly neglected those which would result to them."[13]

In that same letter he again asks Darius to get him a situation on the Ohio Canal. "I am in fact getting tired of this canal and this place. All the small and difficult jobs about the locks, devolve upon me and my business is more like an overseer than an engineer." He says he wrote their father asking again that he might go work with Darius, but "told him at the same time I would rely entirely upon his judgement as to whether I go or stay." He says he could come in a wagon that was going to be driven north from Louisville by a "respectable blacksmith," saving money on transportation. He signs his name "Ink."[14]

— | —

On October 7, 1829, the letter came that Increase had been waiting for for more than a year. His father wrote to say he had gotten Increase a situation on the Ohio Canal as rodman. He was to leave immediately.[15] A week later, Increase had arranged his affairs and took his "final departure from Louisville, and without much regret."[16]

Traveling by stage and steamboat and having to sleep in a bed "with a Drunkard at Maderas Hotel," he met his father in Chillicothe.[17] Seneca had just taken on a job at Dry Run, which turned out to be five miles above Chillicothe. After visiting his family, on October 20 Increase went to his job site at Circleville. The site was about ten miles north of Yellow Bud, where the Lapham family lived, and five miles north of Dry Run, where his father worked—all of them along the north-south line of the lower Ohio and Erie Canal. Darius was only twenty miles away in Carroll, Ohio, northeast of Circleville at the junction of the Ohio and Erie. Increase was finally out of Shippingport. He was only a rodman, but he was near his family. He was in that "new country" he longed for. Temporarily.

At Circleville in the bluff above the canal site were new rocks to observe. Increase could walk for more than two miles along these bluffs looking at the strata "beautifully exposed" by the erosion of the Scioto River. He noted a deep blue clay containing "many fragments of a dark colored *argillite*" in the gravel. And he saw a yellow clay containing fragments of unworn limestone and sandstone.[18] This Ohio canal, like the others he had known, was a rich source for Increase's study of geology.

The Ohio and Erie Canal that now occupied all the Lapham men was an ambitious project. It would connect Lake Erie at Cleveland to the Ohio River at Portsmouth, 308 miles south, opening the rich central agricultural land to markets to the north and south. Begun in 1825 upon completion of the Erie Canal, it was built by private contractors who worked under state-employed engineers and assistants under the supervision of the Ohio Canal commissioners, one of whom was Micajah J. Williams—

The canals and waterways of Ohio, which provided work to the men of the Lapham family

soon to be important to the career of Increase Lapham.[19] Though most engineering problems had been solved by the builders of the earlier canals, the Ohio and Erie was delayed by the usual diseases, floods, and problems with contractors and laborers. The first segment, from Portage Summit to Cleveland, opened on July 4, 1827.

Increase Lapham was hired as rodman on the Ohio Canal at Circleville, where an aqueduct was being built to carry the canal over the Scioto River. Unlike the Erie Canal aqueducts, which were made of stone, those on the Ohio and Erie were built of wood and stone. The piers and abutments were stone, but the trough that carried the canal was built of wood.[20] Increase began working on the aqueduct the day he arrived, but he doesn't say what that work entailed. It must have been routine or uninteresting or he would have mentioned it in his journals and letters. Yet the geology was interesting. Right away he wrote to Darius, Dr. Clapp, Thomas Taylor, and Mr. Henry back in Louisville to tell them about the "more re-

markable geological appearances observable in this part of Ohio."[21] More
significantly, he wrote to Benjamin Silliman at Yale suggesting that he and
Darius write an article about the "geology, Mineralogy and botany" of that
part of Ohio.[22] He didn't want to just observe and think about his observations. He wanted to write about them. He was certain that what he and
Darius would observe would be worth telling other scientists about.

He and Darius, he tells Silliman, "intend to pay considerable attention
to the examination of the boulders of primitive and other rocks scattered
over the surface of the country, and we mean not to suffer the imputation
cast upon us by our neighbors of a 'disposition to indolence' to apply to
our case."[23] Increase was responding to a comment by Benjamin Tappan
in an article entitled "On the Boulders of Primitive Rocks found in Ohio,
and other western states and territories."[24] Tappan refers to several theories
that might account for the erratic boulders found across Ohio—boulders
we now know were dropped by glaciers. Tappan says that the theories
recently put forth were not "advantageous to the cause of science" because
they were not well founded in observation and evidence. The next sentence is the one that seemed to get a rise out of Increase: "So great in most
men, is the inclination to indolence, that a plausible theory is likely to be
preferred to a laborious investigation of facts, and to render the person
adopting as well as the one inventing it, satisfied and contented, with so
cheap and easy substitute for knowledge."[25] Though this criticism was of
course not aimed at Increase, he seems to have taken it personally; he in
any case knew that he did not want to be counted among the indolent; his
work would be founded on a "laborious investigation of facts."

A few days after he had written to Silliman, he wrote to Darius proposing that the two of them "resume our regular, reciprocal correspondence
which was commenced two years since for mutual improvement."[26] Darius
quickly wrote back that "with pleasure I renew the reciprocal correspondence."[27] Each of the next four letters between the brothers contained a
hypothesis for the existence of the great scattered boulders on the rolling
land of Ohio. Each hypothesis was successfully argued away by the other
brother. In response to Darius's "theory of attraction," which held that
some gravitational force—a comet perhaps—pulled the boulders across
the land to where they lay scattered, Increase said, "You ascend rather too

high in the scale of *speculative geology*, when you attempt to account for the origin of the deluge by the theory of *exterior attraction*." He continues, saying that their object should be to observe the facts that prove that a deluge once stood over this part of the country and that the primitive boulders are the result of its action; they should, he says, leave to "more distinguished" geologists the business of accounting for the origin of the deluge.[28]

Silliman wrote with good advice on December 24. He wanted a letter on erratic boulders from the Lapham brothers, but he advised them, when they looked at the boulders, to look for the following:

1st. Their size, number and variety in given places.

2nd. Any appearance of being deposited by the effect of running waters and what direction or directions.

3rd. Any scratches or furrows &c. in the rocks in place over which they may be supposed to have passed, the number, directions, and depths of the furrows, &c.

4th. Whether more abundant on the borders of lakes and rivers and in the valley than else where.[29]

Silliman himself seems to be directing Increase and his brother to look for evidence that moving water deposited the boulders in their odd and mysterious positions—the diluvian premise. He doesn't direct them to look for furrows *on* the boulders but *behind* the boulders. The real evidence for the force that moved the boulders was usually on the scratches and polishing and the shape of the boulders themselves. The Laphams and Silliman were looking for evidence of their current hypothesis, rather than letting close and fresh observations lead them to a new hypothesis. Tappan was correct in his article: "How then came these boulders in their present situation? In the present state of knowledge, this question cannot be answered. In the meantime, ignorance is preferable to error, and what is unknown may be examined."[30] Yet Tappan, a federal district judge and member of the Ohio Canal Commission who had an interest in shells and minerals, also goes on in *his* ignorance to speculate more about the origin of these boulders. It was irresistible to the curious with an interest in natural science to try to solve this mystery; they knew that if they could determine how those boulders got pushed or dropped or made in the middle of open fields they would know a great deal about the history of the earth. Yet the answer to this mystery, Louis Agassiz' theory of continental glaciation, which first was published in 1837, was still years away.

Increase and Darius did write an article: "Observations on the Primitive and other Boulders of Ohio." Silliman published their four-page article, not as a letter but prominently as the seventh article, in the winter 1832 issue of the *American Journal of Science and Arts*. In the article the brothers say they will not enumerate the various theories for the erratic boulders, but will "proceed to state such facts as have come under our observation and the conclusion to which they have 'irresistably' brought us."[31] They describe where and in what quantity the boulders are found on the bed and banks of the Ohio River and then at the bluffs near Circleville. They note that the boulders become more numerous the farther north they go. From these limited facts the brothers assume that "we must look beyond the Great Lakes for the origin of our primitive fragments."[32] If they knew where the boulders came from, they write, the question would be settled and "we should then know everything concerning them."[33] The brothers say that the boulders are not only primitive but secondary. And they say that there is a connection between the boulders found on the surface and those, for example, at the bluffs at Circleville "in the various and very curious windings and turnings of the different layers of clay, sand and gravel" they have dug up. Large boulders on the surface are nothing more than gravel on a larger scale. "Is this not conclusive as to the origin of the boulders?"[34] The Laphams are correct that the same force that deposited the gravel, sand, and clay deposited the boulders, but they are incorrect as to what that force was. This article was probably drafted primarily by Increase since it contains *his* "irresistible" theory and the evidence comes primarily from Circleville.

In late November Increase visited Darius on the line of his canal between Lancaster and Carroll, Ohio, then he returned to Circleville. Increase probably expected to be in Circleville until the aqueduct and that section of the canal were complete; however, his time at Circleville was coming to an end.

On January 1, 1830, Micajah Williams arrived from Columbus and announced to Increase that he wanted him to go to Portsmouth to work on the canal sections between Deer and Paint Creeks. He was to be an engineer again. (Neither Increase's diary nor his correspondence mentions that that day a large number of free black people were thrown out of Portsmouth by racist white citizens.) The new appointment would not start immediately, however. On January 5, before Increase was out of bed, Erastus Chapman, a contractor on the Louisville and Portland Canal, arrived with letters from Benjamin Sayre and Judge Bates requesting Increase to go with him immediately back to Louisville to be a witness in contractors Peter Carney and Benjamin Sayre's suit against the Louisville and Portsmouth Canal Company.[35] This necessity overrode even the canal commissioner and his work at Circleville.

—⊢ | ⊢—

Increase left for Louisville with Chapman the next day. After riding south on horseback all day—nineteen miles before breakfast—they arrived in Portsmouth, Ohio, at eight o'clock on the evening of January 7. Increase immediately called on family friend and engineer T. G. Bates, another son of Judge Bates, who introduced him to Francis Cleveland, the resident engineer of the whole Ohio and Erie Canal. In a letter the next day to his parents, he told of meeting these two men and being "much pleased with them." He continues, "They are quite scientific; & Mr. Cleveland is a subscriber to Silliman's Journal, and has several other scientific works of value. I think I should have been much pleased with my situation if I had been stationed here."[36] Increase is not so much interested in their engineering abilities, as Darius might have been; it's the science that interests him.

The next day shows the kind of energy and drive Increase was capable of. Between breakfast and 5 p.m. when he and Chapman caught the steamboat *Robert Fulton*, Increase explored Portsmouth with the younger Bates, who told him he wanted Increase to take charge of his line when he got back from Louisville so he could visit friends in New York. Increase took a good look around Portsmouth noting and describing and drawing three distinct geological formations where the Scioto connects with the Ohio. He noted many types of trees in the area: "Opposite Portsmouth I saw

the American Pine tree upon the hills."[37] In the engineer's office, he wrote
to his parents telling of his journey so far and adding an underlined note
to Darius: "There are no boulders of primitive rocks in the hilly country
north of Chillicothe."[38] That same evening on the steamboat, Increase
talked to a bridge builder who was in the area to find a place to build
bridges over the Ohio at Louisville and at Cincinnati. Increase didn't seem
to have a need to just stop and rest for a bit.

The next day they arrived in Cincinnati but it rained, which Increase
said in his journal prevented him from "making such observations and excursions through the city of Cincinnati as I would otherwise have done."[39]
He was a tireless traveler and observer.

On January 10, he was back in Louisville. At Allen's Tavern he saw
Judge Bates and nearly all the contractors he had known on the canal. He
saw John Henry and Benjamin Sayre and rode out to see Peter Carney on
his farm west of Louisville. It must have been a bit awkward to be back
among all the same people he had left in October, some of whom he
would have to testify against in the lawsuit brought by Carney and Sayre,
who were suing the Louisville and Portland Canal Company for money
they believed was owed them. But he thought he would testify in a few
days and then go back to the Ohio canals. Again, circumstance intervened.
The proceedings were delayed for many reasons, primarily the absence of
the arbiter, Samuel Forrer, and a director of the company, one Canvass
White.

By February 8, when he had been in Louisville for nearly a month and
still had not testified, he received a letter from his parents that must have
said to go to Ohio and get back to work.

Increase wrote to them that same day telling them that the proceedings
probably would not happen soon, but that he would not "take the 'decided step' and leave now as you suggest" because Judge Bates was absent,
and he didn't want to leave without seeing him, and because the river was
frozen over and it would be very expensive to travel. Judge Bates was an
old friend to the Laphams and it was partly to help rescue Bates's reputation that Increase wanted to testify. Increase continued to Seneca and
Rachel: "I have since my arrival in this place had considerable leisure time
as it is called, i.e. I have not had any regular and constant employment,

but it has not been wasted in 'idleness and dissipation' as you will perceive when I enumerate the objects which have occupied my attention."[40]

Increase was not yet nineteen years old. He had been working since he was thirteen. He was a little over five feet tall, probably still growing. With his slight build, fair complexion, and pale gray blue eyes, he must have looked a bit more fragile than he was, but he had been working hard for years. Unexpectedly he had a little time off, and though the relationships among his friends and acquaintances in Louisville must have been touchy and tense, he was having a good time. He was back in Louisville and Shippingport, which he had been anxious to leave for more than two years, but he wasn't working for Mr. Henry. Increase told his parents what he had been doing. He copied plans of the wooden locks and sold them for five dollars each. He copied and studied plans of the canal. He did some surveying for five dollars a day—not a small amount of money in comparison to his rodman's salary. He made plans and estimates for a railroad around the Falls of the Ohio and told his parents that if a railroad were built there he had a good chance of being the engineer. He also spent time preparing for his testimony. He wrote a letter to Judge Bates explaining his observations of the mistakes Mr. Henry made in laying out the lower curve of the canal.[41] He took dimensions of the locks, made calculations and measurements of the bridge, and drew a plan of the locks and bridge for Carney and Sayre, showing defects of the work. He set stakes and took the levels to estimate the quantity of excavation for Mr. Howell's railway from the steamboat landing to his warehouse.

By the eighteenth of that snowy February, Increase had been in Kentucky about six weeks and still the principals for the arbitration had not been gathered together, so the two parties—the canal company and Carney and Sayre—agreed on four new arbiters. They were Samuel Forrer, important to Increase in his coming six years in Ohio; John R. Henry, Increase's Mr. Henry, for the canal company; and two businessmen from Louisville, William S. Vernon and a Mr. Meriwether—"a nincompoop," Increase wrote in his journal.[42]

On Sunday, February 21, after he went to church, Increase again wrote to reassure his father that "the arbitration is now in progress, it is supposed that it will occupy but a few days"—he was very wrong about that—"and

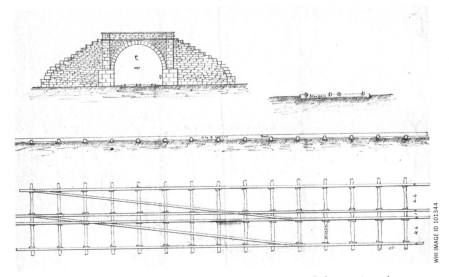

WHI IMAGE ID 101344

Increase Lapham's plan of a railroad, drawn in 1830 at age eighteen. Before most canal engineers, Increase was developing an interest in railroads and foresaw their eventual transcendence over canals.

then I may be expected home as soon as my legs can carry me." He again told his father he was using his time to make a little money, and he added an eloquent paragraph that gives a hint of the complexity of human relations around Louisville at that time:

> You have frequently advised me of the importance of a knowledge of human nature, that the "greatest study of mankind is man" and here I have an excellent lesson. Placed in a circle of about one hundred of the most respectable people of Louisville, and more immediately in connection with a set of contractors, planning and devising how they can most effectually attack their opponents and Judge Bates assiduously endeavoring to retrieve his reputation, I say, placed in the midst of all this, who could neglect to learn some useful hints for his benefit through life?[43]

After spending a day with both Mr. Henry and Judge Bates making out an estimate of the work done by Carney and Sayre, Increase finally went before the arbitration on February 24. Judge Bates was testifying, but occasionally a question was asked of Increase. From February 24 until

March 4, Increase was either helping Bates with calculations or he was in the arbitration room in the courthouse waiting and listening and observing. On February 27 he was finally asked for his testimony. "It was only a few questions in relation to the deposites, floodwood shantles &c. &c."[44] He was so well prepared that he must have been a bit disappointed at the triviality of the questions. On March 3, however, he was asked to explain an extra charge for the changed curve. In his journal he said, "It had been laid wrong by Mr. Henry & the contractors were obliged to excavate upwards of 3,000 yards of earth in consequence of it."[45] At noon on March 4 the case was submitted to the board of arbiters. No one knew when the decision would be made.

The next day, March 5, Increase and Judge Bates left Louisville for Portsmouth on the steamboat *Robert Fulton*, both probably thinking they were through with the "tedious business."[46] After a stop in Cincinnati, where he noted in his journal his nineteenth birthday, Increase arrived in Portsmouth on March 8 and then left on one of the stage horses for an all-day ride to Chillicothe, where he met up with his father at Madera's Hotel. The next day, after depositing fifty dollars in a bank in Chillicothe for the expenses of moving the family to the farm, he and his father rode home to Yellow Bud through the rain. His father told him he had bought for 450 dollars a farm on King's Creek in Champaign County, Ohio, which it was agreed Darius and Increase would help pay for. The farm was rented to another farmer for the coming season.[47] Increase stayed two days with his family and then took a stage down to Portsmouth, just missing Darius, who arrived for a family visit the day Increase left.

Increase didn't go back to his job in Circleville. As assistant engineer— the job promised to Increase by Micajah Williams directly before he was asked to testify in Louisville—Increase worked with Judge Bates on the canal in Portsmouth until the end of May. He laid out stations of excavations, ran test levels, and laid out curves—all in rainy weather that made the roads slippery and the Ohio River nine feet above the highest recorded watermark.[48] In spite of the bad weather, Increase had a great deal of work to do on the canals and his journal entries thin out. In early May Darius wrote a long letter about geology, but it was six weeks before Increase wrote back, saying that he "scarcely had a thought" on geology for the past month. "My duties on the canal are so constant that I can hardly think of any other subject."[49] Yet Increase's duties—and his attention—were again about to be divided.

On May 29, Mr. Daniel Carroll, a canal contractor from Louisville, arrived with two lists of questions for Increase to answer in writing—forty-three questions from Carney and Sayre and forty from Goodwin and the Louisville and Portsmouth Canal Company. Carney and Sayre's lawsuit was continued until November 1830.[50] For more than a week Increase worked on his answers to the questions, presumably along with his work on the canal. His answers took up twenty-five closely written 8-by-13-inch pages and about ten days of writing.

This deposition, even if it had been written by a person twice Increase's age, is exceptionally clear and precise. Carney and Sayre were claiming that they did more work than they were paid for because the engineers, Judge Bates and Mr. Henry, didn't provide them with adequate plans in a timely manner and did not keep accurate records of the amount of stone the contractors placed in the lock walls. Daniel Carroll's and Mr. Goodwin's questions for Increase were about who made the plans and when, who drew the plans, who measured and kept records of the work done, and how those records were kept. Much of Increase's information, then, is about Mr. Henry and his incompetence. Yet his answers are not acrimonious and he seems to have been careful with his wording. Here is one crucial question and its answer:

Question 23. Wasn't Mr. Henry very careless in making his estimates [of daily work done]?

Ans.—Estimates as they are usually made, and as they were originally intended on the L. & P. Canal are only an approximate statement of the quantity of work done within a certain period, made for the purpose of showing the commissioners or directors how much money they can advance to the contractor with security. It is therefore not important that they should be exact. Considering Mr. Henry's estimates in this light they are generally made with sufficient care.

The "final estimate" is an exact statement of the whole of the work done by the contractors and is the basis of the final settlement between them and the directors. It is therefore important that this estimate should be made with care and accuracy. Taking Mr. Henry's estimates in this light & considering that each weekly estimate was "final" of the work done in that week it may be said that he was too careless in making them.

Frequently as in the case of the embankment at the locks he would only guess at the quantity of work done. At other times he would get the number of men engaged and allowing a certain quantity for a days work would calculate in that way. That portion of the work which he measured weekly was generally in such irregular & indeterminate shapes that it was almost impossible to measure them with accuracy.[51]

Increase goes out of his way to be clear and accurate. When asked how accurately Mr. Henry had measured for the stone bridge Carney and Sayre built across the canal, Increase answers, "The errors which Mr. Henry committed at the bridge were not so much in the measurements as in the calculation."[52] He then explains that Mr. Henry does not know how to figure the amount of masonry in an arched bridge. He tells how Mr. Henry roughly guesses this amount and then includes a drawing with his, Increase's, own measurements and calculations and the admission that Mr. Henry did make a slight error in measurement of the abutments.[53]

In his cross-examining questions, Mr. Goodwin asks a fair question: "If errors were made in the measurement of the [unclear] or calculations, did you know it at the time, and if so, was it not your duty to make it known and cause it to be rectified?" Increase's answer:

It is the duty of an assistant Engineer to correct his superior in errors of simple multiplication, division &c, which the best calculators are liable to commit and even to correct errors in the manner in which the calculations should be made to a certain extent. But I had done this so frequently for Mr. Henry that it became troublesome at last. There are however only a few instances of which I was at the time aware that I did not correct him.[54]

These instances were most likely at the end of his work in Louisville, when he seemed worn down by the problems of the job and the incompetence of Mr. Henry.

This testimony—very supportive of Carney and Sayre's case and supportive of the reputation of Judge Bates—was taken to Louisville by Daniel Carroll, where it was undoubtedly read by some of the most prominent men on the Ohio canals: Micajah Williams, Samuel Forrer, Judge Bates, Benjamin Tappan. These twenty-five pages probably did as much as any of his articles and all of his actual work on the canals to elevate among canal men the reputation of Increase Lapham.

For all this work on the deposition, Daniel Carroll gave Increase three books and $2.50, that being the "lawful allowance" for his time as a witness in the case.[55]

—|—

On June 24 Increase got word that his presence as a witness was once again required in Louisville. He left Portsmouth on June 26 on the steamboat *Nile* "accompanied by some acquaintances, among them Miss Eliza McCoy of Chillicothe and Miss Rachel Peebles of Portsmouth."[56] On the twenty-seventh in Cincinnati, he "went up to the city with the ladies." This is Increase Lapham's first mention of "the ladies." And it isn't his last. Increase was a handsome young bachelor, a respectable engineer on the Ohio Canal. In Louisville, he put up at Allen's Tavern and looked up his friend and canal contractor, Mr. Carney, and then looked over the work done on the canal since he had last seen it. But when he went to the courthouse the next day, he found that the trial he'd come all this way for was postponed until November![57]

That day Increase did a remarkable thing. He showed the deposition to Mr. Henry. Increase and Mr. Henry had a long conversation over tea at Miss Ming's boardinghouse near Fifth and Main in Louisville. There is no record of bad feelings in Increase's journal, but this must have been a difficult conversation, yet one that Increase apparently felt confident in. He must have been very sure of his accuracy and the fact that much of what was in the deposition was not new to Mr. Henry.[58] The next day Increase Lapham's deposition was resumed. Questions and answers were slightly revised. The last day of June, Increase mentioned that General Andrew Jackson, the president of the United States, passed through Louisville and that he went to see a mechanical exhibition of "automatons and of the conflagration of Moscow." July 1 he left Louisville for Portsmouth. He had no reason to linger.[59]

But on the second of July in Cincinnati, he bought some books and drawing paper and went to the "Gardens." He was invited to take tea with a Mrs. Lodwick, which he did. There were "ten or twelve ladies present." He said he learned to play the harp that day. And then that evening he and the Misses McCoy and Peebles whom he came down to Cincinnati with took a steamboat back up to Portsmouth, arriving the next afternoon.[60]

Increase Lapham had reached a man's estate from one summer to the next. He had tried in imaginative ways to get a college education, and when these efforts failed, he went on with what he had to do. He wrote with his brother a respectable article for *Silliman's Journal*. He continued to collect shells and minerals and correspond with prominent men who also were collectors. He contributed both plans and money to the well-being of his parents and younger brothers and sisters. He worked hard on three canal sites. He wrote a deposition that no doubt helped win Carney and Sayre's suit against the Louisville and Portland Canal Company. He made new and influential connections and maintained connections with old friends—friends who could have but didn't become enemies. And he discovered "the ladies."

CHAPTER 8

Acquainted with Everybody but Intimate with None

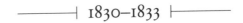 1830–1833

Dᴜʀɪɴɢ ʜɪꜱ ᴛɪᴍᴇ ɪɴ Pᴏʀᴛꜱᴍᴏᴜᴛʜ, Oʜɪᴏ, Increase Lapham exhibited not only a man's character but also a man's ability to supervise other men—though not with the authority of an iron fist: He was small, quiet, and nineteen. His authority came partly from his friendship with Judge Bates, until March 1829 the principal engineer of the Ohio and Erie Canal; but more important, it came from detailed knowledge of how a canal should be built and his close and honest observations of what was being built. Alongside his hard physical and mental work on the canal, Increase took up a new line of research: agriculture. While precocious in his professional and intellectual life, in Portsmouth Increase did some catching up in his social life. He discovered girls.

On July 3, 1830, Increase Lapham arrived in the placid valley of the Scioto River at the point where it emptied into the great Ohio. On July 5 he was invited to celebrate "American Independence" with "a party of twenty or thirty persons, half of whom were ladies" in the Kentucky hills across the Ohio River from Portsmouth.[1] While he ate the refreshments provided by these friendly strangers, he could see Portsmouth, a town of about thirteen hundred people less than ten years older than Increase. (In 1830 Milwaukee was an Indian village and fur-trading post, and Chicago was just being laid out.) When he wrote to Darius a few weeks later, he didn't tell him that Portsmouth was "a straggling and rather an untidy village," with widely scattered frame and log buildings surrounded by gardens.[2] He didn't mention that the town bull wandered the weedy streets, that the intersections were pitted by hog wallows, or that a dozen or so wolves were killed nearby every year.[3] He didn't tell Darius that some of the white people of this peaceful town in this free state, months before he arrived, had driven the free black people out of Portsmouth to settlements farther up the Scioto River.

For the three years he would live and work in Portsmouth, Increase Lapham seemed to float above the unpleasantness, the conflicts and catastrophes of southern Ohio. He told Darius about another reality, a careful Increase Lapham reality, but a reality nonetheless. He wrote Darius that his situation as an assistant engineer on the Ohio and Erie Canal was a good one and very pleasant and that he expected to remain on the line until it was completed. He was pleased with the town as well as the job. "The steamboats which navigate the Ohio river are constantly passing bringing information from almost all quarters. . . . The inhabitants are sociable, agreeable and friendly. The surrounding scenery is extensive, beautiful and pleasing."[4]

Important to Increase were the unspoiled natural aspects of the Portsmouth area and his work on the canal there. And without his family and old friends nearby, the inhabitants of the town became important as well. Though he says that he values "information from all quarters," over the next three years Increase does not mention in journals or letters the information he undoubtedly heard and read about the growing abolitionist movement in the country. Nor does he mention the Underground Railroad active in Portsmouth. He draws and visits Indian mounds, but he does not mention the Indian Removal Act signed into law May 26, 1830, by President Andrew Jackson, which was to remove all Native peoples east of the Mississippi from their native lands. He does not mention the 1832 Black Hawk War in Michigan Territory, in which his contemporary, Abraham Lincoln, had a small part. He mentions little about the presidency of Andrew Jackson. At this point, Increase's life was—except in natural science and engineering—local. And his life was not political, except as personal relations and work are political.

Resident engineer Francis Cleveland, Increase Lapham's immediate boss, seems to have been Mr. Henry's opposite—"a man of some science

...perfect in mathematics and other branches of engineering."[5] Cleveland, who was soon teaching algebra to Increase, shared his interest in minerals and natural history. Increase told Darius that Cleveland was having "a cupboard with glass doors made to display them in. There is room in it for our collection also."[6]

Cleveland was the resident engineer on the Ohio and Erie Canal from the Ohio River north to ten miles above Chillicothe. Increase was the assistant engineer under Cleveland for the lower twelve miles of the canal, from about Dever's Run down to where the canal joined the Scioto River at Lock 55. As assistant engineer, Increase was helped in his surveying by a rodman, Mr. Bradford, and an axeman to clear brush, Mr. Garrison. Little is known about the rodman and axeman, not even their first names, and, more surprising, little is known about Francis Cleveland, a man with a prominent position on the canal and in the Scioto Valley. He was apparently an educated and well-read man who was a founder of an informal literary society in Portsmouth, as well as of the public library with loans from his own library. He may have been the man who named Waverly, Ohio, after the Walter Scott novel he was reading when a name for the town was called for.[7] He laid out the southern division of the Ohio and Erie Canal, so he must have known surveying. After the canal was built he became city clerk and city engineer of Portsmouth, recording secretary of a resurrected Scioto Agricultural Society, and later the owner of a newspaper in Portsmouth. He was called Captain—but this was a common honorary title of the time.[8]

We do know that he was a good boss for Increase—knowledgeable, not overbearing, trusting of Increase's abilities. He apparently saw Increase's abilities when others doubted that this slight nineteen-year-old could supervise about eighteen contractors on twelve difficult miles of canal that would include the building of five culverts, two bridges, eight locks, and an aqueduct.

Increase's duties as the assistant engineer were to lay out the work on the ground, make accurate measurements of each kind of work to be done, estimate the quantity and description of all materials required, and oversee every part of the construction in his charge.[9] Well into this new position, he admits to his mother that when he first started the job in

Portsmouth, "it was objected to my appointment that I would not have sufficient firmness and resolution in giving directions and in other dealings with the men."[10] It is likely that Increase himself was unsure of his ability to supervise all of these contractors and their work.

For the remainder of 1830, until the cold closed work on the canal and Increase got several weeks off to visit his family, he wrote only ten brief entries in his journal, though he kept up his letters to Darius and to the naturalists Isaac Lea in Philadelphia, Dr. Fitch in Louisville, and S. P. Hildreth in Marietta, Ohio. He mentioned no social events, though he occasionally looked for mineral and fossil specimens. He told Darius, but not his father, that he "had four fits of fever and ague [chills]" in September.[11] He just worked.

Much of Increase's early work in Portsmouth was surveying and drawing. On August 28, not quite two months after his arrival, Increase mentions in his journal that he "Drew a map of Portsmouth, O for Mr. Williams."[12] Micajah Williams was the canal commissioner responsible for the southern portion of the Ohio and Erie Canal. This map is not in Increase's journals nor in his papers, most likely because he based it on the earlier surveying, plans, and information of Francis Cleveland for the Ohio Canal Commission, so he didn't think of it as his own. But there does exist a beautiful, graceful map of Portsmouth labeled "Topographical Map of Portsmouth and its Vicinity with the Southern Termination of the Ohio Canal, Drawn by F. Cleveland, Engineer."[13] Since this very accomplished map bears close resemblance in detail, style, and interests to Increase's later maps; since it could only have been drawn in 1830[14]; and since a map for Micajah Williams would not have been needed had Francis Cleveland drawn this earlier, it is very likely that this map is the one Increase drew for Williams—who later told Increase that his drawing was so good he could go to work in Washington, DC, for the United States Land Office.[15]

Between late August of 1830 and early December of that year, though there is little mention of his work on the canal, Increase must have been kept busy learning his new job and getting to know the contractors and the ways of Captain Cleveland. This was a demanding job that required that he learn not only the lay of the land but also how to oversee the work of many projects and many men—all of whom were likely older than

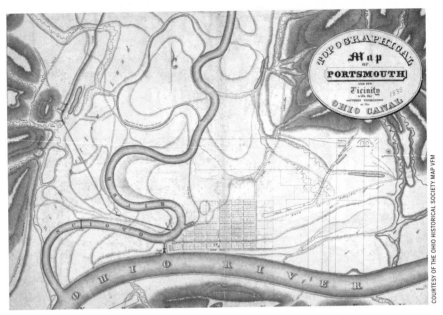

This map of Portsmouth where the lower Scioto River joins the Ohio River was likely drawn in August of 1830 by Increase Lapham, based on Francis Cleveland's earlier surveys and plans.

Increase. As usual, his letters to Darius reveal what is on Increase's mind: he is thinking about farming.

In August of 1830 Seneca Lapham left canal work to move his family from Yellow Bud to the farm outside West Liberty in Champaign County, Ohio, that he had bought the previous year. About thirty-five miles northwest of Columbus, the farm was difficult for Increase and Darius to get to from their work sites. Though Increase had not seen the land, the fact that his father and mother and younger siblings were now settled on a farm opened a door in Increase's imagination. Not only did he immediately refer to this unseen piece of land as "home," but he began to read about agriculture. In August he wrote Thomas C. Clark in Philadelphia for a copy of his *Farmer's Library*. He bought a book of essays on raising silkworms, which many American farmers were trying. He wrote to a man in Cincinnati for mulberry seeds and silkworm eggs. These canal men were going to grow silkworms on the farm in West Liberty, Ohio. At the end of a few months of this reading, Increase even wrote an article about the culture of silk for the *Portsmouth Courier*.[16]

That fall, Increase sent one hundred dollars saved out of his quarterly payment of two hundred and fifty dollars to his parents to help pay off and furnish the farm. Rachel wrote thanking Increase, praying that her kindhearted son would be blessed both in "basket and store."[17] To Darius Increase proposed that they make the farm a place with an orchard, a kitchen garden, and a hothouse, "a place on which it would be agreeable to live." And he proposed that they grow not merely useful, edible plants but also "such as are ornamental and pleasing to the sight." And they should put in a summer house and fountain.[18]

Increase Lapham's aesthetic sense had up to this time been apparent only in the beauty of his maps and drawings. But it had been educated not only by the images he had seen in books, magazines, and homes in Portsmouth and Cincinnati, but also by the beauty of the natural world as he traipsed up and down the canal. Increase is romantic. Darius is more practical. He writes back that all are agreed about the orchard and vegetable garden, but as to the summer house and fountain, "you are too refined." After they have what is necessary to eat, drink, and wear, Darius wrote, then they can think about a summer house and a fountain.[19]

In December of that first year in Portsmouth, Increase spent fifty-nine dollars for a mare with one white foot to take up to the farm, when he could have used a horse himself to ride up and back the twelve miles of the line of his canal—a trip he walked many times a week. When work on the canal shut down for the hard freeze of winter, Increase bought a pair of saddle bags and set out on the mare to visit his family and the new farm—"home." The trip of about 150 miles took him eleven days. Just before he left he had mentioned in a letter to Darius that he had been thinking that after he finished work on the Ohio Canal he wanted to take a trip to the eastern states to see the canals and railroads there and perhaps to get a job on one of them, to see the country and collect minerals and plants and to meet engineers and scientific men of the country.[20] Instead of this daydreamed trip to the East, Increase was heading northwest on the mare. And though it's not an exotic trip, he would see new country and old friends and family, making the most of his travel.

After traveling up the route of the canal and staying a night in Piketon, Increase rode on to Chillicothe where he saw his brother Pazzi. Then to

Bloomfield where he spent the night with his newly married sister, Mary, just fifteen, and her husband, Dr. William Jones. The next morning he rode up to Carroll where he stayed three days with two old canal friends, waiting for Darius to return from his work on his line of the Ohio and Erie Canal. Finally on the sixteenth of December, Darius returned and the brothers set off toward the farm, Darius only going five miles with him as he had work to do on the canal. Increase had waited three days for only a few hours with Darius.

Then, on a below-zero day, Increase rode on alone to Columbus where he saw the House of Representatives in session, "put up at Watson's Hotel," and saw again in a bookstore Thomas Say's *American Entomology*, "a very fine work with colored plates."[21] He still coveted that beautiful book. He could have bought it; he had the money. But his first priority was helping his family buy and furnish the farm. After Columbus, he headed northwest for two very cold days through country new to him, looking closely along the Scioto River and Darby Creek at the limestone, which he drew with cold hands and later described in a letter to Darius. On December 19, after many cold days on horseback, he arrived at the log house on the Lapham farm.

Papilio turnus, or tiger swallowtail, ca. 1824, from Thomas Say's *American Entomology*

COURTESY DEPT. OF SPECIAL COLLECTIONS, MEMORIAL LIBRARY, UW-MADISON

After he had been at home with his family, Increase wrote to Darius that "we have got a first rate farm, excellent soil, fine water." He told Darius of the level bottom land, the gently sloping hills bordering the south tending valley, and the level ground where they would build a stone house. He described a mill that Seneca would build to grind the corn they would grow—cob and all—for cattle and hog feed. And he told of Seneca's plan for a hexagonal orchard, each tree to be surrounded by six others in order to place a larger number of trees on a smaller piece of ground. The mathematical Laphams had figured this out.[22]

The day *before* Christmas, after only a five-day visit, Increase left his beloved family and set out for Portsmouth, this time by stagecoach, leaving the white-footed mare behind. By January 1, 1831, he was back at Wilcoxon's, the tavern west of Portsmouth where he stayed and where the canal company had an office. Soon he was hard at work again on the canal.

—I⊢—

At the time of Increase Lapham's arrival in Portsmouth two major canals were being built by the state of Ohio, which had been separated from the Northwest Territory by statehood in 1803. The westernmost canal, called the Miami and Ohio, connected Dayton to Cincinnati through the valley of the Great Miami River. The Ohio and Erie Canal connected the Cuyahoga Valley with the Scioto Valley and Cleveland with Portsmouth. These canals were designed to open up Ohio to the New York markets, which paid better prices than the often glutted New Orleans markets. Through the Ohio canals farmers could ship wheat, pork, corn, and other products north to Lake Erie and the Erie Canal, then to the Hudson River down to the port of New York. And products like iron and manufactured goods could be imported from the East more cheaply.

This canal work was very different from Increase's jobs in Louisville and Shippingport. For one thing, the Ohio canal system was one of the best managed and funded in the northwest. There was little scandal or mismanagement in the building of the Ohio and Erie and the Miami and Ohio and the other feeder canals, as there had been on the Louisville and Portland Canal. And on the Ohio canals, the engineers and contractors did not use as many cranes for lifting and machinery for stump pulling. The Ohio canals were cleared and excavated by men with picks and shovels and wheelbarrows.

On this canal, Increase had much more responsibility for more complex work than he had on the Louisville and Portland Canal—twenty-five half-mile sections of canal—which involved contracting for a variety of work. The width of the canal had to be grubbed and cleared of trees and vegetation to twenty feet on either side of the canal. The canal banks were mucked and ditched—cleared of all wood and rubbish so the bank would rest on solid ground. The canal bed was excavated through rock and soil or by digging under water. Embankments were raised above the natural surface of the earth. Locks, bridges, culverts, and aqueducts were constructed of stone or wood.

While his work in Louisville took place over about a five-mile stretch of river, his work in Ohio took place over a more than twelve-mile length of the canal. Increase had to spend much of his time walking to his sections, which were all above Portsmouth. Occasionally Increase borrowed a horse, but, most often and in all kinds of weather, he walked. He frequently spent nights at taverns or inns or with friends who worked on the canal. He was allowed $2.50 a day for expenses.

His work in late August and early September of 1831, recorded in his journal, was typical of his work on the Ohio Canal. After writing Darius on a Sunday that the rainy summer had caused much damage to the canal and delayed its completion for a season, Increase went nine miles up to Brush Creek with his axeman, Mr. Garrison, where he stayed for two nights. Then he went three miles farther up the canal to take estimates at McLane's sections where McLane was building two culverts and a bridge. Leaving McLane's, he walked the twelve miles back south to Portsmouth where he met canal commissioner Micajah Williams, who was going up the canal. He stayed in Portsmouth for two days making out estimates. Then he went five miles up the canal to Old Town where he examined John and William Dickey's three sections, saw that they were almost complete, and spent the night at "Old John Williams." Back to Portsmouth where he did "considerable work." On Monday morning he went back to have another look at the Dickeys' sections and spent the night. He continued the next day up to Eddy's Section 231, where he set some stakes on the lower end of the section in the morning, and then traveled three miles down to Section 236, where he laid out the Sheep Run culvert. After spending the night on the line, he went on up to McLane's in the rain where he met Captain Cleveland, who had just come down from Camp Creek. The following day he reset the stakes for the Sheep Run culvert—no doubt corrected by Cleveland. Increase went up to Brush Creek the next day and stayed at Lucas's—probably a tavern. On the way he collected some shells.[23]

All of this coming and going is dizzying to read, but there were for Increase long quiet stretches to look around at the valley and the woods, the plants, the shells, the rocks. Long stretches of thinking and figuring and making plans in his head that he would write about to Darius or his father on Sundays. It was hard physical work in good and bad weather. He must have been a strong and fast walker. He must have had good shoes.

—|—

For three years Increase worked like this, from the time the ground thawed in the spring until the middle of July when malaria closed down work on the canals, beginning again at first frost and working until the hard freeze.[24] In 1831, he didn't go home to see his family as he usually did in December. He says vaguely in a letter that he couldn't get away.[25] Perhaps it was because of the unusually cold winter that year, so cold the Erie Canal froze solid and was closed for the month of December. But perhaps the main reason he didn't go visit his family was his intense and enjoyable and *new* social life and his close connections with friends in Portsmouth: the "Ladies!"[26]

Though his relationships to the young women of Portsmouth were no doubt most interesting to him, Increase was friends with many of the leading citizens of the town as well, at whose homes he would meet eligible ladies. He spent evenings with the family of J. A. Bingham, a businessman who also owned a newspaper, with Mr. Glover who worked for and then owned a newspaper, and with Dr. Hempstead, then Clerk of Common Pleas. He was friends with a Mr. Voorhees and a Mr. McVey. He was friends with a woman named Eliza Dupuy who later became a prolific novelist. It was the friends of Miss Dupuy who started the first public library in Portsmouth—"a very respectable library." Eliza Dupuy "drew around her a choice circle of literary friends, among whom were Dr. Hempstead, Judge William V. Peck, Edward Hamilton, John Glover, Francis Cleveland and others."[27] Increase Lapham was one of those others; all of these people are mentioned in his journal in 1831 and 1832.

But there are also mysterious mentions of Miss O, Miss G, Miss B, and Miss Y—especially in his giddy year of 1831 when Increase went to tea parties, skating parties on the Scioto, even a ball for Washington's birthday at the McCoy Hotel. "Danced!" he wrote in his journal.[28] On many Sundays he attended three churches and then went home with some young women. He went paw-paw hunting and to the circus with his female friends. He went with them to an exhibition of a man eating fire and swallowing a sword—"a vulgar offering" he says, but he went back the next day without them and "helped the clown off the stage! Made as much fun as he did!!"[29] This Quaker boy was having a wonderful time. He took flute lessons and went to singing school and singing parties. He was "Quilting!"[30] He went to the theater and an Irish wedding. He went skating and rigged a whirligig for the girls. He went to a large party on the new steamboat *Senator* where he "waited on Miss B. but paid the most attention to Miss Jennison."[31]

It all sounds casual and relatively innocent, but it was giving him ideas, ideas that upset his family. In June, Increase wrote to his parents saying, "Would you be surprised if I should get married before Fall?" This set off a chain reaction of terror in his family. His mother, reported Pazzi to Darius, didn't sleep a wink the night after she received the letter. So Darius wrote to Increase, "Pshaw! Pshaw!! boy!!! Be careful how you tie yourself to an apron string. . . . Will you let me deeper into this subject?"[32] Increase immediately wrote to his parents. "I am grieved to hear that what I said on the subject of marrying should have caused you uneasiness. My only object was to present the subject in a general way and elicit your opinions. I have not the most remote idea of doing anything contrary to your wishes or without your consent, but I think the subject may be discussed now with advantage."[33] As far as the journals and letters tell, it wasn't discussed at this time. Yet his sister Mary had been married with apparently no discussion—among

A token saved by Increase Lapham from the whirl of his most social time—1831 in Portsmouth, Ohio

WHI IMAGE ID 102041

the brothers, at least. It seems his parents and Darius had expectations of Increase that they didn't have for a daughter, expectations they didn't have for wandering Pazzi, either.

From the tone of the journal entries, it seems unlikely that Increase had a "serious" relationship with any of the young women. It is more likely that he fell in love with female company. This was another "new country" to Increase—the talk of "ladies," their voices and laughter, the music and theater and parties, the clean scents and rustle of dresses. He found he was attractive to the girls and he was a healthy young male having fun— probably not as much fun as his parents worried about. "You need not be alarmed for my safety," he wrote Darius.[34]

For most of that summer of 1831 Increase wrote little in his journal. On August 6 he explained in the journal that his repetition of phrases like "went up the canal" and "remained in town" were becoming too tedious. And besides that, he added,

> it would occasionally have been [his] painful duty to write such sen-
> tences as these; "Sunday went to singing school in Old Town and then
> went home with the girls and spent the evening." "Thursday took a
> jaunt to the hills in Kentucky with a party of ladies and gentlemen,
> walked myself with the prettiest girl in the country." "Saturday eve-
> ning took another walk to the mouth of the Scioto river with a com-
> pany." Such outlandish and abominable notes as these, I should abhor
> to write, and I am well convinced that no reasonable person will be
> disposed to censure me for omitting them.[35]

His fun with the ladies must have cost him money because he sent less money home to his parents than Darius expected him to. And Darius asked why. Increase explained by saying that his expenses were higher, perhaps, than those of Darius, and, well, a few dollars "have gone away foolishly" but he said he could account for every cent he received.[36] He then wrote seriously to Darius about the benefits of making plans in a way that only a twenty-year-old can do:

> He who never looks ahead and forms some plan is tossed about by
> every change of circumstance, as the rolling waves are by every change

of wind; and will be apt to turn out a worthless, trifling, restless fellow.
... By forming a plan we can regulate our future lives... so as to enjoy
happiness as far as is practicable in this world and gradually improve
our fortunes.[37]

He added that he had almost come to the conclusion to make Portsmouth
his permanent residence. When it was no longer necessary for him to send
money home, he wanted to buy land near Portsmouth.[38] A month later he
wrote his father that he wanted to quit canalling—and the main reason
was that on the canal they had to move around so much. He hated being
transient and was perhaps for the first time experiencing what it might
feel like to belong on several levels to one place. "We do not more than
get acquainted in a neighborhood before we are required to leave it and
take up our abode with a new and strange set of people," he complained.[39]
He liked this "set of people" in Portsmouth and could see himself buying
land and perhaps setting up housekeeping with a wife there. The rest of
the year of 1831 was the same whirl of work and social life. His last journal
entry of 1831 was: "Lovers Quarrel."[40]

The next year began with a letter from his mother who was worried
about his extensive social life. He wrote to Rachel in response, "You suppose I have become tainted with vanity by 'mixing too much in what is
called fashionable society.'" He defended his Portsmouth friends, first by
saying he doesn't want his mother to think they are "light headed, trifling,
good for nothing kind of people." Rather, they are "mostly good, common sense, sort of folks, who regard industry a virtue and skill at work
an accomplishment." As usual, Increase's second reason is the most true
to what he thought: "If I do not go into *this* company, where shall I go?"
His third justification for his social life—the one he had to reach for—was
that he needed to "mix in the world and become better acquainted with
human nature" so that he could deal more effectively with those whose
work he supervised. And to reassure his mother he added that he taught
Sunday school and attended church regularly.[41]

He strongly defended his social life, yet from this point on his journal of
1832 would take a very different tone. Most entries were still about work as
they were in 1831, but there was very little mention of "the ladies." In 1832

he mentioned his friend Mr. Voorhees and other male friends as much as he did girls. Either he was not writing things down or he had decided that his life *was* a bit "trifling." He did attend to more serious matters. He went to several lectures on grammar and a lecture on political geology. He attended meetings with about thirty other men who were forming a lyceum. Ten or twelve of them would give lectures on subjects well known to them. Increase, who was to lecture on botany, worried in a letter to Darius about public speaking. The tone of this whole year is much less giddy than that of 1831, except for one delirious entry written the day after his twenty-first birthday, which reads in its entirety, "Returned to town—Maple sugar!—high water!—girls!"[42]

In April 1832, Benjamin Silliman published Darius and Increase's article on erratic boulders.[43] The article appeared as Increase sent it to him—with Darius as first author. Under the influence of his literary friends, Increase tried writing the same material as fiction. He called the resulting dialogue, which ran in the *Portsmouth Courier*, "The Stranger."[44] The piece begins with two friends walking along the bank of the Ohio River. They see another man "engaged in picking up and examining the *gravel stones*, as he walked thoughtfully along the shore." Then the narrator reports, "'Those pebbles,' said I sneeringly, 'seem to have some magic charm to him—let us overtake him.'" This gives the "stranger" the chance to make the same points about the origin of gravel and boulders that Increase made in his article and in almost the same words. The piece was not a success, but it shows Increase experimenting with form and writing for a popular audience.

—| |—

In 1832, Increase's life went on much as it did the previous year with hard work on the canal, going to lyceum meetings and lectures, seeing "the ladies" to some extent, collecting botanical specimens and keeping his botanical diary, writing to and missing his family, and thinking about the family's farm and how it should be run. Yet it was a dramatic year in Increase's life—a year full of events he could have no control over. First there was a great flood. Then Darius and Pazzi surprised him with big changes in their lives. And there was the terror of a cholera epidemic.

The worst flood that Portsmouth had seen to that date began in January of 1832 when the Ohio River rose thirty feet in a week. Increase excitedly recorded in his journal the details of this great flood over the next few weeks. February 5: The river was rising rapidly. February 7: Increase ferried a mile and a half across the "noble river." February 9: The river was forty-eight feet above the low water level. The weather was warm, the mud deep. February 12: Increase wrote Darius that the Ohio was now sixty-five feet deep: "Still rising, still raining!" February 14: "River higher than ever known before! Many frame houses floating down—with hay, rails, wood, & almost every thing else. Many small houses & stables upset in Portsmouth." February 15: The river was still rising an inch an hour in the street. Increase "went to a party in a skiff—fell overboard going home." The flood was a catastrophe for the town, for people's lives, and for the canal, but Increase floated above it. February 16: Increase couldn't get over to the canal. "Water in most of the houses. 1½ feet deep in my tavern & the passage leading to the [canal] office."[45]

February 18: The water began to fall in the night. "Water 5 feet higher than ever known before & 63 feet higher than low water of 1830." Increase went out on the flood with friends in a sailboat. February 19, 20, 21, 22: "River falling slowly." He sounds a little disappointed. But on the twenty-second he celebrated Washington's birthday with friends who fired twenty-four guns and "by drinking Egg-nog & apple-toddy!" On February 23 he was back at work, heading up the line of the canal in a snowstorm. Increase Lapham the engineer of course knew how much damage the flood had caused to his canal and to his friends' houses, but the boy naturalist thoroughly enjoyed the flood of 1832.[46]

On May 11, Increase wrote a letter to Darius who was at the farm with the family, his job on the Ohio and Erie Canal having ended. He was responding to a letter Darius wrote on April 9 when he was apparently in an uncharacteristic "desponding" mood, having no work and wanting to give up canalling to live near his parents. "Surely you have no ground for such feelings," writes the young Increase. "Ought we not rather to be thankful for what we now enjoy, a good farm, tolerably well stocked, is a little fortune to any one." He also tells Darius what he isn't telling himself—don't give up canal work, which was to Increase more clearly

Darius's "profession" than his own.[47] He reminds Darius of the many railroads and roads that are to be built and that he could look for work on one of those. If Darius needs money for traveling expenses, Increase promises to send him some. But Darius has more reason to "despond" than he is telling Increase.

The next thing Increase heard about Darius, on May 20, was that he was married to Nancy L. Fallis in Wilmington, Ohio, and that the couple had moved to Cincinnati, where Darius was assistant engineer in charge of nine locks on the Miami and Ohio Canal: one problem solved, another complication introduced. About Darius's marriage Increase was speechless in his journal until the end of July. He finally wrote to Darius in early June after Darius wrote to him with *some* information. Unfortunately we don't have that letter, which doesn't tell Increase what he wants to know: "I wish you would be a little more *verbose* in your next and tell me how it ever happened to enter your head to be married."[48] It is impossible to know for certain, but it sounds like Darius's wedding was a hurry-up affair. Innocent Increase doesn't suspect that his adored older brother could have gotten himself in this sort of situation. He deals with the hurt caused by the secretive and surprising Darius with jokes. The rest of his letter is the jocular Increase who has forgotten that he surprised his family a year earlier with the idea that *he* might marry:

> Perhaps you may wish to know my views with respect to the "future state," but as you are now a married man you must not expect it. You no longer belong to the tribe of bachelors. . . . Well I must tell you that *I am married already.* My wife is twelve miles long and has twenty five children. (You may know that I am speaking of my line.) I have watched the growth of each, have experienced as much pleasure in observing the growing beauty and development of the perfect parts and proportions of my Brush Creek Aqueduct as you will a few years hence from your favorite son or daughter."[49]

Increase is hurt and perhaps jealous, yet he sent a box of books to Darius with a "token for Mrs. Nancy" and one hundred dollars to help them set up housekeeping.[50] He even went to Cincinnati to call on the newly married Darius and Nancy, but Nancy was away visiting her mother.

And on August 1 Increase noted in his journal: "Pazzi also married! At home to Ann Marie Bennet—of Logan County."[51] There are very few journal entries for the next months.

The marriages of his brothers and their intimacy with their wives excluded Increase and highlighted to him his own loneliness. He wrote to his father after he had heard of Pazzi's marriage, "All that I love are far from me; on this account I am growing cold and indifferent to every body. I have no one here on whom I can rely with confidence so that my better feelings toward my fellow man are becoming slack for want of exercise." He realizes that his social relationships are not of the depth he sees possible. He wants to hear from his family more often, he says, particularly from his mother. "I am," he writes, "acquainted with every body but intimate with none."[52] He wishes Pazzi, who is at the farm, much happiness, and he gives his respects to his new sister, Ann Marie. Increase is growing in depth and self-consciousness. And sadness.

The "sickly season" of 1832 made it all worse. An epidemic of cholera—the first in this country—swept the United States and its territories, after arriving first in Quebec on a ship from England. In late summer the disease arrived in Ohio on the improved transportation from the east. Cincinnati was hit hard, as were other Ohio cities. In Portsmouth they feared its arrival daily. Cholera, a disease spread through contaminated food and drinking water, affects people with severe diarrhea, vomiting, and cramps, sometimes killing them with dehydration within a few hours after they are hit with the first symptoms. The epidemic was terrifying.

Increase was very afraid for Darius. In the middle of October he told Darius to "write me often during the prevalence of the cholera in the city, it is the only evidence I can have of your being alive."[53] Though there was as yet no cholera in Portsmouth, the planned celebration for the opening of the canal was postponed for fear of the disease. By November 8, Darius had not written to Increase who lived in fear that Darius was dead. He wrote chastising him for his "cruel" silence. Yet somehow he must have received word that Darius had had cholera but survived. Illness, work, and marriage had cut Darius off from his younger brother. "Give some account of the complaint and how you felt while it was raging," he asked

Darius.[54] He then reported that "my line stands well and leaks but little"—work, as always, goes on.

About that same time Rachel Lapham wrote to her son Increase that she was disappointed she hadn't seen him that winter—it had been almost two years since he had been home. She is obviously worried about him as Increase is worried about the rest of the family. She tells Increase that her Redeemer "is able to deliver us out of all trouble if we put our whole trust in Him."[55] But the scientific Increase asks Darius "what precautions you adopted to escape and what you would recommend to me, for we live in daily expectation of its breaking out here."[56] Cholera never did arrive in Portsmouth that year.

—| |—

On December 14 Increase left Portsmouth on horseback for his winter break and a visit with his family. Riding up the canal and then through creek valleys in the snow and rain, losing his way several times, he arrived in Urbana, Ohio, where Pazzi lived with his new wife. After sleeping on the floor—"Froze last night"—he rode ten miles the next day through three inches of snow to the family's farm near West Liberty in Champaign County.[57]

Oh, they were glad to see him! His mother "threw her arms around my neck and thanked the Lord for preserving us to see and enjoy this happy meeting." His sisters, Hannah, Lorana, and little Amelia, each gave him a little shy kiss. Then William came in, now fifteen and much grown from the last time Increase saw him. "I could hardly realize him, there appeared to be little of William left in him. . . . Next came the little pup and the enormous Random [the family dog] to show their joy. Father returned from West Liberty at night."[58] Increase was home.

Thanks to Darius who had described it in a letter to Increase the previous March we can see the inside of the Lapham home on King's Creek near West Liberty. The house was a double log cabin "one-half made by father" connected with a passageway. One of the cabins was the kitchen, in a corner of which was a workbench and tools where in the evening the family "made small artifacts for about the house, meat tubs, bread trays, etc." The passageway between the houses was divided into a bedroom—probably where William and visiting siblings slept—and a porch. In the main room was the big armchair "facing south" where Darius sat writing this description. Next to it stood a writing table for Hannah and Lorana. On the other side was Darius's bookcase containing books and papers. Hanging on the wall was a basket full of newspapers and letters. In another corner of the room was their parents' bed with the girls' trundle bed under it. He even detailed what was hanging on the walls: "father's picture in chalk," a picture called "St. Clotilda Reine," a "looking glass surrounded by a variety of ornaments," a vignette of "Ovid's Metamorphoses," and a picture of "a dying Christian repeating to his soul the lines of [Alexander] Pope beginning 'Vital spark of heavenly flame.'"[59]

On the other side of the armchair was the table "on which lie three numbers of the New York Spectator received last evening." This is where their mother, the "schoolmistress," collected the little girls about her every day after the work was done to read their books. Nine-year-old Hannah, Darius tells Increase, is a "capital scholar" who reads "quite fluently and has a remarkable tact for committing to memory." Lorana, seven, can read words of two syllables and improves rapidly. And even five-year-old Amelia can spell words of four or more letters. William, fifteen, however, "hates his book almost as much as he hates snakes."[60]

Darius goes outside to the hilltop site for their planned new house from which they will be able to "take a survey of that little speck on the surface of the earth which we have the pleasure of calling our own, and where we can sit under our own vine and fig tree." On the side of a west-facing hill will be their orchard. And over there will be a new barn and a corn crib; there a root cellar and a spring house.[61] Increase later added that the farm was eighty-five acres, seventy of which were cleared. And there was "a horse, two yokes of cattle, 50 hogs, plenty of potatoes, beets, rutabaga etc. but little corn—Large quantity of clover hay, stacked & salted." Not only that but Seneca had "raised a large quantity of silk & mother has made some sewing silk."[62] The clever Laphams were building a prosperous and happy life in one dear place.

While at the farm that December, Increase helped his father lay out the fields using a clothesline three rods long. Using a board level they set stakes for the stone wall that would be in front of the house not yet built, Increase making the wall curve to "conform to the shape of the ground."

And Seneca walked Increase around, showing him where his white mulberry hedges and his hexagonal orchard would be.[63]

On the twenty-second Seneca, Rachel, and Increase went to West Liberty where they saw Increase's sister Mary, now sixteen, with her husband, William Jones, and their baby "Little Cecelia, who just begins to talk." The three of them also visited Pazzi and his wife in Urbana. From Pazzi's home, Increase went on alone, heading back toward Portsmouth, on December 29. He traveled first to Columbus on the National Road, not yet covered with stone, where he went to see the legislature. Seeing canal friends along the way and stopping in Waverly to attend a tea party, Increase arrived back to Portsmouth "safe and sound" on January 3, 1833.[64]

March 15, a week after his twenty-second birthday, Increase wrote in his journal that his "engagement on the canal at Portsmouth expires this day, having lasted a few days more than three years."[65] The Ohio and Erie Canal was complete. Where would he work next? He had no idea. He had thought he might settle in Portsmouth, but what work could he do? He had considered working on the new railroads near Lexington, Kentucky. Mr. Micajah Williams, now the head surveyor of the entire Northwest Territory as well as Ohio Canal commissioner, had said he might work in Washington as a draftsman at the land office, but nothing had come of that. The life of a farmer appealed to him, but he knew he still had to help support his family and the farm; the farm was supporting too many people as it was. He had written and published several articles, but there was no money in that, as there was no money in his first loves, geology and botany. He had daydreamed of a trip east after the canal was finished, but when the time came he didn't have the money. He had even discussed going into the pork business with a canal friend. He didn't know what he would do next. What he did know was that he did not want to work on a canal again.

And he wanted to live near Darius. "I want to live in Cincinnati and be with you. . . . The advantages of a city life I conceive to be great, perhaps I overrate them; but the pleasure and benefit which I should derive from your society *never can be overrated.* . . . I hope you will assist me in attaining my end. I do not care if it is not engineering, there are many other situations which I think I could fill with credit."[66] This dream was not to be.

Canal commissioner Micajah Williams had a job for Increase Lapham, but sadly not in Cincinnati near Darius. It was a situation that had never occurred to Increase, a situation in Ohio's capital, Columbus, that he would fill with credit.

CHAPTER 9

In the Capitol

⊣ 1833–1835 ⊢

THE INCREASE LAPHAM WHO LEFT PORTSMOUTH, Ohio, in March of 1833 just after his twenty-second birthday was no longer a young prodigy but a confident engineer. He also was more experienced socially and he had finally learned how to spell. In his next venture, Increase would deepen and widen his connections with influential scientists, engineers, and politicians, and he would begin to be a writer of influence. He would also survive a deadly cholera epidemic.

On March 18, 1833, Increase left Portsmouth, traveling down the wide brown Ohio River on the steamboat *Alota*, which was towing a scow of stones. These stones would be used by Darius Lapham in the Miami Canal locks near Cincinnati. When he left Portsmouth, Increase was still hoping to get work in Cincinnati, the seventh-largest city in the nation and the home of Darius and Nancy on Sycamore Street. He did not yet know his fortunes would shift in another direction. For a short time, he did work with Samuel Forrer leveling a line for the Miami feeder canal— but this turned out to be Increase's last fieldwork on a canal until he arrived in Milwaukee three years later. Instead, when the board of Ohio canal commissioners met in Cincinnati in early April of 1833, they appointed Increase Lapham to work as a part-time secretary for the Canal Commission for four hundred dollars a year. The appointment was spearheaded by canal superintendent Micajah Williams. This part-time work would allow Increase time for his natural science activities, for writing, and for other work that might come along. Other work did come along. By the end of his stay in Ohio, he'd been offered four additional jobs, three of which he took. The only disappointment with the Canal Commission work was that it was in Columbus rather than in Cincinnati.

So, on April 29, 1833, Increase left Cincinnati on the steamboat *Guyandotte* heading southeast for Portsmouth again. But thirty miles above Cincinnati, the steamboat hit a snag and wrecked—not an uncommon occurrence for the period. Increase wrote Darius that he was "not scared out of more than one or two years growth." He climbed up a post to keep himself above water until the passengers and crew were all rescued by the steamboat *Rambler* and taken upriver to Portsmouth. Of his baggage, only his box of minerals got wet.[1] He seemed to enjoy the experience—at least in retrospect. From Portsmouth he took a packet boat north along his own sections of the completed Ohio and Erie Canal to Columbus, Ohio, his home for the next three years.

Once in Columbus, Increase arranged to take his meals, along with a Mr. and Mrs. Mills and Mrs. West, a few blocks from his work at the home of Mr. Medberry, the superintendent of the prison. "Columbus," Increase wrote Darius, "appears to be very pleasant."[2] West-facing Columbus, on high ground above a bend in the Scioto River, had been designed in 1812 to be the Ohio state capital with a grid of wide streets and a capital square. By the time Increase arrived, Columbus, with a population of about four thousand, had a capitol building, a state office building, and almost a thousand other structures, only a few of which were still log buildings. A busy market sold farmers' produce. Three impressive churches and the growing temperance movement attempted to counteract the effects of the many taverns, grog shops, and bowling saloons. Common schools had not yet been established, but there was a private high school for boys and a Female Academy where girls from the country could board. Columbus had several newspapers and four printing companies, as well as three bookstores, one of which sold Sir Walter Scott novels, Strickland's book on canals and railroads, and the latest edition of John Kilbourn's *Ohio Gazeteer*. The new National Road coming from Cumberland, Maryland, passed through Columbus along High Street and then crossed the bridge on the Scioto,

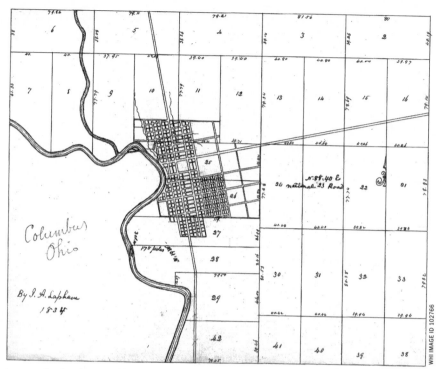

Increase Lapham's map of Columbus, Ohio, ca. 1834

heading west for Illinois.[3] But Columbus was still a frontier town in the 1830s. A traveler wrote in 1835 that it "would be a cleanly town were it not for the pigs."[4] Increase didn't write about the pigs; he wrote Darius that "[w]e have the prettiest little church that I have seen in the Western country and it is filled with the prettiest girls!"[5] This is Increase's only mention of girls during his very busy and productive time in Columbus.

A building over from the one where Increase worked, the Ohio State House was an imposing, square, two-story brick structure on the southwest corner of the public square. Increase had visited the State House twice before as he passed through Columbus—quietly observing the House of Representatives on the ground floor and the Senate on the second floor. Atop this State House was a steeple with a "good toned bell" and a balcony with railed walls where one could stand and look west across High Street and view "the whole town as upon a map," as well as "a most pleasing view of rural scenery in every direction" and the "town of

Franklinton, one mile to the west, and the intervening meandering of the slow winding Scioto."[6] Not to mention the cattle that grazed below on the mostly empty ten-acre public square, making "an awkward appearance to strangers."[7] It's hard to think that Increase would not have availed himself of this overview of the world he now lived in. From inside the State House and from atop it, Increase was widening his view of the world.

Increase worked just north of the State House in the two-story State Office Building, which held almost all of the state offices—governor, secretary, treasurer, and auditor of the state as well as the Canal Commission and the state library. Increase must have quickly developed at least a nodding acquaintance with most of the men who ran the state of Ohio, including the Democratic Governor Robert Lucas.[8] And since the Canal Commission office was one room, Increase would have known everything that was happening on the Ohio canals. His secretarial job did not carry with it much power, but Increase's small desk was near the center of Ohio government. He would use this proximity well.

Increase's trustworthiness quickly got him both another job and a bizarre but free place to sleep. Henry Brown, the Ohio state treasurer, saw in him a young man who might be useful. He offered Increase a job for one hundred dollars a year guarding the state's funds at night. He was to lock himself into the treasurer's office with the state's money every evening. So

The first state buildings at Columbus from left to right: the Courthouse, State Office Building, and State House

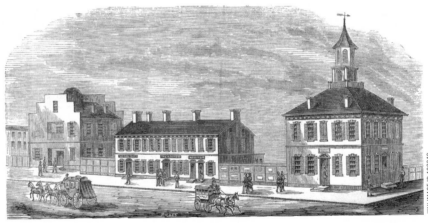

WHI IMAGE ID 102249

Increase bought a bed and rigged it up in Brown's office "whose door, window and shutters are faced with thick sheet iron" that he locked, barred, and bolted, sleeping with a loaded pistol at the head of his bed. He was an unlikely security guard, but Brown had him do a few other clerkly tasks as well.[9] Increase explained his morning routine in a letter to Darius:

> Perhaps as I am now in a new business it would interest you to give an account of the manner in which I spend my time. First, then, four o'clock a.m. fast asleep secured as aforesaid [by the bolted ironclad windows and doors]. Five o'clock I get up wash my face and hands and prepare for breakfast; I open the doors and windows to admit some light into my dungeon and here I discover the wisdom of Providence in creating twilight, for when I first open the door the sudden glare of light even at that early hour is quite painful to my eyes. Half past five I take my breakfast, but in order to get it I must walk up High street to the National road down that towards the bridge one square to Front street, up that street three squares to Mr. Medberry's, in all about ⅓ of a mile; here I meet mine host and landlady and also Mr. and Mrs. Mills.[10]

—| |—

In 1832 cholera had eaten around the edges of Ohio, arriving in the north with soldiers who died along Lake Erie on their way west to fight the Indians in the Black Hawk War, floating and steaming down the Ohio River to Cincinnati where hundreds died and some, including Darius Lapham, survived. In 1832 the citizens of Columbus waited and dreaded but the city was spared the cholera that usually arrived wherever transportation eased its way.

By 1833, because of the new National Road, the time distance between cities was greatly shortened. Mail arrived from Wheeling and from Cincinnati in about twenty-four hours and from as far away as Washington, DC, in about eighty hours. Seventy crowded stagecoaches arrived each week in Columbus. And packet boats arrived every day coming south on the Ohio Canal. Because each house in Columbus had its own well and disposed of its sewage into a pit on that same lot, it took little time for seepage to contaminate all water.[11]

Cholera arrived in Columbus on the canal in July of 1833. On July 14, Increase recorded in his journal that there was "a death in town from cholera."[12] On Tuesday morning, the twenty-third of July, Increase had breakfast as usual with the people in his boardinghouse whom he called "the family"—the Medberrys, the Mills, and Mrs. West. All seemed well. Then he walked a third of a mile back to the State Office Building through streets emptied of farmers and traders and travelers by fear of cholera. After work that day Increase walked back to his boardinghouse for his evening meal. Mrs. Mills and Mrs. West—who had seemed fine at breakfast—had died during the day of the disease. After that day three or four people died every day in Columbus and then there were "cases unnumbered." Among those dead were friends and acquaintances of Increase's—a well-respected Dr. Horton Howard and his beautiful young daughter Ann, and two grandchildren of Dr. Horton who were the children of a friend, H. D. Little.[13]

Cholera is a monstrously ugly and frightening illness that can kill in hours, the violent dehydration draining the body and spreading the disease. Writes one historian,

> Those who die of this disease are a gruesome sight. It attacks the bowels and causes a stupendous loss of bodily fluids.... The whole body becomes covered with dank moisture. Cheeks become hollow, nose pinched, eyes sunken, voice husky. Death's rigor sets in quickly. Muscles literally become hard as a board. Sometimes a stiffening corpse jerks about; it may kick out a foot, wave an arm, flap its jaws or roll its eyes.[14]

The curious Increase very likely knew these details. Alfred Kelley, one of the canal commissioners he worked with, was also prominent on the Columbus Board of Health and he certainly knew.[15] Kelley and the other members of the board knew to try to clean the water supply and the trash and rotting animals in the streets, but they knew no more than anyone else how to treat the disease. Treatments for cholera at the time were as useless or exacerbating as emetics, calomel (mercury), laudanum (opium), heat (bags of hot sand or ashes placed on the body), mustard plasters, copious blood letting, or castor oil. Some of these treatments could kill a healthy person.

On July 26, Increase wrote Darius that "I have not yet been able to drive away from me that dread and fear which you may readily suppose has got fast hold of my mind. I awake every morning after an unquiet dreamy sleep and am perfectly astonished to find myself alive and free from the disease."[16] After only a few months in Columbus, Increase may have been acquainted with a number of people, but he had no close friends or family near him. "The idea," he wrote, "of being in a strange place, where it is probable... that I should receive but little attention if I should be attacked does not tend towards composing my mind." He reported to Darius, eleven people in Columbus had died of cholera, most in the last few days. The citizens were "very much panic struck." Most people refused to go into a house where someone had died of cholera. His fellow boarders—and perhaps Increase, though he doesn't mention it—were shunned; one of them went into a house where every person in the room got up and left and asked the person to leave. Some who had fled cholera-ridden Columbus to rural areas were beaten.[17] Most upsetting to Increase was sleeping alone in a room that was a locked vault. If he got sick in the night, who would know and who could get in to help him?

So, in spite of his work obligations and perhaps leaving Ohio's treasury unguarded, on July 27 Increase, along with many other frightened Columbus citizens, left town. The rural areas were spared the worst of the epidemic because the water was cleaner, so crowded stages and canal boats left every day that summer crammed with people who hoped to escape the contagion. In West Liberty on the farm, Increase found, to his relief, that his family was well. He stayed with them about ten days and then returned to Columbus to find the cholera epidemic was even worse than before. Yet he stayed, picking up his work again, sleeping with Ohio's money in the treasury.

Oddly, his room and his work in the Ohio State House might have saved Increase's life. The water for the State House didn't come from the usual shallow-dug well near a seeping cesspit but from a large freshwater spring near Broad Street.[18] But he did eat the same food as those who had died in his boardinghouse.

Cholera in Columbus continued through September, tapering off with the colder weather. In September, Darius wrote Increase a letter proposing rightly that one solution to the filthy water problems that spread the disease was deep-water bored wells instead of the shallow hand-dug wells that were easily contaminated. But he wrongly objected that the deep water might be too salty.[19] The Lapham brothers referred to their knowledge of geology gained years before from Dr. Clapp's borings in New Albany, Indiana, and to their readings in *Silliman's Journal* and to their own observations to try to solve a serious public health problem. At a time when cholera was attributed to causes like excessive eating of celery and too much alcohol, the scientific knowledge of the Laphams was to be used to improve the lot of the people.[20] On the American frontier, science was not for its own sake; it was to promote the happiness of development and economic progress.

Not only did the brothers Lapham discuss in letters what could be done to get cleaner water, but Increase also suggested to Darius how to disseminate the information in a way that shows he was quickly picking up a canny political sense. He told Darius not to put the idea of deep wells all at once before the "good people of Cincinnati." He said, "First, write a short notice for one of the daily papers, just merely hinting the thing; then after a week or so another, stating some of the principal facts and advantages; and finally, come out with your essay in the magazine giving the whole subject at large and in full. This would lead them gradually to thinking of it, whereas, otherwise they will read your essay and it will soon be forgotten."[21]

In Columbus that summer and fall of 1833, more than two hundred people died of cholera, a little less than one in twenty.

— | —

During the ten days in August of 1833 when Increase was in Champaign County with his family escaping from cholera, he wrote to Naaman Goodsell, the editor of *The Genesee Farmer*, a monthly magazine with short articles on such subjects as the culture of silk, walnut trees, insect pests, killing rats, and harrowing versus plowing. Increase's letter to Goodsell was in response to Goodsell's general request in *The Genesee Farmer* for articles on the agriculture of Ohio. Increase included an article that was well written but contained no new ideas. However, it shows us what Increase was able to think about in a time of great trouble, how he was able to distill in a short period of time his reading from agricultural journals and his observations from his travels about Ohio. In the article he recommends crop rotation and the use of manure on fields—both practices widely known to be useful because "the soil is the real capital of the west"—a good phrase but not likely original to Increase.[22] On the frontier, however, they were not the usual practice. Most farmers of the time would clear the forest from the land, farm it for several years, wear out the soil, and then move on to clear and wear out other cheap land. Increase's article included the fact that the Ohio Legislature had just passed a law establishing an agricultural society in each county.[23]

At the same time he sent Goodsell the article, Increase wrote him saying that for many years the experience of agriculturalists was recorded only in "short and desultory articles" in many agricultural journals. Increase had just described Goodsell's magazine in an insulting manner. He does say the articles are valuable and well written. But this accumulated knowledge is difficult to use as it is "dispersed through so many volumes and entirely disconnected and unarranged."[24] So this twenty-two-year-old canal engineer with no farming or editorial experience suggests to Goodsell that the subject of agriculture be divided into about fifty subjects and an authority assigned to each to write a treatise on that subject—"each treatise...to be full and complete in itself, written in plain style and well digested and to give the best and most approved views of the subject." The project is to be "wholly *American*, adapted to our climate, soil and condition." Not having a clue how much money such an undertaking would involve, he suggests that it could be done "if suitable exertions were made."[25]

He also suggests that Goodsell promote the "diffusion of useful knowledge" in *The Genesee Farmer* not only by publishing a list of useful agricultural books but by reviewing agricultural books the way books are reviewed in scientific journals. Only these efforts will "remove the prejudice which now so generally prevails against reading agricultural books." "I reckon," Increase quotes farmers, "I know about as much of farming as the printer does."[26] His undertone of contempt for willful ignorance is as palpable as his arrogance and ambition and the orderliness of his mind. This is why he is a Whig, not a Jacksonian Democrat.

This proposal to Goodsell demonstrates some of the strengths and weaknesses of Increase's mind. He can see very clearly how scientific information can be classified and disseminated in order to be useful, but he has no idea of the practical hurdles of such an immense project. It's also likely that he thinks everyone has the level of energy and dedication that he possesses. Goodsell apparently ignored Increase's agricultural castle in the air but did publish the article on the agriculture of Ohio.[27]

In the privacy of their letters and journals, Increase and Darius for several years had been making fun of Andrew Jackson—calling him "the Gin'rul," and like others mocking his "kitchen cabinet," his backdoor advisors.[28] They opposed his spoils system, which gave government jobs to his political friends rather than to the best-qualified men. Increase Lapham's Whiggishness and opposition to President Jackson became more and more outspoken during his years in Columbus.

The Whigs were a confusing and short-lived amalgam that originated in the 1830s to oppose President Andrew Jackson's policies. Both southern slaveholders and abolitionists opposed Jackson—the first for his Tariff of 1828, which helped only northern industries, the second because he was a slaveholder who advocated expansion of slave territories into the Northwest. Businessmen opposed Jackson because he did not support banks and didn't renew the charter of the Second Bank of the United States. And many farmers and industrialists and canal men, like Increase Lapham, opposed Jackson because he did not believe the federal government should support internal improvement projects like canals and roads. All of these opposers called themselves Whigs. Ohio in the early and mid-1830s was split nearly evenly between Whigs and the Jacksonian Democrats, who emphasized the rights of the states and of the common man.

Increase Lapham not only opposed Jackson; he supported (with reservations) a party that he saw as *for* the federal support of canals and highways and western expansion. He was *for* banks that would provide capital for improvements by the states and private enterprise. He was *for* being governed, in every sense of the word, by men who were smart, educated, and competent—the sort of men he was used to having dealings with—not by appointed political hacks. Increase was an elitist. He probably didn't even know the name of an actual canal worker or a real farmer—other than his farmer father. Being governed by Andrew Jackson and men like him must have felt to Increase like working for the underqualified Mr. Henry back in Shippingport. Yet many men he respected and worked for, including Micajah Williams who became so influential in Increase's life, were Jackson men. And Increase, though increasingly outspoken against Jackson, usually got along with them. Later in life, his political beliefs would severely limit his choices.

In February of 1834, Increase made his feelings about political parties clear in a letter to Darius. The brothers were discussing their younger brother Pazzi's new enterprise. With a loan from Increase, Pazzi had bought a newspaper in London, Ohio. Darius and Increase felt he was doing very well, but Increase was afraid that he would be swayed to take sides with the Van Buren Democrats who were strong in that area. He wrote Darius that this "would be against his honest convictions, and I think if he cannot sustain his paper there without joining a party who put *self interest* above the *interest of the country*, he had better by all means abandon the project entirely."[29]

Increase of course acts out of self-interest but his interests are parallel to the interests of science, the state of Ohio, and his family. It is these parallel interests that cause him to write a long geological description of Scioto County for the Historical and Philosophical Society of Ohio, which in 1834 may have helped get him appointed to be city surveyor for Columbus—now officially a city.[30] His own interests and the wider interests of science drive his correspondence with prominent Lexington physician and botanist Charles W. Short and with S. P. Hildreth, who wrote to Increase that "you have a talent as well as the inclination to accomplish something that will be permanent and useful."[31] These parallel interests give Increase access to J. L. Riddell's herbarium in Worthington and the library of naturalist and son of the founder of Franklinton, Ohio, Joseph Sullivant. It's also this parallel interest that sends Increase out to the prairies searching for the new *dodecatheon* (shooting star now called "The Pride of Ohio"), which Darius and his botanist friends had been seeking.[32]

—| |—

All through 1834, Increase led a vivid and multilayered life—sending seeds and money to the family farm; writing small signed and unsigned articles on a variety of subjects for local papers and journals. The following year will be much the same but with more layers.

In January of 1835, Increase wrote to Darius saying he was reading Charles Lyell's *Principles of Geology*, but he did not like it much because he said some theories were not supported by facts. And he asked Darius about his daughter, Phebe, who was born the previous summer. "Has she grown much? Is her hair black, white or red? Her eyes ditto and so on?"[33] Darius wrote back in a few days, which was unusual for Darius who wrote about one letter for every two of Increase's. Darius told Increase about all the mistakes he had heard in a lecture on geology he'd attended. "To hear this man talk you would think he had all the geology himself and nobody else had a particle of it.... I have some notion of giving him a wipe or two in the papers." Then Darius wrote a lovely paragraph about his beloved little Phebe:

> In answer to your queries respecting the specific character of the *Lapham Phebe*, I will give you the following zoological description taken from nature.
>
> Genus   Homo.
> Species   Lapham
> Var.   Phebe.
>
> Female, age 6 months; height 1 ft. 10.5 in; horizontal circumference of cranium 1 ft. 3.3 in.; from center to center of ears over the occiput 0 ft. 10.3 in.; circumference of body below axillae left 4.1 in.; circumference of leg at origin 0 ft. 8.4 in.; circumference arm at origin 0.ft. 5.5 in.; color of hair dark brown; color of eyes blue; color of skin very fair; beauty superlative; capacities in order developed, cry, laugh, oo oo, ago &c. We are now all well.[34]

This scientific and imaginative description by Darius incited Increase to one of his own. "I wish you would exert your imagination a little," he writes, "and see me in the canal commissioner's office sitting at a small desk, appropriated to the use of the secretary of the board, placed at a

point 40 degrees west, distance six feet from the center of the fireplace."[35]
And we see it too, all these years later: in front of his desk the large case of
drawers and shelves with glass doors that held the cabinet of the Historical
and Philosophical Society of Ohio, an organization authorized in 1831 by
the Ohio Legislature and first headed by Judge Tappan, who was there
in the room that day Increase wrote in 1835. Increase had recently been
elected as a member of this society, and he was chosen to be the "curator"
of their cabinet of natural curiosities. So he was in charge of cataloguing
and storing the society's petrified rams' horns, calves' horns, honeycomb;
curious stones, rocks, and minerals; mussel, snail, and sea shells; bugs and
butterflies and dried plants—an activity he must have delighted in.[36]

At Increase's right was Mr. Byron Kilbourn, thirty-four, whom Increase
had known and worked with for years on the canals, standing at a high
desk writing a long report (probably the annual canal report). Kilbourn,
from a well-to-do family in Worthington, Ohio, will play a major part in
much of the rest of Increase's life. Already he knows Kilbourn very well:

> Observe the dark frown on his brow, and you will be able to anticipate
> something of the nature of his composition. Perhaps it relates to some
> wolf-scalp vouchers which are missing; or perhaps to the $10,000 of
> 3 per cent money drawn from the U. S. treasury without authority
> and paid into the State treasury when it suited the convenience of the
> one who drew it; or perhaps it may relate to the $504 paid to a certain
> printer for work which could have been done by others for half that
> sum. At any rate, you may imagine that Mr. K. is writing at the high
> desk.[37]

And there, Increase points out to us, is "Mr. R. [probably Mr. Riddell]"
writing "a chapter on the effects of the July rains and floods on the canal
for the annual report." At that time of year they were probably all working
on the annual canal report—except Increase, of course, who was describing the scene to his brother.[38]

If we, and Darius, "exert our imaginations further" we will see on
Increase's left "Mr. Samuel Forrer writing to the honorable committee on
canals."[39] Forrer, previously a primary surveyor of the routes of the Ohio
canals and resident engineer on the Miami and Ohio Canal, was then on

the board of canal commissioners. Forrer, forty-two, a mechanically apt man who learned engineering working in Virginia as a millwright and machine builder, was a longtime supervisor and friend of the Laphams. Unlike many of the men around him, he was apparently neither political nor caught up in speculation fever. Forrer was pure engineer, spending almost his entire and long adult life working in some manner for the Ohio and other canals. His "sound judgment, practical sense, and high reputation for probity and fairness" made him the final reference for professional questions. "His decision and advice were usually the end of a controversy."[40] Forrer had been on the panel in the dispute between owners and contractors on the Louisville and Portland Canal. The next year, when Increase and Darius had to make an important professional decision, it was Samuel Forrer they first called upon. The Laphams never speak of him except with high regard and respect. He was probably the easiest man Increase ever would work with.

Forrer, unlike most of the men Increase worked with, was not angling for some other opportunity: he worked for the canals of Ohio. Byron Kilbourn also worked on the canals, but he was at that time angling for a contract to work for Micajah Williams in his speculative land ventures. Williams himself was at that time secretly dealing with New York backers to form the Ohio Life Insurance and Trust Company. Judge Tappan, also a naturalist, was also dealing with Williams to own a piece of the Ohio Bank. What we now call conflict of interest, the men of the Old Northwest called the main chance. Canal men regularly surveyed for canals and also scouted for good land to purchase for a mill site or a town site. Forrer must have been by comparison to most of the men around him refreshingly straightforward and easy.

And over there is Judge Tappan "with his cross eye studying some abstract question of law or politics, or possibly reading the 'Globe.'"[41] Benjamin Tappan, a Democrat and abolitionist, was on the Canal Commission and a federal judge at the same time. He was born in Northampton, Massachusetts, in 1773, and studied painting briefly with Gilbert Stuart and law in Hartford, Connecticut.[42] Later he was a United States Senator from Ohio. He shared Increase's interest in conchology and geology, but Increase had little respect for him as a geologist. As early as 1828, before

Increase knew Tappan, he criticized Tappan's ideas about Ohio boulders published in *Silliman's Journal of Science*. Tappan later criticized the appointment of John Riddell to the state's geological survey committee because Riddell couldn't tell graywacke from limestone, when graywacke is a type of limestone. Increase had a lot of fun with this in a letter to Darius.[43]

"Imagine," Increase continues, "that the fire is so hot as to render my back uncomfortably warm, while the air which enters through the cracks of the windows and partitions renders me uncomfortably cold, and you will perhaps have no difficulty in imagining the reason why I wrote 1834 for 1835 in my last letter."[44] And so we imagine, at his urging, Increase in his dark jacket finishing his letter to Darius, and then taking up his work—minutes or report writing or copying letters. The room is quiet except for the rattle of Mr. R's newspaper, the scratch of pen nibs on paper, the shifting of coals in the fire.

CHAPTER 10

Leaving Ohio

—| 1835–1836 |—

B Y THE TIME HE WOULD LEAVE COLUMBUS for Milwaukee in 1836, Increase Lapham would be an influential writer; a politically savvy and well-connected citizen; a geologist, conchologist, and botanist valued by fellow scientists; and a trusted former employee of the Ohio's Canal Commission board and state treasurer. Yet, in spite of all the successes of his time in Ohio and in spite of the fact that he wanted to stay in Ohio, Increase would go to Milwaukee because he had no choice. He was pushed out of Ohio and pulled to Milwaukee by forces that were out of his control. And these main forces were the land speculation let loose in the Northwest by the Indian removals of the 1830s and the rise of Jacksonian democracy.

Black Hawk's defeat in 1832 and the Chicago treaty of 1833 under which the Potawatomi were to leave Illinois and Wisconsin opened the dams to further immigration of white people from the East. The possibilities of cheap land and the fast buck in ill-sited towns and well-sited Chicago and Milwaukee pulled a frenzy of speculators west. Increase Lapham would be caught—only partly unwillingly—in this frenzy.

In early 1835 in addition to his work for the Canal Commission board and the treasurer of the state of Ohio, Increase was made treasurer of the Historical and Philosophical Society—for no pay, of course. In July, Increase was appointed to two more city jobs: street commissioner and city fire engineer. He fired off a letter to Darius asking how log conduits are joined in Cincinnati; he had to superintend laying a mile of logs and he thought he had a better idea.[1] That same month he was made deputy surveyor of Franklin County and was elected a corresponding member of the Western Academy of Natural Sciences based in Cincinnati.

In spite of all this, Increase continued his botanical studies, exchanges, and correspondence with botanist, geologist, and lecturer John Riddell and Lexington physician Charles W. Short. Increase was making a catalogue of the plants of Ohio. Short writes that "your specimens are beautiful and I have rarely seen a collection so well preserved."[2] Increase exchanged shells with John Milton Earle in Worcester, Massachusetts, Charles J. Ward in Chillicothe, and Jared Potter Kirtland in Poland, Ohio. He exchanged and discussed fossils with Cincinnati businessman and president of the Western Academy of Natural Sciences Robert Buchanan; canal commissioner Benjamin Tappan; and naturalist and physician Samuel P. Hildreth.

How did he do it all? This question will be asked over and over as we look at the life of Increase Lapham. He had to have extraordinary powers of concentration, plus an ability to compartmentalize his endeavors. It seems he wasted no time, and he had a strong constitution. During his years in Columbus he mentioned no illness. But he did take some time off.

From October 15 to November 2, 1835, Increase was away from Columbus—traveling and visiting. And it's a good thing, too, because when he got back from this trip his life got even more complicated. On October 15 he took the mail stagecoach down to Chillicothe where he stopped and saw Charles Ward's and canal commissioner and Ohio legislator William Price's cabinets of shells. On to Portsmouth where he took the steamboat *Fairy Queen* to Cincinnati, where he must have seen Darius and Nancy and their baby, Phebe. After four days in Cincinnati, he rode up to Dayton to see botanist friend John W. Van Cleve, and then up to Springfield and the farm near West Liberty, where he stayed a week with his family, including (again) Darius and family.[3] Though travel in those times could be grueling, this must have been a time for Increase to just look around and think about things. He was at no one's beck and call on the road; he was on his own.

When he went back to Columbus on November 1, he took his twelve-year-old sister, Hannah, with him. Hannah, a good student, was to go to

Early pages from Increase Lapham's botanical diary, Columbus, Ohio, 1835

a girl's boarding school in Columbus, at least until a school was built in Champaign County near the farm. Increase would look after her; he very likely paid her school tuition and expenses.

On November 13, 1835, John Riddell wrote a letter to Increase from Cincinnati College.[4] It was the beginning of a letter-writing campaign intended to set Increase to work on another important and unpaid job.

Riddell, born in Massachusetts in 1807, was only four years older than Increase, but he had attended the Rensselaer School (later the Rensselaer Polytechnic Institute) at Troy, New York, and studied with the famous Professor Amos Eaton, so Increase may have been a bit intimidated by him. Promoted and sponsored by Dr. Samuel P. Hildreth, Riddell had been delivering a series of lectures in Ohio, some on botany and some on geology. Riddell wrote to Increase that he wished the Historical and Philosophical Society "would influence the legislature to prosecute a geological survey of the state! You will be there at headquarters:—suppose you work the wires a little and see what will result."[5] December 9 there was another letter, this time from Robert Buchanan, founder of the Western Academy of Natural Sciences in Cincinnati, who asked Increase to "[p]lease say to my friend Judge Tappan if he will start a resolution of a geological survey of this state... he shall have full and effective aid from this section of the state."[6] The next day Dr. Hildreth in Marietta wrote Increase saying that he was pleased to hear that Governor Lucas recommended the geological survey. It would be a credit to the legislature if they were to go along with it. And it would "open a field to display the practical skill of some of our young engineers like yourself."[7] That same day an official letter came from George Graham at the Western Academy of Natural Sciences. He was directed to request that Increase cooperate by memorandum or letters addressed to the members of the Ohio Legislature "urging the necessity of a law authorizing a geological survey of this state." He continued, "The importance of this subject will induce you to exercise your influence with your friends in the legislature."[8]

But Increase was already working the wires. The same day Dr. Hildreth and George Graham were writing to him, December 10, he sent a letter to John H. McCreed, who was chairman of a legislative committee exploring the idea of a geological survey. Increase suggested some of the reasons Ohio should follow the example set by the federal government and five other states. What follows in this introduction is a masterful piece of rhetoric, enumerating six reasons for the survey, with each reason documented by references to geological authorities, common sense, and useful examples. The argument is in clear and nontechnical language, obviously written to legislators with little or no knowledge of geology, but whose interests are the development of the mineral and agricultural resources of the state. Briefly, he states that a survey would show the extent of known deposits of coal, iron ore, salt, and gypsum; show where minerals, to date unknown, exist; show the probabilities of other valuable minerals; prevent the useless waste of capital looking for minerals and ore in the wrong places; diffuse the useful geological knowledge among the citizens of the state; and aid agricultural interest in their knowledge of soil types.

He concludes by saying to McCreed, "If these remarks which I have hastily drawn up...give you a deeper impression of the importance of such surveys, my object is accomplished."[9] Little could Increase know that he would be making these arguments with mixed success for the rest of his life.

In the meantime, on December 28, the Historical and Philosophical Society of Ohio met in Columbus. During that meeting a committee was formed to "make a catalogue of the animals, plants, and minerals that have been found in the state of Ohio."[10] The members of the society involved were Robert Buchanan; Joseph Dorfeuille, curator of the Western Museum in Cincinnati who, years before, had told Increase that his mineral specimens were worthless; John L. Riddell; Samuel P. Hildreth; William H. Price; J. P. Kirtland; and Joseph Sullivant. And Increase Allen Lapham, twenty-four, self-taught botanist, geologist, conchologist, engineer, writer, and lobbyist. His abilities made him completely accepted among such men, most of whom had formal educations. It was probably only Increase who thought his lack of advanced schooling put him at a disadvantage.

As spokesman for these men and in the cause of the geological survey of Ohio, Increase wrote in December a long and clear essay on the nature of geology for McCreed and the legislature. Geology may be defined, he writes, "as that science which teaches the natural history of the earth." He tells the derivation of the term and then tells McCreed that geology "is not...as was formerly supposed a mere speculative science; but one of facts and observations and legitimate deductions from them."[11] He then enumerates the seven types of rocks or formations currently studied and the current geological theories.

Increase also wrote, probably at McCreed's request, a plan or proposal for how the geological survey would be carried out. It's a simple two-page plan for four geologists, a botanist, and a zoologist to work four years for a thousand dollars a year. There was money for preservation of specimens and for publishing three thousand copies of the report. And he gives simple directions to the geologists and the naturalists. The total cost, he estimated, would be thirty-one thousand dollars. It is likely that this was not just his work; he told Darius he had written the "greater part" of a report on the survey that McCreed gave the legislature.[12]

Increase must have been thinking that he was writing himself a job description, but he somewhat coyly wrote Dr. Hildreth in Marietta asking him who would be competent for the geological survey. Hildreth wrote back saying that Hitchcock in Massachusetts would be good, but that, in Ohio, John Riddell and Increase were "in every way fitted for the service."[13] Just what Increase wanted to hear. But now he had to wait for the legislature and the governor to act.

—|—

In March of the previous year, Increase had written Darius mentioning that the legislature had "ordered an examination into the whole business and conduct of the canal commissioners and commissioners of the canal fund."[14] An ominous note, but the subject was not mentioned again until January of 1836. Increase wrote Darius that there was still uncertainty over whether the Canal Commission would be reorganized.[15] The proposal was from the Democrats in the legislature to reorganize the Canal Commission into a Board of Public Works, which would deal with roads, highways, and bridges, as well as the canals. Though this was primarily a move to replace Whigs on the board with Democrats, it does seem to make some sense. Though Increase got along with the Democrats, he was a Whig and he was worried about his job. Late in January he reported to Darius that the legislature had sent a bill to create a five-member board to be considered by a joint select committee. Judge Tappan, he said, thinks it won't pass; Mr. Kelley, he told Darius, thinks it will. "I think with Mr. Kelley," he wrote to his brother.[16] Anticipating he might be pushed out of his job with the impending decision, Increase reluctantly began to look for another job. He couldn't be a Whig secretary to a "Jackson" board. And all his city jobs went the way of other patronage jobs.

But he had a job offer—C. J. Ward wrote urging Increase to take a job working for a Colonel Medaria with the Hydraulic Company of Chillicothe. Mr. Ward wanted Increase's company on nature "rambles" and he says, counterproductively, that "the internal improvements which the state is about engaging in together with your standing as an Engineer will *at all times* secure you employment."[17] But the offer didn't thrill Increase; he didn't want to be an engineer again. Increase made it clear in a letter to Darius that this was a last resort. "The Chillicothe Hydraulic Company offer[s] me $50 per mo. & $3 per week to do their Engineering—If the [Board of Public Works] bill should pass I want to accept this offer—if it does not I do not. I have concluded to risk it and decline going to Chillicothe. But if [the bill] should pass and the Board there by sink into a confused mess of democracy, it may not yet be too late for me to accept the proposed situation."[18]

For two and a half months Increase waited and watched as the "Jackson Board bill" and the geological survey bill ground through the Ohio Legislature. Increase and Hannah, who was in boarding school in Columbus, got a joint letter from their mother, Rachel Lapham. Rachel told Hannah, "I hope thee will duly appreciate thy present advantages for which thou art under great obligations to thy brother Increase." And she told Increase, "Thee hast well requited the pains and anxiety endured in raising thee."[19]

Darius wrote from his work north of Cincinnati at Reading, Ohio, sending Increase some shells he had picked up in a road excavation. Some of the shells, he said, "are as perfect in all their delicate lines and processes as if they were the original living shell." The various shells, some "exhibiting the hinge and connection of the valves, very perfect, perhaps quite so," all came from different layers of "greenish friable slate" and limestone "not four rods in length." "A book," the amazed Darius wrote his brother, "might be written on this simple fact."[20]

—⊣ | ⊢—

Events were already taking place that would shape Increase's professional future. Though he either didn't know about it or didn't pay attention, on May 5, 1835, an agreement was signed between Micajah Williams and

Byron Kilbourn. Each would put up eight thousand dollars to buy land in Milwaukee, Kaukauna, and Green Bay in the largely unsurveyed western part of Michigan Territory. That summer Kilbourn took surveyor Garret Vliet with him to Chicago and then north into the territory west of Lake Michigan. The two began the process of buying land for 10 percent down, land that was not quite authorized for sale. But this "[l]imited completion was deemed an acceptable practice in an era when business ethics were ill-defined."[21]

Increase was not opposed to land speculation. He had seen that it had made his friend Joseph Sullivant rich enough to spend all his time with his studies of the natural world. And in January of 1836, he had impulsively suggested in a letter to Darius that the family sell the farm and buy land on the Mad River Feeder of the Miami Canal where Darius was working. A railroad was rumored to be built across the feeder. "Had we not better sell the farm and purchase there?"[22] They might get rich.

And Increase did get a break of sorts. On March 21, 1836, the secretary of state of Ohio appointed a four-person committee to report to the next legislative session the best method to go about a complete geological survey of Ohio. Samuel Hildreth, John Locke, John Riddell, and Increase Lapham were appointed to this committee. The four members in different parts of the state began coordinating their operation by mail. Hildreth was chairman of the committee and would examine the southeastern quadrant of Ohio; Locke, the southwest; Riddell, the northeast; and Lapham, the northwest. They were to look especially for lead, iron ore, and coal in their appointed areas and then meet in December to make out the report.[23] So it looked like things were set for Increase for a while—except that the members of this four-person board were not to be paid for this *preliminary* report on a geological survey. The other three members of this committee had other means to make a living—as physicians and a lecturer. Increase could only daydream about striking it rich buying land along the Mad River Feeder.

Finally, in April, the day of the board's decision came. On April 7, 1836, Increase wrote a long letter to Darius as he waited for word to come down from the Board of Public Works. In his letter he told Darius in detail about the plan for the geological survey—who would examine what part of the state. Then he asked Darius a very puzzling question: "Ought I to comply with this arrangement—or to modestly leave the whole to be examined by the other three members?"[24] It is not clear if his concern was the uncertainty of where his income might come from, income that would support his work on the scientific survey; or did he think he was not as qualified as the other three older and better-schooled members? The latter is likely though never stated.

For Darius, April 7 brought good news, which Increase appended to the letter he'd been writing throughout the wait that day. That day, unconnected to the board's decision Samuel Forrer procured for Darius an allowance of seventy dollars a month retroactive to December 1 in addition to the usual allowance for subsistence and keeping a horse. But this news was superseded by general raises for canal engineers so that, as a senior resident superintending a canal, Darius's pay would go up to fifteen hundred dollars per annum "and [he] will not be troubled with anything like an allowance for subsistence or Horse keeping."[25] Darius would stay on the Miami Canal.

For Increase the news was not so good. The board had put its decision through and, just as he expected, the Canal Commission was to be reorganized into a Board of Public Works with Democrats taking the key positions—as Increase called it, "the Jackson Board of Public Works."[26] Late on the afternoon of April 7, Increase wrote his brother rather plaintively that all day "his cogitations have been in favour of entering again into the [canal] Engineer service, more especially as it is evidently the wish of the new board that I should make room for some more Democratic person as Secretary. They would probably continue me in my present situation if I insist on it—but they would rather I would not." Earlier that morning he had written that he could go into the canal service with the rank of senior assistant at the pay of a thousand dollars a year. But, he told Darius, "I have declined doing so; not only because it would be harder work & require my whole time, but because when the expenses are considered this salary is not more than 750 or 850 here. And the command of ¼ or ⅓ of my time is worth the difference between either of these later sums and my present salary." He asked Darius, "Am I wise?"[27]

Increase had good work to do on the geological survey, but it didn't pay and he might not think he was ready to do it. He felt he couldn't insist on the job he enjoyed as secretary of this new Democratic Board of Public Works. He had the offer of a very unglamorous job for a hydraulic company in Chillicothe that he didn't want to take. He could be a canal engineer again, but he knew how hard and physical the work was and how little time it would allow him for his science and his writing.

What did he do next? He traveled down to Cincinnati to see Darius and to see if he could get a canal job. The answer was not long in coming—but it wasn't what he expected. The offer made to him in Cincinnati would be one that had the potential to shift his direction away from everything and everyone that he had known, and from the state where he had made his home for many years—into a new territory, and profession, altogether.

—| |—

While in Cincinnati Increase met with Micajah Williams's partner, Byron Kilbourn, who had just returned from Milwaukee to recruit workers for his new town. The timing couldn't have been better for Kilbourn. Kilbourn asked Increase, his former rodman, to come to Milwaukee to be his business agent. He would be able to choose between a salary of "$800 per year or a quarter of the profits to be derived from his new land agency."[28] Increase accepted the offer, but it's not clear under which terms.

It is clear, however, that Increase had little choice. This work for Kilbourn was a solution to his immediate problems, but it came at great cost to Increase. If he followed through, it would take him away from his brother and family and away from his strong connections with Ohio scientists and the geological survey that was just getting started. He may have been flattered at the offer and curious about the new territory and he might have thought he could make a lot of money, but his second thoughts must have been immediate. Increase's work as secretary of the Canal Commission had centered him in Ohio engineering, government, and science. He had learned in his little more than two years in Columbus watching the Canal Commission at work how smart men worked together, how bills were written and passed, how to keep public records, how private enterprise and the state meshed. To Increase it must have seemed that this offer of Kilbourn's would narrow his world and send him off to the edge of it besides.

But his decision was made. He went back to Columbus and prepared to leave, sending Darius his books and the flower bulbs Hildreth had sent him because he didn't have a place to plant them. Increase also sent Darius a drawing of a cabinet he might make "if you take charge of my collection."[29] Things were settled. Increase was leaving Ohio for Milwaukee.

But Darius, too, had second thoughts. He went up to Dayton to see Samuel Forrer. It seems he wanted to work on the arrangements of the canal north of Dayton and wanted Increase to work with him. But when he got to Dayton, Forrer had already completed these arrangements, "so of course," he wrote Increase, "my project is knocked in the head and you will go to Milwaukee...a place which sounds to me as far off as if it were beyond the rocky mountains."[30]

Then Increase voiced his reconsiderations; the decision had been too hasty. He wanted to get out of the Milwaukee job. His parents too didn't want him to go so far from home.

By May 13, Increase's doubts about going to Milwaukee had solidified into a determination to get out of his agreement with Kilbourn. He wrote Darius that day that their father had expressed "dissatisfaction and regret" at the plan to go to Milwaukee. By this date Forrer must have come through with an offer of a canal job for Increase because he wrote Darius that "to please Mr. Forrer and to dispel father's dissatisfaction and regrets, I have fairly come to the conclusion that if I can get out of the scrape handsomely, I will relinquish my Milwaukee project." And he asked Darius to "apply to Mr. Williams for me...but get his permission to substitute another person in my place. I suppose if I send Mr. Kilbourn a suitable person (with the advice and consent of Mr. Williams) at the time I was to have gone, he can not have any just ground of complaint."[31]

Increase waited again—this time for word from Darius. Perhaps to calm his mind, Increase left Columbus and traveled around northwest Ohio taking notes for the geological survey for the next five days or so. He bought a mineral hammer and a chisel. He saw interesting plants, but no shells.[32]

On May 23, ten days after he'd written to Darius asking him to intervene with Williams, an irritated Increase wrote Darius asking him to "please have the goodness to let me know what you have done in relation to the subject of my last letter."[33] The next day he rode or walked ten miles to Jefferson looking at limestone quarries, noticing a "fine grained quartzose sandstone...very pure white" that was used to cover the bridge over the Scioto at Columbus, "to prevent the Yankees from whittling it!"[34] Again he could find no live or fresh shells.

Finally, Darius wrote Increase on May 26; Increase probably received the letter May 28 or 29 on the family farm. Here is Darius's report:

> I called on M.T.W. [Micajah Williams] on the subject of your Milwaukee project; to get his "advice & consent" to send a substitute in your place. We went to the [surveyor general's] office to see who it was that would go. Mr. Warren is the only one that he thought Kilbourn would accept in your place & he gave no satisfactory answer but left it entirely indefinite whether he would go or not. Mr. Williams is anxious that you should go, says there is no one that Kilbourn would prefer to you, none that he could have the same confidence in. And says if you have any enterprise about you that you would not fail to go for a year or two at least, & if you did not like [Milwaukee] when you got there; why then you could come back, you know. Nevertheless Mr. Williams will not let you go until the committee get through with their examinations of the Canal Commissions accounts which I suppose will not be until the meeting of the next Genl. Assembly. Mr. Kilbourn I understand engaged you subject to the duty of the settlement of the account, so that he cannot complain if you do not go there until the committee get through. Mr. Williams advised you to write to Kilbourn immediately & let him know the state of affairs, etc.[35]

So Increase had to go to Milwaukee. And soon. Right after the meeting of the Canal Commission board in early June.

Increase and Darius had failed to figure out a way to keep Increase in Ohio where his family was, where the geological survey he promoted would take place, and where most of his naturalist friends lived. Increase could have just not gone. He could have taken that job with Forrer and stayed. But he went, in spite of the doubts of the whole family. However hastily, he had *agreed* to go. So Increase resigned himself to going to Milwaukee.

He would see new territory, new rocks and plants and shells. He didn't know he was making a decision that would shape the rest of his life.

But in the Lapham way, once he knew it was inevitable, he made the best of the situation and enjoyed the ride—a full twelve hundred miles in a stagecoach and steamboat from Cincinnati to Milwaukee. As the crow and the interstate highways fly, Cincinnati is about four hundred miles from Milwaukee. But 1836 was before railroads, and the roads were so bad or nonexistent that the trip was quicker by water even if it was eight hundred miles farther and you had to travel the lengths of Lake Huron and Lake Michigan and a third of Lake Erie.

On June 12, 1836, Increase said good-bye to his brother Darius in Reading, Ohio, just north of Cincinnati and climbed on an overnight stage to Columbus—muddy roads making the twelve-hour trip two whole days. He spent a day in Columbus with his brother Pazzi, packing up his books and rocks and specimens and his few clothes. He said good-

Increase Lapham's journey from Cincinnati to Milwaukee in June of 1836 took over two weeks.

UNIVERSITY OF WISCONSIN CARTOGRAPHY LABORATORY

bye to his brother and his friends there—his sister Hannah was probably
back on the farm—and left on the morning of June 15 on another stage
for Sandusky 113 miles away. Halfway to Sandusky, he later wrote Darius,
the stage stopped for the night at Bucyrus, a "miserable hole—full of idle,
good-for-nothing folks."[36] The next morning they "crossed Sandusky's
Ridge... in the midst of the beautiful prairie," arriving in Sandusky on
Lake Erie after twenty-six more hours of jouncing and rocking on bad
roads, crowded in with six or more strangers, bugs, and bad smells. We
hope he had a window seat.

At Sandusky he immediately boarded the steamboat *North America* and
traveled east toward Cleveland, where he had business and could catch a
boat for Detroit. Ah, we think. He's on a boat on Lake Erie, a cruise. He
can rest his weary body. But it wasn't like that, he wrote Darius. Though
he said he couldn't describe "the sensation... when the boat is rocking and
pitching and is heaved and tossed about by the waves," his disclaimer almost makes us seasick. He suggests that, since he can't describe it, Darius
should recall "*any time* when you were two thirds drunk, you will have a
better idea of it than can be given on paper." Darius should "only recollect
how you put your foot down, and the ground was not there! Or, how in
putting your foot down you found it strike the ground some six inches
too high."[37] Increase had had the experience of being drunk. There's no
denying it, though he *did* deny it in later life.

Whatever his business was in Cleveland, he finished it by 2 p.m. of the
day he'd arrived and then caught the steamboat *Robert Fulton* for Detroit,
where he would wait for the steamboat *New York* to take him around
Michigan and to Milwaukee. This was a smoother trip and he got a little
sleep on a "settee." In the evening, he wrote Darius, "there was not a breeze
[breath] the blue wave to curl," waxing poetic and misquoting the poet
Thomas Moore's "Canadian Boat Song." The next morning he enjoyed
seeing the islands as the boat came into the St. Clair River, but he didn't
much like the windmills of the old French settlement—he'd never seen a
windmill before—calling them "old-fashioned, outlandish, ugly looking,
clumsy and ill contrived contrivances."[38] Seems a bit over the top for just
windmills. He arrived in Detroit on the morning of June 17, 1836, mad at
windmills, though Detroit pleased, or at least, amused him.

He reported to Darius the basics of Detroit—135 years old, eight thousand inhabitants, fine four-story brick buildings, and numerous fine churches. He went to one of the fine churches one morning and, on coming out, was surprised to see a line of one-horse carts waiting outside—carts just like the ones they hauled dirt in on the canals. He hung around to see what that was about and was surprised to see that fine ladies and gentlemen came out of church and got into them, "sitting flat on the bottom, having only a mat or some hay under them!" Then the driver got in and "took his *stand* [his emphasis] among them" and drove off in what Increase supposed they considered fine style.[39]

And Increase told Darius of a market house in Detroit "as large as an ordinary Presbyterian church and built much in the same style!"[40] The market was a religion in the Northwest Territory. Increase stayed a few days in Detroit and wrote Darius on June 21: "I am now, and have been since I arrived at Sandusky in, what might very properly be called, the world of speculators; everybody you meet is engaged in some speculation; everything you hear has some speculation at the bottom. The Hotel where I am now writing has suspended on the walls of the barroom plats of 8 new towns; I have added the ninth."[41] He had put up a plat of Kilbourntown in Milwaukee. Increase also reported excitedly to Darius that he had talked to a man who had just gotten off a steamboat that had stopped in Milwaukee. The gentleman informed him that "the town and its immediate vicinity numbers 1000 inhabitants! Eighteen months since the number was about 10!"[42] Increase was catching the fever around him.

But he was still grounded in his love of science. In Detroit he called on Dr. Douglass Houghton, a physician one year older than the twenty-five-year-old Lapham. Houghton, having had the benefit of a formal education at Rensselaer Polytechnic Institute and a wealthy father, was further along in his career than Increase.[43] He had also served as the physician and botanist on Schoolcraft's famous expeditions in the Northwest Territories that Increase had read about extensively. They must have had a good time together, these two young naturalists who had plenty to talk about—a geological survey of the Michigan Territory that Houghton was to take part in and the one in Ohio Increase had planned. Houghton gave Increase "many fine plants from the northwest."[44] But even more important, Houghton's life must have served as a model for Increase, encouraging his fever for speculation. He wrote Darius that Houghton "has made a fortune here by Speculation—has now relinquished his profession & intends to devote his time to Natural History."[45] This was Increase's dream and here was a man who was living it.

On the evening of June 22 Increase boarded the steamboat *New York*, heading deeper into the world of speculators where his new bright dream might become a reality. Increase noted the current and width of the St. Clair River, visited Fort Gratiot, and soon was on Lake Huron and out of sight of land for the first time in his life.[46] What he didn't notice, or at least didn't mention, was the scene "worthy of the pencil of Hogarth" on the deck of the steamboat—which another traveler has described. Here the passengers

> lie stretched in wild disorder and promiscuous confusion like the slain on the field of battle in all shapes and positions both sexes and all ages. . . . Here is crying and scolding, and snoring and groaning. Some in [berths] and some on chairs and trunks and [settees] and the rest on the floor. Some sitting and some lying, some dressed and some undressed, some covered and some uncovered and naked, some are stretched on beds and others on mattresses and cushions and cloaks, and not a few are trying to find the soft side of the hard floor.[47]

Increase was probably the quiet young dark-haired man with the pale blue eyes leaning on the rail, gazing at the gulls and the water and the western sky.

At the top of Lake Huron the boat stopped at the "ancient" town of Mackinac on the island and Increase ran up to the famous Schoolcraft's gate to see him, but the bell for the boat departure rang, so he just left a letter on the *New York* to be delivered with its attached plat of Kilbourntown on the return trip of the boat. Then for almost two days the boat slowly slid along the low sandy eastern shore of Lake Michigan, passing Grand Haven and St. Joseph and Michigan City, heading southwest now, then west along the southern shore of this lake he would live near and study for all the rest of his life. North along the western shore to Chicago where he went ashore to look up old friend William Gooding, but Gooding was out on the Illinois and Michigan canal line and Increase missed him. In Chicago he heard that the canal commissioners were selling town lots for huge amounts of money—feeding his speculation fever.

That day, July 1, 1836, he boarded the *New York* for the last ninety-mile northward leg of his journey, Chicago to Milwaukee. He must have stood at the rail watching the sun set over the low wooded hills and bluffs of what would in three days officially become the Wisconsin Territory, not Michigan Territory or a big hunk of the Northwest. He arrived in Milwaukee "heartily sick of steam boat traveling." He wrote Darius that "[a]t night we arrived at Milwaukee and glad enough I was, to get home."[48]

"Home." It's an odd word for a young man to choose after traveling twelve hundred miles away from his beloved family and all his previous "homes." He meant, of course, that he was home in the sense of not traveling, that he'd arrived at his destination safely. What he couldn't have known on that first day of July was that he was truly home. Wisconsin would be his life's study and in it he would make a happy family life. Increase Lapham was home.

CHAPTER 11

Begin We Then at Milwaukee

———————| 1836–1840 |———————

I N JULY OF 1836, INCREASE LAPHAM'S TRUNKS, packed to the full with
his books and journals and specimen collections, were unloaded from
the steamboat *New York* to a smaller boat that could pass over the sand-
bar at the mouth of the Milwaukee River. Then Increase and his worldly
goods were rowed up the river, past Solomon Juneau's trading post, up
to Byron Kilbourn's wharf on the west side of the river at Chestnut,
later Juneau, Street. Byron Kilbourn's house was on higher ground at
Chestnut and Third on the edge of the maple and beech forest. Increase
would stay with Kilbourn, his sickly wife, Mary, and their two young
daughters until the Leland Hotel, also on Chestnut Street, was finished
a few months later.

From Kilbourn's house on Chestnut Hill, Increase could look south
down the river and see smoke curling up from wigwams and a canoe here
and there gliding among the wild rice marshes.[1] He could see the cedar
and tamarack swamp and Juneautown as well as the Menomonee River,
which emptied into the Milwaukee River. The town, he wrote Darius, was
beautifully situated on a plain elevated from five to thirty feet above the
lake and enclosed by hills about one hundred feet high.[2] Increase's temporary home was a good vantage point that looked over a beautiful and
promising land:

The ridge of the hill upon which [Kilbourn's house] stood, ran off gradually for nearly a mile, where the spur was lost in the level ground. Beyond was the location of the future Milwaukie,...which, for want of a better name, might be called bluffs; the blue stream, winding between rice-covered banks and spreading out into a marshy lake before it reached the sea [Lake Michigan]. Indian planting grounds and oak openings marked the level country, and beyond, the white surf of the lake gleaming, as it broke on the almost desolate shore.[3]

In a few years, on this vantage above the blue river and the white surf gleaming, Increase would build his own house on Chestnut Hill next door to the Kilbourns'.

On his second day in Milwaukee, Increase got right to work laying out and leveling streets in Kilbourntown on the west side of the river so that the high places could be graded off and the swampy places filled in.[4] Chestnut Street was a strip of hard ground rising from the river, passing between two tamarack swamps—"dense and impervious"—to the bluff on the west side. It was the only way to take a team of horses out to the country.[5] Crews of men and horses and carts moving earth swarmed the street behind the surveyor as he worked westward on the hard ridge between swamps and into the timber, laying out on the complex topography the simplicity of level streets and town squares. He leveled up about ten blocks of Chestnut, west from Kilbourn's warehouse on the Milwaukee River, past Kilbourn's house, to the site of his new hotel and his land office. At Ninth and Chestnut the country was so wild and remote that one Yankee settler wrote home that timber wolves jumped over the fence into the yard and that the town would never reach that far west.[6] This was newer, wilder country than Increase had ever lived in before. Increase the curious naturalist, the surveyor, the engineer must have been thrilled at the country and the work he saw ahead of him. Increase the family man might have been a bit lonesome.

In two days Increase had surveyed north too, finding that the hill north of Chestnut rose 107 feet above the lake. He made a profile of the Milwaukee River, as far north as a little settlement called Humboldt where two Yankees had built a sawmill called Bigelow's Mills—a mill where lumber was planed so rough that at least one carpenter complained it was almost unusable.[7] But it *was* used, because the settlements were short of lumber; many buildings were "framed and raised" and waiting for the lumber to enclose them.[8]

In 1836, Milwaukee was made up of three fast-growing villages: Kilbourntown west of the river, Juneautown east of the river, and Walker's Point to the south where the Menomonee River joined the Milwaukee. The day Increase arrived, Milwaukee's "new and interesting country" was still Michigan Territory.[9] But three days later, on July 4, 1836, it became Wisconsin Territory when Michigan became a state.

Between the tenth and the nineteenth of July in this brand-new territory, Increase Lapham had made a very optimistic map of Milwaukee

W. Haviland engraved this 1836 map of Milwaukee, based on Byron Kilbourn's 1835 map. This was a map of the future city as Kilbourn and Lapham saw it.

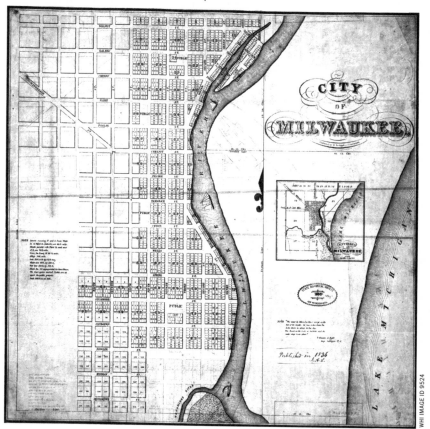

based partly on Kilbourn's earlier map, which was itself based on surveys by Garrett Vliet, another Ohio Canal engineer who had settled in Milwaukee and built a house on Chestnut Street just west of Kilbourn's. This optimism, though based on shaky or at least swampy ground, was well founded. From 1834, when Byron Kilbourn first arrived to scout for a town site, to 1836, the area had evolved from Indian territory populated by only a few dozen white people to one of the fastest-growing cities in the Northwest, from a sleepy to a bustling place where many vessels anchored in the bay, where the beach was crowded with storekeepers and goods, and where adventurers arrived from the east every day. Streets were being built up out of swampland and several were lined with stores. A flagpole had been raised on a bluff where a bright flag fluttered.[10]

—| |—

The three villages that made up Milwaukee had several hotels, taverns, stores, a baker, a tailor, and a population of about twelve hundred people—many of them laborers boarding at taverns. Milwaukee, for the settlers, was a town of about fifty houses—log cabins and mostly unfinished frame buildings. Lots in town were selling and reselling for between five hundred and five thousand dollars, but the sale of land outside the city was held up until January when the federal government would allow the farmlands to be put on the market. Increase wrote to his friend Charles W. Short in Lexington that August that the town was improving much faster than the countryside, making provisions expensive and scarce.[11] While Milwaukee was in a state of fevered growth, the territory around it languished in a state of suspended animation—at least officially. Though the land was not yet on the market, it had been surveyed and settlers were staking claims, felling trees, plowing the virgin land, building cabins, and planting crops.

Increase was at the center of Byron Kilbourn's booming town lot business. Besides his surveying work, as Kilbourn's business agent Increase prepared tract books at the new land office and kept his accounts. Both on paper and on the face of the land, he was helping Kilbourn turn this beautiful country into a city. He became a register of land claims and transfers, a job that kept him busy but did not pay well. By October 1, 1836, Increase reported to his father that Kilbourn had sold lots in Kilbourntown worth about $220,000.[12]

The *Milwaukee Advertiser*, backed by Byron Kilbourn, owned by D. H. Richards, and edited by Colonel Hans Crocker, published its first issue July 14, 1836.[13] This "advertiser" for Kilbourntown lots and improvements was an immediate outlet for Increase's writings, his observations of the natural world, his speculation about phenomena like black sands and the fossils found below the peat at the draining of tamarack swamp. And Increase used copies of the *Advertiser* as inserts in his letters to his father, brother Darius, and men like Charles Short so he didn't have to retell everything that was happening in Byron Kilbourn's town.

Byron Kilbourn, born September 8, 1801, in Granby, Connecticut, to James and Lucy Fitch Kilbourn, was now the most important person in Increase's working life. Before Kilbourn was two years old, his father moved the family to Ohio, where the visionary James founded the town of Worthington, north of Columbus, to which he attracted other New England families. Under his father's direction Byron received a good education compared with the usual meager frontier schooling, and at age sixteen, having studied mathematics, history, law, and music, Byron left home to make his own way. Where he acquired his intensely focused ambition is unknown, but his energy was boundless—he could summon the vigor to work all day all week. A big man physically, he began as a surveyor and soon became a canal man in Ohio, working on both the Ohio and Erie Canal linking Lake Erie with the Ohio River at Portsmouth and the Miami and Ohio Canal from Cincinnati to Lake Erie.[14]

For three months in the winter of 1825–1826, Kilbourn's rodman had been fifteen-year-old Increase Lapham. At the time of their first meeting in Ohio, Kilbourn had been promoted to assistant resident engineer and soon would become resident engineer with responsibility for planning the construction of locks, bridges, and aqueducts. A leader of the Ohio Canal projects and Kilbourn's superior was Micajah Williams of Cincinnati, the man who had helped Increase secure his job in Columbus and who would encourage Kilbourn to move to Wisconsin and act as his agent.

Kilbourn, impressed by his father's success as town builder, businessman, and public servant, was more aggressive than his father. The mentally

WHI IMAGE ID 62389

Byron Kilbourn had a reputation as a hard-driving, not-always-honest businessman, but his relationship with Increase Lapham was enduring and mutually respectful.

quick Byron was too impatient for diplomacy; his impolitic directness created not only admirers but many bitter critics as well. Few who encountered him forgot him and all seemed to form an extreme opinion. He left no doubt that he had come to Wisconsin with a goal: to make himself and Micajah Williams rich. And Kilbourn knew from his father's example that the road to wealth was through land speculation and town building.

Traveling by boat from Ohio, Byron Kilbourn had arrived first in Wisconsin (then Michigan Territory) May 8, 1834, in Green Bay, where he bought surveying equipment and moved south to survey towns in what was to become Manitowoc County. But his more important task, directed from Ohio by Micajah Williams, was to evaluate lands at the mouths of rivers flowing into Lake Michigan for their potential as sites for port cities. He looked at all of them on the west side of the lake and determined that the land west and north of the estuary formed by the conjoining of the Milwaukee, Menomonee, and Kinnickinnic Rivers was far superior to the others.

Using both his own and Williams's money, Kilbourn bought the land that would become first Kilbourntown and later the heart of west Milwaukee in August 1835 at the land office in Green Bay. He hired surveyor Garret Vliet to lay out the town, and he began selling lots at great profit, building his own house on the river bluff. In this, his timing was propitious; Milwaukee, relatively easy to get to with the advent of the Erie Canal and the Great Lakes, was Wisconsin's fastest-growing, most promising settlement. But Kilbourn wasn't the first speculator in Wisconsin Territory. Chronologically, Kilbourn was the third of Milwaukee's founding fathers. Solomon Juneau and George Walker had preceded him, founding Juneautown to the east of Kilbourn's holdings and Walker's Point to the south.

Like the speculators, Indians, settlers, laborers, and idlers in the young territory, Increase went almost daily to Solomon Juneau's trading post on Wisconsin and Water Streets because that's where the mud-splashed postman from Green Bay rode up, threw his reins over the fence, and told the price of land and furs up north and who had died.[15] And it was at Juneau's where the mail coach from Chicago "lumbered up" three times a week, the worn-out horses attended to by Juneau's clerk. Solomon Juneau— affable, imposing, intelligent, generous, trusting—had been a *voyageur* from Montreal who settled the east side of the Milwaukee River in 1818. He was a clerk for fur trader Jacques Vieau, married Vieau's daughter, Josette, and built a successful trading post there. Though most of the Indians had been removed from the area after the Black Hawk War, Juneau still worked in his store, trading and selling with a decreasing number of Indians and an increasing number of whites. In 1836 he and Morgan Martin, his Green Bay business partner, laid out town lots along Water Street and were building a wood-frame courthouse—Milwaukee's first— in what is now Cathedral Square. Solomon Juneau had commenced, almost reluctantly, becoming rich.

The former Virginian George Walker—the second of Milwaukee's founding fathers—had his trading post south of the confluence of the Milwaukee and Menomonee Rivers. From Kilbourntown to Walker's Point was a trip through swamp and over the Menomonee River in a treacherous dug-out boat. There Walker presided and welcomed all. He was a large friendly man only a few years older than Increase, a man well liked by white men and Indians but worried by legal problems proving his claim to his land.

Land was the primary concern for many of the men who chose to make their way in the new territory. But despite the "land frenzy" beginning to boom when Increase came to Milwaukee, it was his love of the land itself, of its rocky forms and the fauna and flora that inhabited its soils, that won his rapt attention.

—I—

As a canal worker at fourteen, Increase Lapham had concentrated his youthful curiosity on rock, its origins, forms, uses, and the stories it told about the history of the Earth. Rock became the base of his spreading interest in all manifestations of natural history. While grasses and flowers may have brought him more pleasure for their beauty, for their importance in agriculture, and for the thrill of discovering a new species in a new land, he never strayed far from the rock that lay at the foundation of his scientific mind.

Like most of his other interests, he valued geology as a practical science: Geology revealed where to dig for valuable ores—iron and copper. Geology led to building materials—clay for bricks, stone, sand, and lime. Geology shed light on the chemistry of soil. Limestone, sandstone, and shale preserved fossils that proved that the Earth was of vast age and that specific communities of ancient animals and plants differed from one epoch to another.

So it was in part a geologist who stepped ashore at the crude settlement of Milwaukee on July 1, 1836. Beyond the Lake Michigan shore spread an unexplored geological jumble whose parts were waiting to be understood and explained. Lapham began to explore his new place immediately. Looking down at the beach, he could see fossils similar to those he found in the deep cut at Lockport. He recognized that the inland surface was composed of "drift," that is, worn boulders, rocks, gravel, sand, and clay that obviously had drifted in from elsewhere, obvious because much of it did not conform to the limestone bedrock. Increase was familiar with drift. His second paper for *Silliman's Journal*, written with his brother Darius in 1832, was devoted to the mystery of drift in Ohio.[16]

Increase and most others at the time explained drift's origins as "diluvial," that is, as having been brought in by a flood. Diluvial at that time had two meanings, one specifically referring to Noah's flood of the Old Testament, a literal belief of Christians; the other evolving as a scientific term referring to any flood. The flood theory of drift was soon challenged by the theory of glacial drift, but it would be decades before Increase gave up on it. Whatever its origin, Increase saw that most of Wisconsin, like Ohio, was covered with drift.

Increase also noted the alluvial composition of the more-than-one-hundred-foot-high hills above the level of Lake Michigan. He noted that the hills on the Menomonee River two miles above Milwaukee were bluish-gray clay and porous limestone that might be useful for brick. He saw on the west side of the Milwaukee River in the swamp that had originally been covered with tamarack and cedar trees, below a layer of loose peat dug through to drain the swamp, a "whitish kind of earth compose[d] principally of clay and sand." In that marl, Increase found "immense quantities of fossils shells similar to those now inhabiting the adjacent grounds," eighteen species of which he identified.[17]

The botanist Increase was also at work and had noted by the first week in August ten plants and four shells that were familiar to him. He immediately began to keep a list of the plants he found in this new country. He wrote to Short that he found a great many plants new to him that he identified from his botanical books, plants that gave him as much pleasure as finding "old acquaintances" from Ohio.[18] About six weeks after his arrival, Increase wrote Short that "[i]n order to inform my friends of what plants are found here and to enable them to indicate such as they want I think of publishing a catalogue of such as I find."[19] And so he did. He published his plant list at the *Milwaukee Advertiser* office as *A Catalogue of Plants & Shells, Found in the Vicinity of Milwaukee, on the West Side of Lake Michigan* by October of that first year in Milwaukee. He sent this book off to his naturalist friends, including the botanist Asa Gray at Harvard, who had asked Increase, in one of the first letters he received in the Wisconsin Territory, to "pay particular attention to my *pets*, the Grasses."[20] This little palm-sized book was called by historian Milo Quaife "the first publication of a scientific character within the present state of Wisconsin." Quaife added that "[i]t would be probably safe to affirm that this was the first scientific work to be published west of the Great Lakes, at least to the north of St. Louis."[21] In his quiet way, Increase had begun to write and make history, building it upon his powers of observation and identification, upon his energy and his increasing command of his language.

And in Wisconsin he found a new interest. Within days of arriving in Milwaukee, Increase had seen both wigwams of Indians who still lived there and the ruins of the wigwams of the Menominee Indians and their abandoned corn hills. He had also sketched eight ancient Indian mounds, some club shaped, some conical, one like a bird in flight, and one his first "lizard" mound. These were on a bluff at Fifth and Walnut, quickly lost

A

CATALOGUE

OF

PLANTS & SHELLS,

FOUND IN THE VICINITY OF

MILWAUKEE,

ON THE

West side of Lake Michigan.

BY I. A. LAPHAM.

MILWAUKEE: W. T.

PRINTED AT THE ADVERTISER OFFICE.

1836.

to the leveling and grading of these streets in Kilbourntown.[22] As soon as he arrived in Milwaukee, Increase became both a recorder of the ancient natural and historical treasures of Wisconsin Territory and a party to their destruction. This would be Increase's life in Wisconsin—using his scientific and engineering skills to improve life for the men and women around him, which also meant the destruction of the Indian mounds, the hills themselves, even the plants he loved so much and knew so well.

By the first of October of that first year, not only had Increase Lapham explored and mapped and described the town of Milwaukee but he had also made several "jaunts" beyond Milwaukee. Alone he walked twenty-five miles south on an Indian trail to the Root River at Racine, then to the Pike River and back to Milwaukee. For seventeen miles of this trip he saw no house, no white men, no Indians. He told his father that he would have been much more afraid of a "gang of white men" than of Indians.[23] He went to Chicago—"a low level muddy place"—where he traveled along the Des Plaines River, which runs through heavily timbered land bordered on both sides by broad, rolling prairies, and "saw some of the finest lands that [he] ever did see."[24] And he traveled with an experienced backwoodsman on horseback fifteen miles north to Menomonee Falls where he camped on a quarter section he had made a claim to. This, Increase said to his father, was his "first essay in camp duty." He told his father how he, Increase, canal man, naturalist, and—we forget—city man, cooked his dinner by hanging a piece of meat on a sharp stick, "fastening the end of the stick in the ground before the fire."[25] The experience of sleeping on the ground under a small cloth tent was new and interesting to Increase, late of Columbus, Portsmouth, Shippingport, and Lockport. *Now* he was on the edge of the known world. But even more interesting to him than the details of camping were the two kinds of rocks he found at the eroded falls, the descent of the water, the lack of fossils.[26]

On October 15, 1836, though he was still excited by the newness and the raw prospects of the place, Increase again wrote to his father telling him that he felt himself much less "out of the world" than he supposed he would in Wisconsin Territory. His father had expressed interest in moving

(left) Title page of Increase Lapham's 1836 *A Catalogue of Plants & Shells*

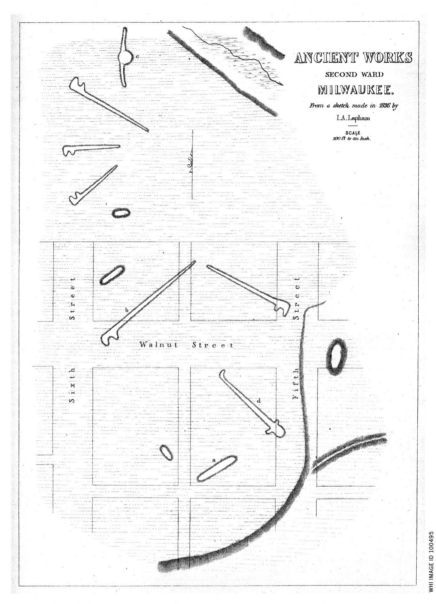

Increase's drawing of the Indian mounds in Milwaukee, based on his 1836 sketch and reproduced in *Antiquities of Wisconsin*. These mounds would have been only blocks from where Increase lived near Third and Poplar.

to Milwaukee to work at "carpentry or other business," so Increase sent him a letter "full of facts from which you can form an opinion for yourself."

Increase told his father who had begun his working life as a carpenter that journeymen carpenters make $2.50 to $3.00 a day, and the work had been constant, though slack in winter. Boarding was $4.00 to $5.00 a week—low considering the high price of commodities like pork and flour.[27]

The rest of this letter from Increase to his father took on a different, less practical tenor. Increase told Seneca the projections of the improvements planned for the coming year. In 1837, Increase breathlessly told his father, Kilbourn would build a dam across the Milwaukee River to power saw and flour mills. Kilbourn would also build a bridge across the Menomonee River, floating bridges across the Milwaukee River, a lighthouse near the harbor, and for himself, a "fine house." Kilbourn was planning a railroad to run west to the lead country. (Interestingly, Increase does not mention the Milwaukee and Rock River Canal, one of the reasons Kilbourn brought Increase to Milwaukee.) All of these plans meant good work for Increase. And the plans of Juneau and other entrepreneurs for churches, schools, houses, and other public buildings could mean work for Increase's father whose farm was always, like most farms, just barely making it. The prices of pork, oats, corn, and potatoes would make farming—if Seneca wanted to continue to farm—"the best business that can be engaged in here and will be for many years yet." The "healthy climate" and the need for houses as well made it such a sure thing that Increase suggested his father sell the farm and come to Milwaukee, where he would make a fine living.[28]

And a fine living indeed it could be. On January 23, 1836, Byron Kilbourn wrote his father, James:

> I will now say to you that my interests at the N.W. [northwest] are worth $100,000 without over estimating them one farthing. My share in the Sales already made are rising of $20,000, about $8,000 of which I have received in cash—enabling me to discharge my bank loan of $6,000—defray all other expenses, and leave a small residue—I have no doubt of realizing a larger Sum than that stated if times remain as they have been.[29]

The land frenzy was still very much under way when Increase Lapham arrived in Milwaukee in July of 1836 to become Byron's business agent. He was twenty-five and single and had stepped into the chaos of the frontier; how could he not be tempted to jump into the land game? After all, it was the community's first economic culture. Under the influence of Kilbourn and the easy-credit habit of the striving frontier village, Increase added to his conservative, studious, and steady identity the riskier one of a land speculator.

Increase had always been susceptible to role pressures. With geologists, he was a geologist. With botanists, he was a botanist. Now he was working for a developer in a land sales office and he took on some of the ways of Kilbourn. In any case, this new Increase was not so readily recognized by Darius and perhaps the rest of his family. With a salary of $1,000, he had bought three Milwaukee lots, risking $5,000, half payable in a year and half due in two years. "I bought them for the purpose of selling again at a higher price," he told his father in a letter on October 1. He also made a claim on 160 acres near the falls of the Menomonee River fifteen miles northwest of Kilbourntown. "This is done by clearing three acres and building a log house, which cost me $50, and entitles me to the land at the minimum price ($1.25) at the time of the public sale, whatever may be its value at the time. The falls... must soon be the site of a town."[30]

He was now a minor participant in a nationwide system in which the federal government sold large tracts of land cheaply and on easy credit, often to nonresidents who had no intention of settling or farming on the land but would resell immediately for huge profits to cash-poor farmers. "The effect of these speculations was greatly to retard and prevent the occupancy of the country by permanent inhabitants," wrote Moses M. Strong, a Mineral Point attorney, decades after he had served as an agent of speculators. According to Strong, total sales of government lands in Wisconsin prior to December 31, 1836, amounted to 878,014 acres, of which as many as 600,000 acres were probably sold to speculators.[31]

But the fever of speculation wasn't to last. The scheme resulted in a national political scandal, implicating several congressmen who were also speculating—all on questionable credit and using questionable bank notes. In 1837 President Andrew Jackson circulated an order that called for payment of land in "specie," that is, coins of gold or silver, commodities that were scarce on the frontier. In addition, the president temporarily suspended sales of recently surveyed lands, including those in Wisconsin.[32] His actions burst the speculation bubble and quickly brought on the depression of 1837.

In Kilbourn's corner of Wisconsin, Jackson's order not only halted the buying and selling of land on credit but also stranded many would-be speculators in the new territory. In an ironic twist, it also landed Increase and Byron Kilbourn in new roles of protecting settlers *against* speculators.[33] With the speculation bubble deflated, Kilbourn must have seen that his fortune now lay in protecting those settlers who would help his city grow. Most settlers were cash-poor squatters who had staked a claim on unoccupied land, hoping to legalize the claim by improving the land through building, clearing, and cultivation, then pay for it with the proceeds from their early crops. Lacking cash, they ran the risk of competing with wealthier speculators for title to land they were farming. Jackson's order did nothing to protect them.

On February 27, 1837, a notice appeared in the *Milwaukee Advertiser* inviting the people of Milwaukee (then including Waukesha), Washington (then including Ozaukee), Jefferson, and Dodge Counties to a meeting on March 13 in the Milwaukee Courthouse "for the purpose of adopting such rules as will secure to actual settlers their claims on principles of justice and equity," and stating that it was the duty of the settlers "to unite for their own protection when the lands shall be brought into market." At least a thousand people attended, an astonishing number. Equally astonishing to some, "It was not a rabble of lawless 'squatters,' but the men who have assisted in laying the foundation and rearing the superstructure of our state."[34]

A committee of twenty-one that included Byron Kilbourn drew up sixteen rules that in essence outlined ways to ensure that settlers who had improved land would be able to buy their claims at $1.25 an acre when government sales resumed. Increase was given the job of registering the claims, and he may even have drafted the rules.[35] It was Increase who stood on a table and read these claim laws to the assembly of settlers.[36]

Under Jackson's successor, President Martin Van Buren, public land sales resumed in the Milwaukee land district on February 18 and March 4, 1839, without competition from speculators. The settlers kept their farms for $1.25 an acre. It is not known whether Increase made or lost money on speculation. His holdings near Menomonee Falls may have been protected by the very rules he helped to enact. And in the end, the wise and cautious family man Seneca Lapham did not sell out and move to Milwaukee. It's a good thing. The boom in Milwaukee—and therefore Kilbourn's and Increase's plans—collapsed with the depression of 1837. The future became as fickle as the Wisconsin spring Increase complained about as being "too short to suit a botanist."[37]

While it had its roots in Jacksonian policy, the depression actually began three months into the presidency of Martin Van Buren and endured for years. Hundreds of banks closed. Cash was hard to come by. Successive regional crop failures in 1836 and 1837 in more settled areas didn't help. "We begin to feel the pressure of the times here in the woods, money is scarce, provisions scarce and dear, very little doing in the way of making improvements, in fact Milwaukee begins to be a dull place and will now jog along at the rate of other towns," Increase wrote to Darius on September 27, 1837.[38] This was very different from the heady tone in which Increase had written to his father a year earlier.

Instead of improvements and growth and a bright future, in the spring of 1837 times began to be narrow and dark. Sales of lots in the towns slowed and stopped. No one in town had cash. Kilbourn didn't have the money to pay Increase. Instead of Seneca and the family moving to thriving Milwaukee, Increase was considering moving back to Ohio.

—∣—

One of the pillars of Increase's life had always been his relationship—often through letters—with his brother Darius. A large part of Increase's pleasure in seeing new country, new plants, and geology had been reporting what he found to his older brother—learning it thoroughly as he described it to the critical, funny, and receptive Darius. Together, over time and the long distances between them, they had figured out the geology of Ohio and the names of many new plants. This demanding and pleasurable correspondence is a large part of what had made Increase an accomplished writer, a master of English vocabulary and structure, by the time he was twenty-five.

Within a week after he arrived in Milwaukee, Increase mailed a long letter to canal engineer Darius, twenty days before he wrote to his parents. This letter, detailing his trip from Reading, Ohio, where he'd left Darius, to Milwaukee, is the only record of that trip. He ended that letter saying that he hoped to hear from Darius by return mail. By October 1, Increase had written Darius three or four times and sent him copies of the *Milwaukee Advertiser* and of his *Catalogue of Plants* and yet he had not heard from Darius: Is he still alive, Increase asked his father—a joke and yet not a joke.[39]

It wasn't until the first week in December that Increase received a letter from Darius when he had "lost all patience and had fairly come to the conclusion that [he] had determined not to write to me." But this was not the usual warm, reciprocal letter from Darius, interested in and commenting on what Increase had to say. Darius uncharacteristically blasts his brother: "You have gone so far to the 'far west' and joined yourself to that outlandish conglomeration of folks called 'land speculators'…that I have been at a loss how to concoct a letter to suit you under the circumstances." Then Darius passes on advice and observations from Increase's friends and former bosses in Ohio—advice that contradicts Darius's opinions. He tells his brother that Micajah Williams suggests that Increase leave the "yarbs [Southern dialect for herbs, plants] and mussels" alone and make himself rich. Samuel Forrer, he adds, is afraid that Increase will forget to get rich in his zeal for collecting shells. Even if he is joking, Darius seems to want his little brother to remain a pure naturalist and engineer, untainted by what he sees as grubbing for money. But he does pass on a compliment: "Micajah says he introduced you to Gov. Dodge [the first governor of Wisconsin Territory] and told him that his government was commenced under most favorable circumstances for internal improvements having two first rate civil engineers in the territory and that you were one of them!!"[40] Darius clearly sees his brother as an engineer and naturalist and finds Increase's even modest speculations distasteful.

Darius had finally written to Increase, but his tone was nasty and he had made no comments about the new country that Increase so carefully reported on. Darius's long-awaited letter contained only contempt for

Increase's endeavors, unwanted advice, and political news. Increase wrote back to Darius in early December expressing his hurt and disappointment:

> I regret very much that you do not answer my letters, for I want to give you an account of my numerous jaunts into the country in this vicinity; the woodlands, the prairies, the beautiful lakes surrounded by regular hills and filled with beautiful islands; our rivers and cascades, the rocks, stones and "yarbs"; the ancient mounds made to represent the turtles, lions (?) bears and other animals, our native copper, our lead ore and our tremendous agricultural productions. I want you to know how I spend my time, what I am doing, how many canals, railroads and turnpikes I have located and a long list of other things which you formerly (or at least I supposed so) took some interest in knowing.[41]

By the first of the year 1837, Darius must have written Increase to say that he would "read with some interest an account of matters in general in relation to this country," because Increase wrote him a long, informative, and somewhat stilted letter telling him what was going on in Milwaukee.[42] Darius wrote on March 15 and April 6 with news that the Columbus, Ohio, Board of Public Works was getting rid of everyone on the board not a Democrat.[43] Increase answered on May 12, telling Darius that he was making a garden for their brother Pazzi, who was to arrive with his wife and baby soon, a short-lived scheme. In this May 12 letter, at the worst of the bad times, in the depression when there was no cash and few prospects, Increase sent Darius a copy of the *Advertiser* wherein Darius could read that Increase had received the "very profitable office of 'Register of Claims'"; "it furnishes plenty of business but no pay," explained Increase.[44]

In his next letter to Darius, on August 8, Darius's twenty-ninth birthday, Increase was so desperate for money and work that he asked if he should return to Ohio. Was there any prospect of his finding employment under the Democratic Board of Public Works? Increase answered his own question: "I suppose not." He hadn't heard from Darius since April 6, when Darius must have made some crack about Increase's Milwaukee boosterism, because Increase closed a lackluster letter to Darius by suggesting he write to Pazzi for "correct" information on Milwaukee: "He has been here [and gone] and therefore he knows."[45] Darius's distance and criticism continued to puzzle Increase through all of 1837. Darius did have his most important and probably demanding job so far—chief engineer of the Cincinnati and Whitewater Canal—but the demands of work had never yet interfered with the correspondence between the brothers.

Finally, on December 9, 1837, Increase was no longer willing to tolerate Darius's silences and sniping; his short, unresponsive letters; his hostility and coldness. This letter to Darius demonstrates Increase's sureness of himself and willingness to defend himself, his ability to obliquely convey information, and his mastery of rhetorical devices—repetition, the use of questions, and irony. In this we clearly hear the strong voice of Increase Lapham:

> Dear Brother,
>
> I received a few days since on my return from the "Mineral Country" your long epistle on the subject of shooting stars, Pazzi's journey, how to get along in the world,...how you are drawing a salary of $3,000, how much you are ahead of the other engineers in Ohio...all of which is very interesting to me, and now it becomes my duty to sit down and write as long a letter in reply. But how shall I do it? What subject shall I select which will be interesting to you? There's the rub! What do you care about our flourishing town, it's all a "Humbug"! so I will say nothing about that!
>
> Shall I tell you about the country between here and the Mississippi which I have just visited for the purpose of studying its features, its productiveness, its mines, its water communications etc.? What are all these to you?
>
> I suppose [you would not want to know] of my descent into a mine 90 feet deep and extending many hundred feet in various directions, or shall I tell you how much I was surprised to find ore lying in horizontal beds instead of veins as had been represented, or...should I tell you of our sleeping in the woods and being awakened by the hooting of an owl instead of the crowing of a cock in the morning, these things I suppose you care nothing about. What then shall be the subject?
>
> If you care nothing about the country I am in, perhaps you will care something about me. Shall I tell you, then, what I am doing and how I do it, how I spend my time, how I amuse myself during my few

leisure hours? Would you like to hear about my private concerns, how I get along in the world? How much is owing to me and how much I am indebted to others? How much of this world's good I have in possession? With these you are doubtless pretty well informed already, and from the fact that I have not sent home anything to assist in building the house [their parents' house] you may infer that I am not in possession of much of the "root of all evil" an abominable thing! Which I never suffer to accumulate on my hands!

If I should tell you how often I go "a courtin" how many pretty lasses we have here and all about our parties and sleigh rides, you would probably think I was consuming time, pen, ink and paper foolishly, and you would think right enough too!...

If I should detail the plan proposed to procure the immediate construction of a canal from Milwaukee into the "Mineral Point Region," should tell you what steps have been already taken to bring about that object, or should tell all that has been done in the way of exploring the route, you would perhaps think it was only a visionary speculating project which would never be carried into effect.

In view of all these difficulties I can only conclude by asking,

What shall I write about?

Yours affectionately,

Increase[46]

Darius's response to this letter is "Yours received. No news, but war and rumors of war, which I suppose you care nothing about. The Seminoles have killed 130 of our troops and eight officers, and the Canadians have captured a steamboat, killed fifteen or twenty persons and sent the boat in the night blazing over the Falls."[47] Darius's (inaccurate) reference to Colonel Zachary Taylor's rout at the hands of the Seminoles in the Battle of Lake Okeechobee and his reference (also inaccurate) to the Canadians' capture and burning of the American steamship *Caroline* seem an attempt to make Increase's concerns appear small in the scheme of things. And it is true that Increase seldom if ever refers to events outside of Milwaukee. Darius, on the other hand, is taken up by inflammatory politics in their tabloid versions. And Darius accuses Increase—who has just walked twenty-five miles through deep snow and cut and infected his hand—of being a dandy and wearing gloves!

Increase, in the middle of these dark times, refused to give up on Milwaukee. Yet as he told only his mother, he suffered from the "hypo"—depression.[48] Rachel wrote back to tell him to rely on "the enjoyment of true and substantial friends"—something Increase had yet to find in Wisconsin.[49] In the meantime he willed himself to make this place *home*. Unlike his brother Pazzi, who moved around constantly, Increase knew that you can't find a home by wandering in search of it; you find a home by staying in one place and connecting yourself to it with tiny threads of observation, work, time, and human relationships. Human roots are not tap roots, but more like the deep thready roots of Asa Gray's "pets," the prairie grasses. Increase had had enough of a wandering life. Though this Wisconsin Territory was not perfect, and he was lonely—he had barely in the first two years mentioned in his letters the name of another Milwaukee citizen—Increase would stay. In May of 1838, Darius wrote saying he could have a job as an assistant engineer on the Mad River Feeder Canal. But Increase wrote back that his "principal objection is a desire to establish myself permanently at some particular place, but I had rather be a principal engineer in Wisconsin than an assistant in Ohio."[50]

In the spring of 1838, Increase boarded at the home of a fisherman and his family, the Blanchards. Mr. Blanchard caught the whitefish Increase loved and Mrs. Blanchard, "kind and motherly," cooked good meals for the family, which included a grown girl who played the piano to Increase's delight and two little girls about the same age as his little sisters.[51] Living with this family lifted Increase's spirits. In his neat, well-carpeted room with his bureau and bookcase well stocked with books on law and natural history, Increase did some of his office work and slept on the little mattress he had brought from Ohio.[52] He reported to his sister Hannah that his income was $1,500 to $1,600 a year—"in promises."[53] But he had no cash. No one in Milwaukee had cash—just "wild cat" money that wasn't worth anything. He told his parents that he could live almost without the "root of all evil"—probably quoting Darius. "Nobody starved *yet* in Wisconsin," he wrote Darius.[54]

In April of 1838, Increase wrote his mother that he supposed finding "a true and substantial friend"—he can't say "wife"—would be difficult in

WHI IMAGE ID 101779

The first known photograph of Increase Lapham was printed in an unidentified newspaper and saved in one of Increase's scrapbooks.

this new and thinly populated country.[55] But it turned out not to be difficult at all.

One day in June of 1838, when the economic depression had lingered for more than a year and when the letters from Darius were coldest and most hostile, Increase, while in the company of three young women of his acquaintance—Miss B., Miss D., and Miss O. as he called them in his journal—was handed a letter. He opened the letter—it was from Darius—and began reading this latest puzzlement. Darius, who had recently chastised his younger brother for writing to him about things he was not interested in, was now telling bachelor Increase about the proper way to raise daughters. Darius wrote that "making delicious pies" is a "better recommendation for a young lady and better calculated to secure her a good husband than all the piano playing that ever was performed."[56]

Increase handed this private letter over to these young women. So they stood together, these four young people, the girls in summer dresses and bonnets, laughing and talking, one of the girls likely reading the letter to the others. They must have laughed at Darius's ideas of the "proper" way to bring up girls, at what he thought would attract a man.

Increase's gesture—the handing over of a private letter from Darius to the three women—is one of exasperation, frustration, and a small betrayal of Darius. It is a tiny point of turning for Increase—away from Darius, toward Milwaukee friends. It is a gesture of trust of the women: What do *you* make of this, he asks them. What am *I* to make of it? He has put something intimate—a puzzle—into the hands of friends.

And one of those women would soon become his longed-for "*dear friend*."

Thinking Increase eligible or charming or handsome or all three, each of these "dear creatures" baked for him, whether or not they agreed with

Darius. One a mince pie, one several cakes, and the other a "delicious pie made of the leaf stalks of the burdock! Or some other similar plant."[57]

Natural scientist that he was, Increase collected data about these three women and reported it to Darius. He knew their weights—110, 112, and 147 pounds—"ascertained by actual experiment." One was "very still among strangers." Two were fond of flowers and the other "rather disinclined to plant them." One attended church regularly, "the others I suppose read their bibles at home—or perhaps novels."[58]

Miss O., likely the woman who loved flowers, weighed 110 pounds, and went to church, was Ann Marie Alcott, age twenty. He had heard her name, but not yet seen it spelled. She was visiting Milwaukee from her home in Marshall, Michigan.

For the rest of the summer of 1838, Increase was silent; there were no letters to his family, though he continued his botanical correspondence with William Darlington, J. R. Paddock, John Torrey, and Charles Short.[59]

Increase Lapham's courtship of Ann Marie Alcott was not committed to paper. He kept it all to himself until things were decided between them. Sometime before the middle of August she had agreed to be Increase Lapham's wife. We know this because on Sunday evening, August 18, 1838, Increase, Ann Marie Alcott, and a Miss Jones, together with some other friends, boarded the beautiful little sailboat *Henrietta* and sailed down the "delightful river, Milwaukee," to the wharf, where their friends left them and the three boarded the steamboat *Rhode Island* for Chicago and then St. Joseph, Michigan. Increase was going to Marshall, Michigan, to ask Ann's widowed mother for Ann's hand in marriage. As Miss Jones stopped in Chicago, the two lovers traveled on alone together, happily.[60]

Finally the two arrived in Marshall—"another delightful place, rendered more interesting by being the residence of the lady in my company." The whole trip happened in a golden light, it seems. Then there was the happy meeting of the brothers and sisters, nieces and nephews, and the mother of Ann. Increase stayed with the Alcotts for two days.

"I don't know as I have any objections," said Mrs. Alcott.[61]

"If thee has found a woman who will share the trials of this life with patience and be a helpmate thee has found a prize worth having," wrote Increase's parents.[62] Increase had found just such a woman in Ann Alcott.

Ann Marie or Maria (both spellings are common) was born in Ballston Spa, New York, on May 22, 1815, the daughter of Amos, a businessman, and Mehitable Simmons Alcott. Ann was the youngest of six children who grew up and attended schools in Rochester, New York. When Amos died in 1824, most of the family moved west to Michigan with their widowed mother, three of them eventually living in Milwaukee, the eldest son settling in Boston.

After making preparations in Milwaukee, Increase traveled back to Marshall, and Ann and Increase were married October 24, 1838—with many members of her family present, but none of his because of the distance. The next day the bride and groom left for Milwaukee, along with two of Ann's young nieces, Elizabeth and Caroline Stone. Their mother, Ann's sister, had died and it was decided that the girls would live with their aunt and new uncle. In Milwaukee, they boarded at the American Hotel until Increase's newly purchased house on Third Street would be ready for them. On November 28, Increase happily "commenced housekeeping" with Ann and the two girls—his readymade family—and Ann Elliot, "the help." He made a rare journal entry noting this important date.[63]

Increase and Ann were homebodies. Ann wrote in an essay when she was sixteen that the very word *home* is fraught with "pleasing...and tender reflections, of childhood years."[64] She, too, had been separated from the dear places of her youth. In December, Increase described his house and home on Third Street for Darius. First, Increase the surveyor tells him that it is on Lot 9, Block 39, one half block north of his office. So besotted is he with this place and his life in it that he even tells Darius that to get into the house you go up three ten-inch steps. The parlor, he says, is nicely carpeted and furnished with a center table on which is a beautiful lamp and a collection of marine shells. In the dining room, Ann usually has something good for Increase to eat. The kitchen has a yellow floor, below which is a cellar and lumber room and a spring. Two bedrooms, plenty of closets and pantries. And a quarter acre for a garden out back.

"Excuse me for writing so much about myself," Increase continued, but "I think getting married has had the effect you supposed it would, of making me feel of more consequence to the world than before, for I cannot even write about anything but *self*."[65] Increase wrote this letter in his "own little family room," probably a study, which Ann had just finished dusting and sweeping, then left him alone there to write without interruption. In this room was likely the one piece of furniture that distinguished his abode from all others in Milwaukee—his "Cabinet." While tabletops and bookshelves might have held pretty things like shells and arrowheads, his cabinet would have been where he stored fossils, minerals, and archaeological items for study, classification, and preservation. Boxes and shelves must have held his preserved plant specimens. A desk or box held his saved, carefully folded, labeled, and filed letters. This first house of the Laphams in Milwaukee is important to imagine because it was not just a family home, but an intellectual home, carefully nurtured now by both Ann and Increase. This little room in this home on Third Street was the first germination site for the ideas of Increase Lapham that would be so important to Wisconsin. Increase's library, described by a later friend, "was a great pigeon hole where all the results of his previous study, and that of others on the same subjects, were methodically filed so they could be found at a moment's notice."[66]

Everything we know about Increase Lapham up to this point tells us that he was a good and dutiful son who had the model of a good and dutiful father to follow. We have seen over the years Increase's frequent expressions of longing to be with his close-knit family, his attention with his time and money to his little sisters' education, and his dutiful, freely given help and cooperation to the family—from peeling turnips to going after the cow to giving them a large part of his income. There is nothing to make us think he was not a good and attentive husband and father. Increase, as he set up housekeeping with Ann, set about re-creating the happiest aspects of his own childhood, minus the disruptive uprooting of the canal life. He would and did make his home safe and ordered with good things to eat and to look at and to read. He would make it a stimulating place by bringing home all the interesting creations of the botanical and geological world that he could put in his pockets, specimen box, or wagon. He and Ann would be warm and welcoming to family members and strangers alike. He would collect and grow useful and beautiful plants and keep a cow and chickens to make his home a little Eden. His mild voice and smiling eye and calm temperament would set the tone for his happy home for his wife and children for years to come. In Ann he had married a woman with apparently the same aims, temperament, humor, intelligence, and warmth. Though not so lucky in his professional life, Increase was among the luckiest of men in his home life.

A year later Increase and Ann's first child was born, a daughter named Mary after Increase's older sister who had recently died. And perhaps after Ann's sister, Mary Jane Alcott Hubbard, who lived nearby in Milwaukee with her husband.

Increase's life had changed forever. He truly was home. Two years before, it had only been wishful thinking.

CHAPTER 12

The Milwaukee and Rock River Canal

$\dashv$ 1836–1840 $\vdash$

O N JULY 27, 1836, ONLY TWENTY-SIX DAYS after he landed in Milwaukee, Increase wrote his father Seneca: "[Byron] Kilbourn and [Micajah] Williams are to make a slack water navigation several miles up the Milwaukee River and thereby create an extensive water power at the upper end of the town. It is to be done next season and I shall do the engineering."[1] This may be Increase's first written confirmation that his career in canal building promised to continue in Milwaukee. Ostensibly, Lapham's first duty in Milwaukee was as a business agent concentrating on land transactions. But the Kilbourn-Lapham relationship began with canals in Ohio and it was nurtured in Milwaukee by their common interest in canal construction. The colleagues would have ample opportunity for canal talk.

Kilbourn and Lapham were not alone in considering possibilities for canals in this settling land. A group in Green Bay was keen to join the headwaters of the Fox River, which emptied into Green Bay, to the Wisconsin River, whose waters eventually reached the Gulf of Mexico. This had long been the route of the fur traders and required a mile-long portage between rivers. A canal linking the Fox with the Wisconsin at Fort Winnebago, today's city of Portage, appeared inevitable. To the south, Chicago promoters were busy with schemes to connect Lake Michigan and the Mississippi River with a canal through the flat prairies of northern Illinois.

The Milwaukee–Rock River canal proposal was under way before Increase wrote his father. On July 6, 1836, the Milwaukee and Rock River Canal Company was organized. The 1840 publication *Documentary History of the Milwaukee and Rock River Canal,* which was likely written by Increase, who served as engineer and secretary of the canal company and editor of the report, begins:

> During the summer of the year 1836, being the first year after the settlement of Milwaukee, public attention was directed to the importance of uniting the waters of Lake Michigan with those [of the] Rock River by means of a Canal; and although the country was then but little known, some general examinations were made by Byron Kilbourn, Esq. (now President of the Milwaukee and Rock River Canal Company) and other experienced persons, which resulted in the convictions that such a canal could be made at moderate expense.[2]

Kilbourn and Lapham, an "experienced person," worked west of Milwaukee that summer locating a route for the canal. Much of it would utilize existing natural waterways—navigable rivers and even the open inland lakes in what was to become Waukesha County—all linked together by a canal a minimum of forty feet wide and four feet deep.

Their first effort was sketchy, perhaps because they were under a self-imposed deadline. The first session of the new territorial Legislative Assembly of Wisconsin was to begin that October and Kilbourn wanted to get his request for incorporation approved so they could go to work. In the petition to the territorial legislature, the canal company described a route roughly forty miles long between the mouth of the Milwaukee River and the Rock River where it is joined by the Bark River, the site of today's Fort Atkinson, "thus affording facilities for the transit of the heavy commodities of the country along the route, and affording a cheap and easy communication by water from the Lead Mine District on Picatolica [Pecatonica] via Rock River to the Lake."[3]

Moving lead east was the main justification. Miners began to migrate into southwestern Wisconsin in 1825, when it still was part of Michigan Territory, attracting speculators and miners from Kentucky, Tennessee,

Missouri, Virginia, and southern Illinois, with scatterings from other regions.[4] They mined lead and a little copper and the area soon emerged as the leading source of lead for the nation. Of course, at the time the only way to move lead from the frontier to markets was by water. The ore was hauled by wagons to local furnaces, where it was smelted into pigs weighing seventy pounds each, which were loaded onto flatboats that moved down the Fever River through Galena, Illinois, or down the Wisconsin River and other streams, eventually reaching the Mississippi River, which provided access to St. Louis and New Orleans.

The Rock River had come into play as a lead route by 1830, before the Black Hawk War of 1832 led to subsequent treaties that settled residual Indian animosity east of the Mississippi River for good and opened Wisconsin for permanent settlement. On June 24, 1830, John Dixon, founder of the town of Dixon, Illinois, wrote to the editor of the *Miner's Journal* at Galena: "The first flat boat built on the Pickatolica [Pecatonica] passed here this day, bound to St. Louis, with one thousand pigs of lead [70,000 pounds] for Col. William S. Hamilton."[5]

Lapham and Kilbourn knew the geography of the newly charted territory well enough to understand that boats moving down the Pecatonica entered the Rock River just south of the Illinois-Wisconsin boundary, near Beloit. If the boats turned downstream, they would have more than one hundred miles to go to reach the Mississippi at Rock Island, Illinois. From there they could make a run to St. Louis or New Orleans. Even at New Orleans, the lead was still thousands of miles by water from eastern US markets. But with the Milwaukee and Rock River Canal, flatboats or small steamboats could turn upstream, navigate a much shorter distance to Milwaukee, and gain access to the eastern seaboard through Lakes Michigan, Huron, and Erie, and the Erie Canal and the Hudson River.

On November 29, William B. Sheldon, a member of the legislature's committee on banks and incorporation, reported a bill to incorporate the Milwaukee and Rock River Canal Company. With the legislature concentrated first on the time-consuming task of forming a government, the canal bill was tabled for the session, an act that so miffed Kilbourn that he organized a public meeting in Milwaukee to condemn the representatives for their failure to pass it.[6]

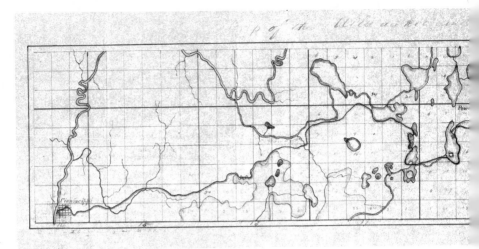

Increase Lapham's plat map of the Milwaukee and Rock River Canal, ca. 1837, as planned by Lapham and Kilbourn. The Rock River is shown at the far left, the burgeoning city of Milwaukee at the far right.

After that setback, Kilbourn and Lapham spent another session in the field to further refine the canal route. "In the year 1837, a preliminary survey was made by Mr. Kilbourn and the Editor [Lapham], by which the entire feasibility of the work was ascertained, and an approximate estimate made of its cost," Lapham wrote in *Documentary History*.[7] He reprinted in his report five articles that ran in the Kilbourn-backed *Milwaukee Advertiser* in May and June of 1837 promoting the canal, describing its route, and extolling the benefits such an internal improvement would bring to Wisconsin. Their style and enthusiasm and description of the countryside are convincing evidence that their main author was Increase, though Kilbourn surely approved them and contributed as well.

These articles were the first public exposition of Increase's response to pristine Wisconsin and his expression of the wonder with which he took in the verdant, water-rich, rolling glory of a Wisconsin spring. Not yet a year in Wisconsin and he had found his lifelong subject of study and appreciation, and he was finding his voice. His was not the hard-edged voice of a hell-bent developer, but the voice of a practical romantic about three idealistic rungs above that of the moneymen. Increase Lapham does not

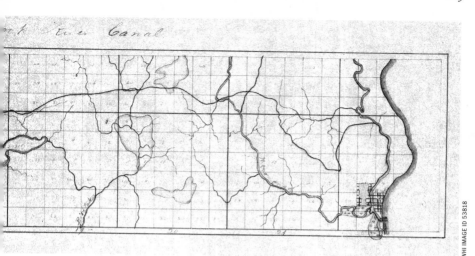

have dollar signs in his eyes when he looks at this ripe country; what he sees is a prosperous and civilized Eden.

The first article described the canal route in detail. In the second, Increase let some rhapsody rise to the surface. In going west out of Milwaukee up the Menomonee River, Kilbourn and Lapham found "a delightfully rolling surface, a rich limestone soil, productive in all the variety of agricultural productions, perennial springs of pure limpid water abundant."[8] In the same article, Increase describes the lake country west of Milwaukee:

> Whoever will take the trouble to look at the family of lakes promiscuously thrown together on the heads of the Bark and Oconomowoc rivers...shall observe the fantastic course of those streams, ridges, and valleys, as they descend, will require but little persuasion to agree with us, that nature has not only pointed out to us a great line of communication, but has almost prepared the work ready for our use.[9]

Next he envisions a future, peaceable kingdom:

> Around these lakes there will be several villages of some importance, which by means of steam boats and pleasure boats of all classes, together with the life and animation inspired by the business and enterprise incident to commercial pursuits, will be interesting for the man of business, and pleasant to the man of leisure, for summer residences.[10]

This was his first view of these beautiful sheets of water, as he was to refer to some lakes, and it left a lifelong impression. His last home was with his children at a farm on the south shore of Oconomowoc Lake in the 1870s.

Lapham described the route of the canal as running in the Menomonee River valley westward from Milwaukee, then across the Pishtaka (Fox) River valley, to the "summit" north of Pewaukee Lake, and then down the Bark River valley into and out of Lake "Namabin" (Nemahbin), and then down the Bark River valley again to the Rock River at Fort Atkinson. A branch canal would connect to Prairieville (Waukesha), which would allow for navigation from the Pishtaka rapids down to the Des Plaines River and into Illinois.

Another branch would run through Lakes Okauchee, Oconomowoc, and La Belle and would eventually connect to the Four Lakes of Madison. The possibilities seemed limitless. New discoveries of iron, copper, and lead, in Kilbourn and Lapham's vision, could bring about the "Birmingham of Wisconsin," emulating England's seat of the Industrial Revolution.[11] As new agricultural districts opened, wheat and other products would be floated into the Port of Milwaukee and farm equipment and luxury goods floated inland to increase production.

And all done more cheaply than anyone heretofore could imagine. With the rivers and lakes already in place, only 50 miles of "artificial work" would be needed to open up 460 miles of navigation. If the lower Rock River, a connection to the Rock and the Wisconsin Rivers, and the Wisconsin River itself were brought into play, the navigational system would be 880 miles in length, "and yet no point in this whole connection would be more than one hundred and sixty miles distant from Milwaukee, and the greater part of it (say seven eights) would fall within one hundred miles of this place."[12]

On December 29, 1837, the second session of the Territorial Council passed the bill to incorporate the Milwaukee and Rock River Canal Company. The bill was signed into law by territorial governor Henry Dodge on January 5, 1838. The act enabled the corporation to sell up to a million dollars' worth of stock. At the first meeting of the stockholders on February 3, 1838, a board of seven directors was elected, with Kilbourn serving as president. Increase was named engineer. "I have been appointed by Governor Dodge 'by and with the advice and consent of the council' a Notary Publick for this county and also Engineer of the Milwaukee and Rock River Canal Co., pay $5.00 per day (or $1,825.00 per annum) when engaged in actual service," Increase wrote his family.[13] But a letter to Darius almost two years later revealed the uncertainty of the economy. "I have a salary of $2,000 from the canal company but it is as yet only 'promises to pay.' No funds have been provided."[14] The canal company directors authorized Kilbourn to go to Washington to lobby for congressional support.[15]

In late March 1839, the canal crew again went inland to perfect the route. "In a day or two I am to start out to survey the Rock River canal, that will take about six weeks. I must 'live in tents' and sleep on the ground! What fine times I shall have!" Increase wrote to his family in Ohio; his exclamations suggest that he was particularly addressing young William and his sisters. "Tomorrow [March 18] Ann will be busy preparing me for this six weeks trip in the country."[16]

It was cold enough that March that ice still covered the inland lakes. On April 5, in a letter to Ann, Increase describes his work. He deeply wants his bride to share and to understand his work:

> Very soon after I closed my last letter to you, Mr. Kilbourn came around the hill riding his favorite Katey; he was soon followed by a troop of others carrying long poles....
>
> We soon commenced work; it was my duty to do the leveling. I have to talk at the top of my voice, such as this, "Up, up—halt—very light up—halt—fast—right." We ran across Nemahbin lake on the ice, then coasting along Twin and Oconomowoc lakes also on the ice, and along the creek to Hatch's Mills, thence across the Big marsh, which extends to the Rock river. We have been more than a week on this endless morass. Its northern shore consists of bays and points of high land running into it which form the first real difficulty in locating a canal. These difficulties are easily overcome however.
>
> Our course led through some immense Tamarac swamps near the south bend of Johnson's Creek and Wingfield's Run. It would have done you good to see us winding about the Tamaracs, where we occasionally, by making a miss-step, would sink into the mud and water three feet deep....

> We ran a Random level down to the river on Wednesday and gave
> three hearty cheers on reaching the waters of this beautiful river....
> We stayed at Jefferson, a town containing *one house*! (Mr. Darling's).
> It is the county seat for Jefferson County.[17]

Meanwhile, Kilbourn's Washington lobbying succeeded. On June 18, 1839, President Martin Van Buren signed into law Wisconsin's first land grant, a strip of land five miles on either side of the proposed Milwaukee to Rock River canal, a total of 139,190 acres. Mapped as a checkerboard pattern in square miles, the odd-numbered sections were to sell for $2.50 an acre and the proceeds would go to the canal company to pay for canal construction. Proceeds from the even-numbered sections would go to the national government. This was twice the price per acre of other government land. Kilbourn was elated. Land sales began along the "canal strip" in July.

The timing couldn't have been worse. Between the heady times when the Milwaukee and Rock River Canal Company was organized in July 1836 and its incorporation in January 1838, the economic boom that had caused Milwaukee to become the nation's fastest-growing community had collapsed. In the depression that followed, the price for the canal land was much too high. Eastern speculation dried up quickly and few settlers had cash. Many who had bought land in the boom with small down payments lost their land and their money. Only 43,587 acres of canal grant land were sold, most at a down payment of only 10 percent, and the company grossed $12,377.29, "an amount much too small to meet even the costs of surveying the canal route," according to Kilbourn's biographers.[18]

Although the company's charter did not expire until 1848, the canal was a dead issue, except for a short segment built in Milwaukee in 1841. This was the "slack water" improvement that Increase had mentioned to Seneca in his July 1836 letter. The canal company paid for a dam on the Milwaukee River just south of future North Avenue and more than a mile upstream from the settled part of town. A canal was dug parallel to the Milwaukee River on the west bank. It took water from the impoundment formed by the dam and ran it southwest into town, providing water and water power to new industries being built on its banks. The canal rejoined the river east of Chestnut Street. Eventually, the new enterprises numbered twenty-five and employed 250 people, "supplying critical employment for scores of German immigrants."[19] Kilbourn built a flour mill on the canal just east of his house on Chestnut. Although he had lost $30,000 on the failed Milwaukee–Rock River canal project, he profited for many years from his other enterprises.[20] In 1888, the canal was filled in and paved as today's Commerce Street in central Milwaukee.

—|—

What are we to make of the unlikely partnership of the modest Quaker and the ham-fisted capitalist? Except for a rare scolding from his brother Darius, one encounters in the written record few critics of Increase Lapham, either while he was alive or posthumously. Those who considered him at all, and there were many, universally admired and praised him. For a time after his death, he may have been Wisconsin's most honored citizen, for whom parks, avenues, university buildings, fossils, grasses, and even glacial moraines were to be named.

Not so with Byron Kilbourn. Many considered him to be, if not criminally corrupt and obsessed with money, at least "overbearing" or "manipulative."[21] During his life, he made bitter political enemies, especially Alanson Sweet, a rival town promoter, and Harrison Reed, editor of the *Milwaukee Sentinel*, Juneautown's counterpart to Kilbourntown's *Milwaukee Advertiser*. No rival was more effective than James Duane Doty, appointed by President John Tyler as the last territorial governor before Wisconsin became a state, who had canal interests of his own, who bested Kilbourn in political elections, and who went out of his way to block Kilbourn's projects.

Kilbourn did not help his own cause. His lack of subtlety was evident after he succeeded in lobbying Congress for the canal land grant. He wrote his Ohio backer, Micajah Williams, that he felt "fully compensated in the reflection that I have been of some service to the country, gratified my friends, and above all served my own best interests."[22] He was politically insensitive, and in later enterprises, especially when promoting Wisconsin's first railroad, was thought by many to have engaged in outright bribery.

Even so, he has contemporary and modern defenders. "In his day he was one of the worst abused men...and the real cause for it all was that he was the leader of the Rock River Canal Company and was determined that...business should grow and flourish on the west as well as on the east side of the river," pioneer Milwaukee industrialist James Seville told the Milwaukee Old Settlers' Club. There were few "whose opinions had more influence in the state at large than Mr. Kilbourn. He could do more with the legislature, governor, etc., than any other man and that too without any seeming effort on his part. He was...sociable, communicative, benevolent and always ready to engage in anything to help his adopted city."[23] And he seems to have been a capable two-term mayor of Milwaukee who put much effort into public works, especially its water supply.

Despite their differing personalities and reputations, Lapham, the quiet scientist, and Kilbourn, the charging bulldog, were close colleagues during the canal project. Both men were obsessed, Kilbourn with building internal improvements and making money, Lapham with building cultural assets and spreading knowledge. They understood each other and each granted the necessity of both internal improvements and cultural assets. They cooperated and supported each other. While their joint business dealings were never again so intertwined as they were with the Milwaukee and Rock River Canal, they remained lifelong friends. Only a whisper of criticism about Kilbourn is found in Lapham's surviving voluminous correspondence, which is a tribute to Kilbourn's intelligence and professionalism. If Kilbourn was inept, Increase would have said so. Late in life, Kilbourn was known to pay for subscriptions to scientific periodicals for Lapham and to favor him with other kindnesses. Kilbourn named Lapham co-executor of his estate along with his son when he drew up his will in 1868. As Kilbourn's biographers Goodwin Berquist and Paul C. Bowers write, "Lapham was loyal, efficient, and dependable, a man of unquestioned integrity. Byron could not have made a wiser choice."[24]

—⊣ | ⊢—

After it became evident that the canal was a lost cause, and especially that the work of canals in the future would be done better at less cost with railroads, Lapham was forced to shed his longtime professional identity of canal man, which he had assumed as a boy and which had sustained him from New York to Kentucky to Ohio and into Wisconsin. If the transition was difficult, he does not say so in his letters or in his pursuits. Even as he was traveling on canal business, he accumulated scientific observations and experiences that would engage him for the rest of his life. He was and wanted to be a scientist, unpaid if necessary, paid if possible.

And he had much to learn about and explore in the new territory. He made his first trip to the "mineral region" of southwest Wisconsin in November 1837 and reported on it in a letter dated November 27 to his young brother William:

> I have been traveling about two weeks through our new country, have visited the famous city of Aztalan, and the new capital of Wisconsin on the "Four Lakes." I have been on the ground which was the theater of the Indian War of 1830, have seen and conversed with persons who were engaged in that war, have been down into the lead mines 90 feet below the surface and followed the drifts for many rods in various directions. All these were new and novel things to me, and I should have liked very much to have had you enjoy with me the pleasure of the trip.... Our cold weather set in about a week ago, and one of our long winters has commenced.[25]

The "long winters" of Wisconsin had their own pursuits. On January 1, 1839, the *Milwaukee Sentinel* printed an announcement entitled simply "Lyceum" inviting "the citizens of Milwaukee desirous of organizing a Lyceum" to a gathering.[26] A week later, on January 8, the *Sentinel* ran a full article under the headline "The Lyceum":

> The organization of a Lyceum at this place is at once an interesting and an important event. It is a strong illustration of the march of mind and civilization, and an evidence that this community has done with the insane rage of making fortunes in a day, and places a value upon the cultivation of mind and those refinements which necessarily attend it. Societies of this nature have been formed in almost every village in the Eastern States, and have proved to be eminently useful. Public lectures and debates, which constitute their chief objects, are

well calculated to excite general interest and highly beneficial to those who are engaged in them.[27]

The Milwaukee Lyceum—the word comes from the garden in Athens, Greece, where Aristotle taught philosophy—was organized on January 10, 1839, with Dr. Lucius L. Barber, president; Hans Crocker, vice president; and Increase A. Lapham, secretary. A bill for its incorporation was passed at the next session of the Territorial Assembly.[28] Members were to contribute cabinets of curiosities, antiquities, natural history specimens, and books.

But after an initial program held January 22, in which J. E. Arnold and H. N. Wells debated "Are Usury Laws Expedient?" the Lyceum apparently languished until October. Then Hans Crocker and Increase announced a revival of Lyceum meetings, to begin weekly on Thursday evenings at the Methodist Chapel. It languished again, and on December 24, the *Sentinel* published its "obituary." But the Lyceum was revived again in January when Increase delivered an astonishing paper on natural science on Friday, January 23, 1840—astonishing for its comprehensive summation of science as it was understood at the time, astonishing in that it was delivered by a twenty-nine-year-old in a rough, unfinished town through which the frontier had just passed, astonishing in that the speaker was wholly self-taught and had to rely solely upon his own library in a town that had no schools, libraries, nor other such resources, and astonishing in its detail, clarity, and elegant organization.

"The natural sciences," he said, "include the three great divisions of nature, the animal, vegetable and mineral." He then defined mineralogy, geology, botany, and zoology and discussed their origins and the scientists in history to whose original insights each discipline could be traced.

> [G]eology also aspires to the formation of a *theory of the earth*, or, to ascertain the various steps by which the world was gradually reduced from *chaos* into its present condition. For the world was created "in the beginning," perhaps millions of years before the creation of Adam, "without form and void." This fact, the great duration of the world, which is clearly proved by the examination of the structure of the globe, and the remains of animals and plants buried beneath its surface, has been adduced as evidence against the truth of the Holy Scriptures. But,

when properly understood, it will be seen that the Scriptures no where say that the world and Adam were created at the same time.[29]

Here is Increase daring to bring modern thought to the farthest edge of a developing nation, relying on his reading of geologists and rationalizing their controversial new ideas with his own religious background, which he assumes is akin to that of his audience. "In the infancy of the sciences," he continued, "before much was known as to the real nature of the earth, many vague fancies were entertained and published which have now been laid aside, and geologists have taken up the examination of *facts* rather than *theories*. It has hence become a practical science."[30]

How far he had come from the teenager who bent down in the Niagara limestone at the Deep Cut at Lockport to pick up a fossil and wonder about its marine origin. Here he is, grown and married and experienced and speaking about geology, while standing coincidentally on the very same formation of Niagara limestone where he began his questioning, but on its western rather than its eastern edge, and here he is years later *understanding* and *teaching*.

Increase talked about fossils and their meaning. He listed them. He remarked that "in our own country (though not yet I believe in this territory), we find the remains of the huge mastodon, an animal much larger than the elephant but resembling it in many particulars. We have the mammoth, which is an extinct species of the elephant." His prescience was uncanny. Mammoth and mastodon remains would be found in a few decades and not far from where he was speaking and they would inform his successor Wisconsin geologists about the age of the glaciers and when humans first entered what was to become Wisconsin. A mammoth skeleton found in rural Kenosha County in the late twentieth century would reveal the clear marks of human butchering.[31] As Increase wrote, "These animals must very recently have inhabited our country as their remains are found in the superficial soil, and generally in the swamps in the vicinity of 'salt licks' where they probably resorted for salt as the buffalo and deer now do, and were sunk in the mud."[32]

He saw so much and had brought many, but not all, of the pieces together, correlating bones with soils and fossils with salt licks and he saw how they relate in a chronology of an old Earth. Mammoths and mastodons were "recent," unlike the ancient trilobites, crinoids, corals, and cephalopods found in the Niagara limestone that underlies eastern Wisconsin, including Milwaukee. He couldn't yet assign accurate ages—no one could—but "recent" is as close as 10,000 years, and Stone Age hunters were already on the continent, stalking migrating mammals at the base of mountains of ice. And the marine fossils? He knew they were much, much older, but science hadn't yet arrived at the scale of thought that includes hundreds of millions of years earlier in an Earth that is 4.5 billion years old, when shallow salt seas punctuated by sunlit shallow reefs spawned countless colorful life forms, where giant cephalopods—squid—roamed the coral gardens like wolf packs. Increase was on the right track. He planted these images in Milwaukee. His audience of young settlers in the new, rough town must have been amazed. How could they not have been changed?

Geology, as he was speaking, had been around for a century, but botany, relatively speaking, was a brand new science.

> But what shall I say of the extent of the objects presented to the admiring gaze of the botanist? Already the number of known and described

Many of the crinoids, trilobites, and cephalopods in Increase Lapham's fossil collection were the products of ancient life in coral reefs found in the limestone bedrock under eastern Wisconsin. This exhibit of such a reef is found today at the Milwaukee Public Museum.

COURTESY OF THE MILWAUKEE PUBLIC MUSEUM

plants amounts to nearly one hundred thousand, each of which has its own peculiar kind of root, stem, branches, leaves, flowers and fruit. How infinite then is this branch of our subject alone! About a thousand species of plants may be found within the range of a few miles from any place in this latitude.[33]

He should know; he's the one who collected them around Milwaukee, catalogued and then published a list of them, and sent specimens eastward to eager botanists like Asa Gray.

And then on to zoology, and how animals are classified:

And when we reflect upon the number of wild animals that range the forests throughout the world, the number of birds that fill the air with their sweet melody, the innumerable fishes that fill the waters of the mighty deep, the myriads of insects, the immense variety of sea shells, who can fail to be convinced of the almost boundless extent of this branch of creation?[34]

And this all-around scientist, this botanist-zoologist-geologist-mineralogist-meteorologist-anthropologist, typical of the scientist of his day, was beginning to understand limits. "To study the habits, manners and economy of one single species of this almost endless variety of animals would require a very considerable time and closeness of observation. Who, then, can become acquainted with the whole?" Who indeed?[35]

"There are persons, and possibly there may be some here," he continued, "who are disposed to ask, 'What is the use of this study?' To such I would say that if we regard it in the narrow light of money making..., then, these pursuits might as well be let alone." After pointing out that a geological survey surely would uncover valuable ores and minerals, he added,

There are other and higher reasons for pursuing these studies. There is a pleasure, a pure and unalloyed pleasure, connected with them that is seldom found anywhere else.... From what I have said of the extent and variety of the pursuits of a naturalist we may suppose that these pleasures, these charms have no end. There is no life so long as to be in danger of exhausting them. There is no condition of life debarred from these pleasures, all may study nature, the poor as well as the rich, old and young, male and female, the ignorant, the learned, all may enjoy the pure and simple pleasures they afford.[36]

In this speech Increase illuminated the difference between Kilbourn the capitalist and Lapham the scientist. Increase had used his canal jobs as a rationale to study his natural surroundings. When the canal project failed some years after his speech, Increase dedicated himself to making a living as a pure scientist. He did not always succeed but he never again worked steadily as a salaried engineer or surveyor. When Kilbourn turned to railroads, Lapham turned to nature. Their acquaintance continued, but their shared careers ended.

The reason was simple: to Increase, the aesthetics and wonder of natural history were more compelling than pecuniary profit.

Increase Lapham's lecture was a paean to the study of nature, his personal hymn to science. At twenty-nine, he was sharing that part of himself that he had recognized and nurtured as a boy, the part that kept his eyes and ears open and in the countryside searching for new wonders. In Wisconsin, he had found an unspoiled Eden, a trove of creation whose special mysteries were ripe to be collected, studied, understood, and communicated. And, generous at heart, he wanted everyone to share these "pure and simple pleasures."[37]

Increase knew his Lyceum speech was a success. On February 2, he wrote Darius: "My lecture before the Lyceum was very interesting! At least my wife says so, and *she knows!*"[38]

CHAPTER 13

The Botanist

⊣ 1841–1843 ⊢

THOUGH MILWAUKEE WAS NOT OFFICIALLY A CITY, in 1841 it was becoming more citified by immigrants from Europe and the eastern United States. Not only did the three settlements of Juneautown, Kilbourntown, and Walker's Point include many new stores, but there were also new churches, hotels, and banks, dentists and doctors and druggists, blacksmiths and carpenters and chandlers. The first brewery for lager beer was built in Walker's Point in 1841 and a temperance society was formed. A handful of jewelers and cabinetmakers was also flourishing. And that year a man named Henry S. Brown set up a shop to take daguerreotype images and Scotsman John W. Dunlop set up a business as a landscape gardener and florist.[1] Milwaukee was beginning to support the finer things of life.

Housing was scarce at the time, but Increase Lapham was snugly established in his little frame house on Third Street with his wife, Ann Marie, and their first child, sixteen-month-old Mary Jane Lapham, born September 15, 1839. Living near enough so they saw each other daily were Ann's mother, Mehitable Alcott, who resided with Ann's sister Mary Jane and her husband, Henry Hubbard, and their four young sons. Darius was hard at work as chief engineer of the Cleves Tunnel on the Whitewater Canal in Indiana, though at home he was "bothered like all fury by two little girls who romp around us as they please."[2]

In the mild, dreary January of 1841 many male citizens of Milwaukee amused themselves in drinking establishments and shooting matches on the frozen Milwaukee River. But Increase Lapham left his warm home, saddled his horse Adelaide, and rode west to see the "famous ancient city of Aztalan," crossing ten miles of parallel ridges that reminded him of the ridges he and Darius had seen in New York when they were boys.[3]

With no leaves on the trees and the ground bare, Increase could more easily see the great mounds at Aztalan and the signs of the ruined enclosure beside the west fork of the Rock River. It would be nine years before he began to seriously study and survey the ruins for his great work, *The Antiquities of Wisconsin.*

In the spring of 1841, the year ahead looked promising for the Laphams as well as Milwaukee. Ann wrote Increase's mother, Rachel Lapham, expressing her delight at Seneca's and Rachel's planned first visit to the Wisconsin Laphams. It would be the first time they met their daughter-in-law. That same day Increase wrote to his father about his recent trip to Madison—all of the travelers covered up in great buffalo robes against the cold.[4] Soon after that February journey, Increase saddled Adelaide and rode 130 miles west to Mineral Point on business, though a snowstorm prevented him from doing any geology and botany on the way. But a Mr. Taylor gave him some geological specimens in Mineral Point and the two must have had some good rock talks.

In early 1841, Increase Lapham's work life was predictable and steady, his wife and child well and happy, his travels in Wisconsin intellectually stimulating, and his botanical correspondence rewarding. Increase was also planning to build a two-story brick building that would include a room for his collections and studies, as well as housing tenants for rental income. The future did look bright.

But in the middle of April, the Wisconsin Laphams heard that brother Pazzi, who was at the family farm near West Liberty, Ohio, was very ill. Increase was at that time packed to go to Ohio carrying territorial bonds for the Milwaukee and Rock River Canal; after his canal business he would go see his family. But at the last minute the Canal Board sent someone else. Then Rachel and Seneca Lapham canceled their planned trip to Wisconsin Territory, no doubt because of Pazzi's illness. Plus, Ann's brother-in-law, Henry Hubbard, and her five-year-old nephew were also ill.[5]

On May 28, 1841, Henry Hubbard died at age thirty-five and the little boy died soon after, leaving Mary Jane Hubbard a widow with three boys, ages nine, seven, and an infant. Two years earlier this besieged family had lost their only daughter before she was a year old.[6]

Leaving Mehitable Alcott to care for Mary Jane and her surviving children, Increase and Ann, probably with their daughter, Mary, left Milwaukee on June 9 for Ohio to see his family.[7] The Wisconsin Laphams arrived in Cincinnati the afternoon of June 21, surprising Darius, his wife, Nancy, and their two daughters. It had been five years since these two brothers had seen each other. Increase was now thirty-one and Darius was thirty-five. We have no record of the reunion from the reserved Laphams but we can imagine the joy. Darius wrote his "Respected Father" the day after his brother arrived, "I shall keep them here a week at least before I can let them go up home. My business is such that I can not leave sooner and to let them go without the rest of us would not answer at all."[8]

It is possibly during this week that Increase and Darius took their oldest girls to a daguerreotype shop in Cincinnati and had their images taken. Both brothers' faces are confident, happy, unlined by loss and disappointment. Though there are other images of Increase, this is the only known image of Darius.

Then the two brothers and their families traveled north together to the family farm near West Liberty, Ohio, where the whole Lapham family was together for two happy weeks: Seneca and Rachel, then in their late fifties; Darius, Nancy, and their two girls, Phebe and Harriet; Increase and Ann and baby Mary; the ailing Pazzi, twenty-nine, and his wife and child; William, a twenty-four-year-old bachelor farmer; vivacious and intelligent eighteen-year-old Hannah, a favorite of Increase's; and the two youngest girls, Lorana, sixteen, and Amelia, thirteen. Rachel's heart must have been almost bursting with happiness to have the family together again, only a little more than all the others of this close and happy family. It was the last time they would all be together.

After this family time—which none of them wrote about since they were already together—Darius and his family left the farm around the eighth

WHI IMAGE ID 100192

COURTESY OF THE OHIO HISTORICAL SOCIETY 0314-2

Increase Lapham with his eldest child, Mary, ca. 1841. It is likely these daguerreotypes of the brothers and their daughters were taken during Increase's Cincinnati visit.

Darius Lapham, Increase's favorite brother, with his eldest child, Phebe, ca. 1841

of July for Cincinnati where they saw a funeral procession for the lately deceased Whig president William Henry Harrison, whom both Darius and Increase liked.[9] John Tyler was now president. Ann and Increase had a long and hot six-day journey back to their "sweet home" in Milwaukee by stage, railroad, and steamboat. Increase, who rarely complained about the hardships of travel, said that in all his travels he had "never experienced a harder ride in a worse wagon or over a rougher road."[10] But poor Ann! She was by the time of their trip home five months pregnant.

Home again, Increase set the canal workers back to work and oversaw the completion of the dam on the Milwaukee River. Their second child, Julia Alcott Lapham, was born November 20 of that year.

Increase worked in the real estate office and for the canal in the early 1840s, but what he seems to have been most interested in is botany. His study of plants meant the collection, identification, preservation, and dissemination of specimens to botanists across the United States and even in Germany and Cuba. On the same day that Julia was born, Increase was occupied writing to John Torrey about revising his earlier plant list and about the identification of an unknown gentian Increase had found. That day he also packed a box with plants for Joseph Barrett, Alvin Wentworth Chapman, William Marbury Carpenter, the Hartford Natural History Society, and the Essex Natural History Society, sent care of Torrey. We can imagine Increase concentrating on botany in his study while, with the help of her mother and sister, his wife gave birth to her second baby, Julia, the daughter who would edit and type thousands of pages of her father's diaries and letters to save for posterity the details of his life.

—| |—

When Increase first arrived in Milwaukee in 1836, he immediately started collecting and listing Wisconsin plants and exchanging them with botanists, many of them old friends from Ohio and Kentucky, but also men in the East like Asa Gray, soon to be at Harvard. So by early 1841 Increase could report to his father that he had "a very handsome collection of dried plants, numbering something over 2000 species."[11] Settled in one place, well cared for and encouraged by his wife, Ann, with room for his growing collection, and with work that was steady but not particularly demanding, Increase Lapham's botanical activities picked up. The canal work was winding down; his real estate work for Kilbourn had settled into a routine. Darius wrote that he was involved in hydraulics at this time, but Increase was doing no engineering, only setting grade stakes for sidewalks and streets.

Increase had been studying plants since he was a boy in Lockport. He learned to preserve specimens while in Shippingport—going out to the woods with his father and brothers to find the early spring flowers and weighting his specimens with the same rock for several years. In Portsmouth he listed in his diary the plants he saw along his canal line. In Columbus, with many experienced naturalist friends to tramp around the woods with, Increase became more knowledgeable about collecting and preserving. But in Milwaukee, Increase was on his own. Amos Eaton's seventh-edition *Manual of Botany* was, he said, his only authority on plants.[12] And apparently at this time no one other than his wife and a few other family members could be commandeered to go collecting with him. No doubt on trips for other purposes he had his botanical eye open and his vasculum—the tin box that prevented plants from withering before being pressed—at his side.

As often as he could, Increase brought others along on his plant expeditions. One May Increase and Ann took their horse Adelaide and a wagon southwest on a two-day trip to Eagle Prairie, a small prairie thought to be shaped like an eagle with outstretched wings, where the ridges and gravelly nodes created many microclimates for different kinds of plants.[13] And one September he took with him Ann and her mother and his niece, Caroline Stone, on a thirty-five-mile collecting trip. The four hardy souls saw ten beautiful lakes, lovely prairie openings, the occasional rude log cabin—and got stuck in the mud.[14]

But most often Increase collected alone. In spite of the cold and damp of early May 1841, Increase went out to the woods and prairies looking for new plants and for flowering plants, with his vasculum over his shoulder to collect them. He found nothing new but among the young bracken and grasses of a bog he spied the "very interesting" *Sarracenia purpurea*, a low-growing carnivorous pitcher plant, the capture of which probably made his knees damp. The purple leaves were tubular, he noted, and standing erect holding water. He brought home specimens of the commonly named side-saddle plant for his growing collection and added it to his list of the plants of Wisconsin.[15]

When it came to recognizing plants new to him, Increase was fluent in the "grammar of botany," which Eaton, a consummate teacher, clearly elaborated in his coat-pocket-sized *Manual of Botany*.[16] To successfully catalogue and collect plants, Increase had to have a sharp eye and a good memory so he could differentiate telling characteristics—leaf shape and veining, stem shape, but primarily the shapes of the fruiting parts of the plants—calyx, corolla, stamen, pistil, pericarp, seed, and receptacle—in Eaton's terminology. For example, Eaton's entry for the side-saddle plant, or *Sarracenia purpurea*, reads "leaves radical, short, gibbous-inflated or cup-form, contracted at the mouth, having a broad arched lateral wing; the contracted part of the base hardly as long as the inflated part. Scape with a single, large nodding flower. In marshes."[17] Increase would have recognized that what he found might be a *Sarracenia*—deep purple among

(right) Lapham's specimen of the carnivorous purple pitcher plant, the side-saddle flower, *Sarracenia purpurea*, found in Milwaukee

the buff grasses of the previous year—because the leaves came up from the root without a stalk, that is, they were radical. It is a short plant, found in a marsh, with gibbous or swollen purple leaves, and there was a broad, arched side membrane or lateral wing with a large nodding flower on the scape or flower stem.

Not only did Increase have to be able to recognize and identify plants, but he also had to remember what plants he already had collected and what his collector friends had requested. Increase the list maker was good at keeping track. He also had to recognize the microclimates where he was more likely to find specific types of plants. He had to carefully collect enough of the plant so the specimen would contain information useful to him and to other botanists. Because plants in flower or in seed were most useful, Increase had to know when the plant would be flowering or seed bearing, which meant timing his collecting carefully. And he had to not collect so much that he would endanger the survival of a rare plant.

To preserve a plant after it had been collected, Increase let it wilt just enough so that it could be carefully arranged on a clean sheet of soft paper, such as old newspapers, to clearly display the type of branching, both sides of the leaves, the flower or seed, the stem, and sometimes the root. Then, in his words, he had to apply "a gentle pressure, enough to prevent the plants from shrinking or crisping, but not enough to crush and destroy them." Then, "[a]s often as the papers became damp, they must be taken out and dry ones put in their places."[18] For the next few days or weeks, he had to check these plants almost daily to make sure they were not growing moldy or being eaten by insects.

When the plant was sufficiently dried, he would arrange the specimen on clean paper, then glue or sew it to the mounting sheet. This sheet would be labeled with the plant's common and Latin names, where and when it was found, and sometimes the type of habitat it came from, and his own name. "By this means," Increase said, "very beautiful specimens are obtained, showing in many cases the natural color of the flowers, leaves, etc., with great beauty and fidelity. The leaves will be spread out so as to show their natural form and positions, and such specimens are nearly as valuable for the purpose of exhibiting the plant as newly gathered specimens."[19]

COURTESY OF THE WISCONSIN STATE HERBARIUM (WIS), UW-MADISON, ACCESSION NUMBER V0141893

This wild phlox was named after Lapham by Asa Gray. Collected in Milwaukee, it is now known as Lapham's phlox.

Increase was well regarded in his time for the care with which he prepared specimens. Among his correspondents, Joseph Barrett was "struck with the beauty and neatness with which your specimens are preserved."[20] Moses Ashley Curtis wrote from South Carolina to say that Increase's specimens are "of the finest style of preserved specimens." In comparison to the Short school of preserving, he wrote, "Yours are just the right thing."[21] And Charles Short in Louisville, Kentucky, from whom he had learned the "Short" method when Short sent him a pamphlet on preserving plants in 1834, again and again over the years commended Increase for the high quality and beauty of his specimens.[22] Not only do Increase Lapham's preserved plant specimens show his attention to the detailed information he wants to convey, but they show his artistic eye and delicate, sure hand. Even today, Increase's specimens stand out as superlative when compared to others from his time.[23]

Increase was generous with his time and the specimens he sent, sometimes sending off packets to as many as eight botanists in one day. His generosity was also financial because paper was expensive on the frontier, as were mailing materials. Increase received many specimens in return, but his position as the only and first botanist in the Wisconsin Territory at the time made his plants very valuable to men in the East. In his 1842 lecture on botany, Increase said that the many botanists all over the United States "appear to be anxious to obtain specimens from our young territory where nature has been so little interfered with by the hand of civilization, and I find no difficulty in exchanging all the duplicate specimens I am able to collect and preserve of the species found in other regions of the country." Taking advantage of this situation, he said, would enable the resurrected

Lyceum to form a collection of plants that would be "very extensive and valuable."[24] But it was Increase, not the Lyceum, who formed the extensive and valuable collection.

—⊣ | ⊢—

Who were these botanical correspondents of Increase Lapham in the 1840s? Most of these men were botanists whom Increase had not yet met, for instance, Asa Gray, arguably the best-regarded botanist of the time. Just before Increase left Ohio, he had sent a letter and package of plants to Gray, the first faculty member of the University of Michigan already working on his massive life's project, now called *Gray's Flora*. Born in 1810 only a year before Increase, Gray had the advantage—much felt by Increase—of having studied at the College of Physicians and Surgeons in New York City—with physician and botanist John Torrey. But the first plants Increase sent Asa Gray from Ohio had "lain in quiet obscurity for several months," thirty-nine types of carex—grasslike plants that bloom in the spring.[25] On his way to Milwaukee, Increase wrote Gray from Detroit on June 18, 1836, asking if Gray had gotten them.[26] Gray's response was one of the first letters Increase received in Milwaukee. He wrote,

> I thank you for your kind offer to collect specimens, and shall be pleased to correspond with you, especially so as I know you will meet with many very interesting plants in the region where you propose to reside. I would like to receive not only Grasses, which never come amiss in any quantity, but also all other plants you collect at Milwaukie. In return I will most readily give you any information, which my metropolitan situation enables me to command and will afford, in every way all the assistance in my power in your botanical pursuits. Let me entreat you to pay particular attention to my *pets*, the Grasses, etc, but don't neglect the others. I will see that you have due credit for any interesting discovery.[27]

Asa Gray's letters to Increase Lapham were informal, generous, familiar, while Increase's letters to Gray were correct and somewhat stiff. Increase must have felt himself the student, the amateur in comparison to the more schooled Gray. But Increase did pay special attention to Gray's pets, the grasses.

By 1842, Asa Gray was chairman and founder of the Botany Department at Harvard and in charge of the new botanic garden. He began to ask Increase to send live plants. "In the fall can you send me live roots of your fine gentian, new species? That I may cultivate it in the Cambridge Botanic gardens, to see if it holds distinct from *G. crinita*?"[28] And, Gray asks, could Lapham send roots of the pressed specimens he sent earlier, a dozen of each, including the new fringed gentian?[29] Increase carefully packed in moss the roots of the plants Gray wanted and sent them in the fall. Gray wrote back the following spring thanking him for the plants, which arrived safely "except that the labels being written on paper, were destroyed by the moisture of the damp moss.... I trust the *Iris lacustris* will flower this spring that I may examine it."[30] *Iris lacustris* (dwarf lake iris) was one of the first plants Increase found in Milwaukee.

Asa Gray also sent live plants to Increase in the fall of 1844. All the plants were doing well, Increase wrote him, except for one *Erica* (heather). All the verbenas were alive in the greenhouse of his friend Mr. R. Woodward. "Such has been the growth of the place, and the taste for the cultivation of flowers is such that we now have a public garden with a green house, in which many of the common house plants are cultivated and instead of putting the tender plants you sent into the cellar I have made arrangements to have them properly taken care of [there].[31] Gray had also sent a Dr. Tuckerman to see Increase in the summer of 1844. Tuckerman, Increase wrote back, was "about the only botanist I have met in Milwaukee, but I am just informed that Dr. Wunderly[,] a young German botanist, a friend of Prof. A. Braun is about to locate here, and if he proves to be a good observer of nature I shall be very glad that he has selected Wisconsin for his future home."[32] Increase was hungry for someone to talk botany with.

Though Increase was working alone for the most part, his plants, thanks to the well-connected Asa Gray, were traveling far. In May 1846, Gray wrote to say that most of the live plants Increase sent the previous fall were planted in Cambridge, but some were sent on to London, to Gray's friend Joseph Hooker at Kew Gardens, and others to St. Petersburg.[33] Increase and Gray would eventually meet in 1847 at a rollicking dinner party of four botanists in Cambridge, Massachusetts.

John Torrey was another of Increase's important correspondents. Born

WHI IMAGE ID 45468

WHI IMAGE ID 45467

WHI IMAGE ID 46202

Three botanists who corresponded with Increase Lapham: Asa Gray, ca. 1864, left; John Torrey, ca. 1865, top; and Charles W. Short, ca. early 1860s, bottom. The images are from Lapham's *carte de visite* collection.

in 1796, Torrey was a physician, a professor of chemistry at West Point, and a professor at the College of Physicians and Surgeons in New York City. He taught botany at Princeton in the summers from 1830 through 1854. Torrey learned botany from the master, Amos Eaton himself, who taught Torrey, interestingly, while he was imprisoned in Greenwich Village, New York, for financial misdealing. Torrey later taught Asa Gray his botany, around the same time Increase Lapham was learning botany from reading his various editions of Amos Eaton's *Manual of Botany*. Torrey and Gray became more than teacher and taught, but colleagues from 1838 through 1843 on their great work on the *Flora of North America*. But first, Torrey published *The Flora of the State of New York* in 1843. With Asa Gray and George Engelman he founded the National Academy of Sciences and began the plant collection at the New York Botanical Gardens. A genus of evergreen trees similar to the yew was named after Torrey. "All over the world," Gray wrote in his memoir, "Torreya trees, as well as his own important contributions to botany, keep John Torrey's memory green."[34]

Increase was exchanging letters and plant specimens with John Torrey as early as the 1830s. In August 1840, Torrey wrote Increase that

> Your plants are in a beautiful state.... Pray, dear sir, continue to explore the interesting region about you and let me have specimens of your rare and new things.... Dr. Gray and I are working hard at Compositae &c. Send us good specimens (with radical leaves) of all your asters and solidagos as well as all your other compositae except the commonest species.[35]

Torrey was especially interested in the "pretty gentian" Increase had found that seemed unlike any other.[36] In September 1841 Increase sent Torrey some of that year's plants, and he wrote that "the new gentian this year is mostly branched, in the same locality where they were last year mostly simple."[37] Increase wrote Torrey again on November 20, 1841, the day Julia was born, asking him to please revise and correct the plant list he had sent earlier and to send it on to Silliman. Then Increase asks what was decided about the new gentian.[38] After that it was four years before he heard from Torrey again. In July of 1845, Increase received a warm letter from Torrey in Princeton. "It is so long since a letter has passed between us that (it being all my fault) you have perhaps crossed my name off your books." But, he says, he is working hard to support a number of people. Torrey wants him to meet his friend Robert Storner who is going to Wisconsin Territory. "You will find him a right good fellow, who, having heard me talk about you was desirous of seeing you face to face." He wants to know what novelties Increase has found and he reports that Stephen Long is off on another expedition.[39] The two botanists pick up easily where they left off and continue for years.

Finally, starting in Increase's Columbus years, physician-botanist Charles W. Short carried on a warm correspondence with Increase for much of his life. Increase had first read Dr. Short on botany in the *Transylvania Journal of Medicine* in the late 1820s. Short was a professor of medicine at Transylvania College in Lexington and then in Louisville. In the fall of 1835 Increase sent from Columbus a large packet of plants. Short wrote a long, instructive letter back, commenting on each plant Increase had sent, on Increase's identification or misidentification of the plants. Short's tone was just right for a slightly insecure young botanist. He always noted when Increase was correct and rather than criticize when he was not, Short asked questions and made suggestions or asked for more detail. "*Helianthus Macrophyllum*—I venture with hesitancy to correct your name, but this seems rather to be *H. decapetcheus?*" He is generous with his praise: "Your specimens are beautiful and I have rarely seen a collection so well preserved." When Increase writes that he is moving to the Wisconsin Territory from Columbus, Short is "not a little gratified to hear of your determination to remove still farther west. . . . As a naturalist you can hardly fail to meet with much of novelty and interest in that distant and untrodden field."[40] He asks lots of questions and wants specimens from this place. Short's warmth and interest elicited long letters of description from Milwaukee, and he became one of Increase's most frequent correspondents. Between these two men existed not only a strong professional accord, with both generous in specimens and in connections to other far-flung botanists, but also a palpable warmth. Short must have been Increase's most satisfying correspondent.

—| |—

On March 10, 1842, Increase Lapham delivered a lecture on botany at the Milwaukee Lyceum that had, since its "obituary" in the *Milwaukee Sentinel,* come full circle.

This was a wide-ranging lecture carefully designed for a popular audience, a talk that encouraged the study of Increase's first love, botany. Increase, always concerned with human advancement, explained that knowledge of plants was useful to the farmer, to the shipbuilder and the lumberman, and to the physician. He mentioned in passing that he was often called by physicians to find in the woods, bogs, and prairies plants needed for their medicines. Increase told his listeners who the authorities were in the field of botany, mentioning Eaton, Torrey, Gray, and Silliman along with his *Journal of Science.* He spoke of the new natural system of classification, which divided plants into groups by their structure. He gave a careful description of how he preserved plants and, for encouragement, told his audience that many botanists all over the country were eager to exchange their plants with those from this new territory. He also warned would-be botanists that they'd better hurry because the native plants of Wisconsin were—already in 1841—disappearing:

> It is important that the plants of Wisconsin should be examined and specimens of them preserved *now* while they are, comparatively, but little disturbed by the hand of man. In a few years many of our most interesting plants, which are now found in abundance, will become scarce and even be driven from the soil. Already I find it necessary to extend my ramble some distance to find plants which within a few years were found within the limits of our town.[41]

Most of the rest of his talk described many of the interesting plants of Wisconsin—from the great oaks to the water lilies and wild rice. "The truth is that no part of the United States is more rich in plants than Wisconsin," he expounds, "rich not only in the number and variety of species found but in the beauty and elegance of their flowers. Our prairies and 'Oak Openings,' may, at some seasons of the year be considered as one continued flower garden."[42] Increase, claimed by some to be the first ecologist in Wisconsin, made his early preservationist concerns known as he extolled the beauty of his adopted home.

His botanical meanderings didn't stop there. Increase had been keeping botanical diaries since his days in Kentucky—little handmade booklets in which he listed the plants he found in the area and often their date of first bloom. Knowing that it would be important to more botanists than just himself, he had published his first list of Wisconsin plants within months of his arrival as *A Catalogue of Plants & Shells Found in the Vicinity of Milwaukee.* He revised this printed catalogue in 1838 and printed an addendum in 1840. His sources for these printed catalogues were his botanical diaries.

His botanical diary of 1838 still exists along with several others in the Lapham archives at the Wisconsin Historical Society—a three-by-eight-inch hand-sewn booklet of twelve pages.[43] Though there are temperature charts in this book for January and July and he notes the dates the river and harbor iced up and then cleared, the catalogue is primarily a listing of when plants are first in flower. And he lists almost all plants with their Latin names. This is truly a diary, with plants listed by the date he observed them or by first flowering date rather than by their family or type.

Increase was interested not only in the plants themselves but also in what these first flowering dates might tell him about the weather compared to previous years and in other places. In the spring of 1837, Increase had placed a short article in the *Milwaukee Advertiser* asking if "persons residing in different parts of the country [surrounding Milwaukee] would send us notes of the time of the flowering of the wild plants, in order to compare the climate at those places with that of Milwaukee."[44] He got one response to this request—from a David Bonham who lived fifteen miles west of Milwaukee "near the source of the Pishtaka river" (the Fox River).[45] Bonham sent Increase the first flowering dates of twelve plants, including blue flag, bloodroot, buttercup, and prairie pink, which Increase then looked up in his own records. Making a chart comparing the two lists of first flowering dates, Increase concluded that spring flowering came about ten days earlier on the Pishtaka than near the lake shore, which he attributed to the cold winds off Lake Michigan.[46]

Increase's early plant diaries allowed him to compare the blooming of plants from one year to another year—when they bloomed, where they were found, what the weather conditions were—and they hinted at his later serious interest in weather statistics and prediction. They include relevant information like "snow 1" deep" or "Hard frost." He notes the date the first steamboat arrived each spring from the south. And sometimes he sounded like the boy Increase: "June 15—Strawberrys ripe!" In 1838, from March 29, when the red maple was in bloom, to September 2, when there was a frost on the asters and goldenrods, Increase recorded the plants he saw on forty separate days. By this time, Increase listed most plants by their complete Latin names. But hog weed is just "hog weed."[47]

We can watch Increase grow as a botanical cataloguer if we compare this 1838 botanical diary to the diary he had kept in Portsmouth and Columbus, Ohio, from 1831 to 1833. In the earlier sixteen-page, hand-sewn book of eight sheets of lined paper, Increase wrote down the twenty-four plants he saw in flower near Portsmouth between March and April of 1831, noting weather conditions and location. In 1832 he listed the plants he found near Portsmouth in February, March, and April by date. Still in Portsmouth, he more formally began on page 7 a "Catalogue of plants found in the vicinity of Portsmouth, Ohio," and he included the "Forrest Trees." When in Columbus, he began on September 15, 1833, in that same book a numbered list of the plants he found in the area. In this catalogue of ninety-four plants, his command of the Latin names is incomplete; he lists many plants by genus only or by common name.[48]

From 1838 through 1842, Increase kept another plant book, a catalogue rather than a diary, a constant update of what he'd found to date: "List of WISCONSIN PLANTS [1838]–1842."[49] This book is more systematic than the earlier ones, though still a handmade and hand-sewn four-by-six-inch booklet. In this he lists more than six hundred plants by "tribe," beginning with the *Exogenae*, rather than by date. Entries in this list are annotated with initials that tell him where he found the plant, whether it is medicinal, and which are wanted by his correspondents. The 1838 printed revision of his 1836 "Catalogue" and sheets of additions and corrections are sewn into the back. His plant names are now complete scientific names and he also includes common names and notes on the date found. This book, more than the earlier books, looks like a true scientific journal rather than an amateur collector's list.

Increase's next catalogue, "Plants of Wisconsin 1845," though still handmade, was a bigger book in every way. It measures a little more than eight by six inches and is 103 pages long. This book too was improved upon, added to, corrected so that it contains plants listed by "tribe," plants listed within genus, common names, where found and when and by whom. Some plants, he notes, are found as far afield as the "Falls of St. Mary, Dr. H.," "Rainy Lake, King," "Chicago," "Shore of L. Superior—Dr. Pitcher."[50] In other words, Increase was trying to list *all* the plants of the Wisconsin Territory. His code tells him which plants are *not* found in

Plants in Flower

March 29th
  o Acer rubrum
April 4 - Trillium nivale
    o Hepatica triloba

" 18 - Snow 1 inch deep
" 25 Arabis rhomboidea
" 26 - 1st Steamboat from Below
29 - Caltha palustris
  o Sanguinaria canadensis
  Claytonia Virginica
  Isopyrum thalyctroides
  Erigenia bulbosa
  + Alnus serrulata
  Salix conoides?
  Ranunculus

May 7 - Snow —
  Anemone nemorosa
  o Ictodes foetida
+ 12 o Erythronium Americanum
    album
  Fragaria Virginica
  Viola cucullata
  Phlox divaricata
  Gnaphalium Americanum
  Ranunculus abortivus
  Trillium erectum
 o Euchroma coccinea
~ 14 Dentaria laciniata
  Mitella diphylla
  Hippophae Canadensis
  Iris cristata?
18 Xylosteum ciliatum
 o Prunus Virginiana
  Panax trifoliata
 o Dirca palustris

May 18th o Thalyctrum dioicum
  Pedicularis canadensis
  Ribes lacustris
  Viola pubescens
 o Acer sacharum
  Uvularia grandiflora
 o Caulophyllum thalyctroides
19 o Thlaspi Bursa pastoris
  Plantago cordata
  Arum tryphyllum
  Prunus americana
  Crataegus sp. ?
20 Actea alba
  Coptis trifolia
 o Aronia botryapium
 o Corydalis Canadensis
23 o Leontodon taraxacum
 o Populus tremuloides
  Floerkia uliginosa
29 Hard frost
 o Veronica peregrina
30 Dodecatheon integrifolium
  Geranium maculatum
  Juncus polycephalum?
  Orchys spectabilis
  Vicea cracca
 o Zizia aurea
  Carex intigerrima
 o Aquilegia Canadensis
 o Convallaria stellata
  Polygala senega
 o Arabis laevigata
 o Lathyrus albidus
  Menyanthus trifoliata
  Vaccinium Pennsylvanicum

Two pages from Increase Lapham's 1838 fourteen-page booklet in which he lists plants by date of flowering in Milwaukee. He also notes how much snow was still on the ground on April 18 and when the first steamboat was able to make its way north after the ice broke up.

Milwaukee County and which are observed in Jefferson, Sheboygan, Iowa, and Rock Counties, whether the plant has medicinal properties, and if he has a specimen in his herbarium. The catalogue included plants he had not seen himself. He was doing what he had told John Torrey he would do

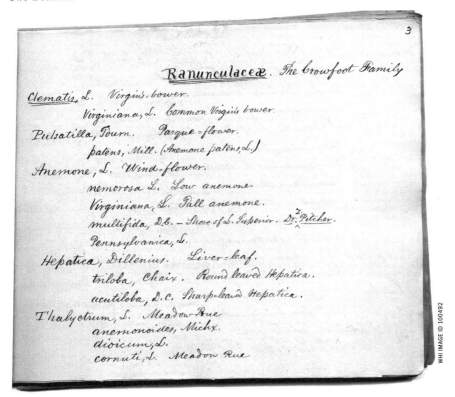

Members of the crowfoot family listed by genus and species with their Latin and common names, from Increase's 1845 catalogue of plants

four years before: prepare a complete botanical catalogue of the Wisconsin Territory from the southern extreme of Lake Michigan to Lake Superior. And he would do more.

In 1849, Increase published "An Enumeration of the Plants of Wisconsin" in volume 2 of Harvard's *Proceedings of the American Association for the Advancement of Science* at the recommendation of Professor Asa Gray. The culmination of his cataloging efforts in what was now the state of Wisconsin, it was "intended to embrace all the plants heretofore found within the limits of the State of Wisconsin. It is, doubtless, very far from being complete, many species remaining to reward the future explorers."[51] Increase, not surprisingly, kept finding and adding specimens even after its publication. How could he not? A lifetime of disciplined habit would not change in later years, nor would his beautifully prepared and generously shared specimens of plants lose his or his far-ranging botanist friends' interests.

—| |—

The Lapham family as well as the families of plants commanded his attention. The year 1841, though brother Pazzi was ill and Ann's sister had lost her husband and son, closed on a high note with Increase's writing of "A Week in Wisconsin," a descriptive piece for *Silliman's Journal*, and the birth of Julia Lapham in November. But early in 1842 Pazzi died. From Milwaukee, Increase wrote home to his family on the farm near West Liberty: "We can now only console ourselves with the hope that we may not only meet him but all other departed friends and relatives in that land where we shall not again be called upon to separate."[52] He wrote to Darius what he could not say to his parents: "Pazzi has not had a very great share of happiness in this world and it is to be hoped that he is…enjoying what was denied him in this world."[53] Increase believed in an afterlife, though his church affiliation at this time was murky.

In May a second blow fell on the family. Much loved, brilliant, and vivacious eighteen-year-old Hannah died of apparently the same ailment that had killed Pazzi. This time Increase was stunned with grief. He had placed special hopes on Hannah—making sure she had at least some good schooling while he was in Columbus, responding to her sprightly and intelligent letters in a pedagogical manner. It was Ann Lapham who wrote Rachel and Seneca when they heard from Lorana on May 22 that Hannah was gone. Increase wrote few letters for the rest of the summer of 1842.

He was working on the canal, he told Darius, and fifty Germans a day were arriving in Milwaukee. The summer was a cold one and crops were failing in Ohio and Wisconsin. Yet Increase seemed to take pleasure and perhaps solace in "farming." He had a plot of land in the city that was planted with fruit trees and seeded with grass. He kept the grass mowed with a scythe so his children and their friends could play there.[54] He was creating little Edens for his family. With the help of a "mad Dutchman" he was clearing another three acres to be a pasture or meadow. It would be planted with native trees he was preserving, among them trees sent to him by Darius.[55]

On their hillside meadow the Lapham family would sit in the grass looking over the town. From there they could see the Great Western Hotel, once Lot Blanchard's old boardinghouse, and Byron Kilbourn's old residence at Third and Chestnut, both of which Increase stayed in when he first arrived in Milwaukee. And Solomon Juneau's old house where everyone gathered and where two bears had been chained in the yard was now a public house with rooms and stables.[56]

The year 1842 ended with Increase worrying about the future of his bachelor farmer brother William, and with Ann caring for her very sick older brother, Volney S. Alcott. Increase was low. He wrote to Darius in December: "Business grows dull here; our canal is probably killed dead; my land agencies will probably not afford me sufficient employment and profit to support my family; so what better can I do than go home and live with you all?" Yet, he said, the sleighing was good.[57]

The next year was no happier, at least at the beginning. Volney Alcott died February 4. He was the second oldest of the six Alcott siblings, born in 1806, ten years before Ann. But the canal work continued, along with the land agencies. On February 20, Increase hitched Adelaide to his cutter and set out for Green Bay on business. Out of this trip he made a fine article often reprinted as "A Winter's Journey."[58] Increase's botanical correspondence picked up and must have kept him very busy, sending not only dried specimens but also live plants, cuttings, and roots to ten botanists. And his "home farm," the garden behind his frame house, was filled with so many fruit-bearing trees and shrubs—pears, plums, currants—and flowering herbs that he said, "it might . . . be called a Botanical Garden or an Ornamental Garden." Increase was especially proud of a "fine large smooth gooseberry bush" he'd found and layered so that in a few years he could supply the area with a fine native gooseberry. He wrote this letter to Darius with the baby Julia on the desk in front of him to be amused while Ann went to her mother's.[59]

This year ends happily with the birth of Increase and Ann's first son, William Allen Lapham, born December 19. Increase is fruitful and multiplying in so many ways.

CHAPTER 14

The Uses of Writing and Publishing

⊢ 1844–1847 ⊣

B Y MARCH 7, 1844, INCREASE'S THIRTY-THIRD BIRTHDAY, the bricks for his new building next door to their frame house were delivered and paid for. By September Increase had moved his office and library to the third floor of the finished building, and though the downstairs store was not yet rented, the other two upstairs rooms were let to a justice of the peace and an attorney.[1] His article titled "Statement of Elevations in Wisconsin" had been published in volume 46 of Silliman's *American Journal of Science and Arts*, sixteen years after Silliman published his first article. Now his writing and publishing career was about to make a great advance.

Since Increase was a boy in Shippingport, his scientific pursuits had not been limited to observation, collection, and classification. His approach to science had never been "pure," that is, science for science's sake. Many naturalists of the time were satisfied to collect and share specimens with fellow naturalists, to deposit their collections in their cabinets of curiosity or in a natural history society. Increase's science had to find a further utility. And it wouldn't be useful without publishing the results, not just for fellow scientists, but more importantly, for fellow citizens. Increase *had* to write about what he observed, collected, and classified.

In fact, it is because he was a writer that we know about him today. And the writings we know him for are not his strictly scientific articles in, say, *Silliman's Journal,* but his writing for and to fellow citizens, contemporaries, and nonscientists. We also know him for the useful Wisconsin maps he revised and republished almost yearly. And we know him through his detailed and informative private letters to his wife, Ann, and to his siblings, especially brother Darius. In each of these areas of communication he exhibits his strong sense of duty to fellow citizens and to the citizens who will come after him.

Besides his correspondence with scientists, which never let up, Increase's writing and publishing between 1844 and 1847 were to impart reliable information to ordinary people so that old, unexamined ways may be supplanted with scientific ways, so that ordinary lives might be more productive. Though not a professional or even a paid writer, Increase Lapham was a skilled writer. Though unschooled, he was well read and valued clarity and correctness. He had an extensive vocabulary and a firm command of the English sentence. And he knew his audience. He knew what they knew, what their limitations were, what superstitions they held. His writing over the next few years imparted an immense amount of useful, accessible scientific knowledge to immigrants, farmers, miners, politicians, city officials, sailors, gardeners. Utility suffuses most of his writing, but he wasn't above slipping in the purely wonderful or beautiful.

Lapham's Milwaukee building at 321-323 Poplar Street (now McKinley) was completed in 1844. Next door to the family's home, this rental property housed Increase's study and his collections on the top floor in back. He gardened extensively in the rear.

WHI IMAGE ID 43414

"I have commenced writing what might be called a Wisconsin Gazetteer, though it is not certain it will ever be published," Increase had written Darius from his bachelor's room at the Blanchards' on February 4, 1838. "When it is done perhaps I shall lay it on my fire! It affords me employment for leisure hours, which I have these long winter evenings, and is a useful exercise of my skill in composition. I shall be able to make quite a large book and perhaps, one that will be useful."[2]

Increase was writing the book off and on during the period of his intense canal work in Milwaukee, while he collected plants and planted gardens, as he met and married Ann Alcott, started a family, bought one house and planned another, wrote articles, collected rents for Byron Kilbourn, corresponded with botanists around the country, and worried and grieved over his Ohio family. Increase Lapham's perseverance in this writing is evidence that he was a writer—a true and driven writer—as well as a scientist.

A gazetteer, perhaps modeled after the *Universal Gazetteer* by Jedidiah Morse—one of the books that Increase bought in October 1827 and pored over in Kentucky and Ohio and had carried to Wisconsin—was far more ambitious than the articles he sent to Benjamin Silliman from Kentucky or the entries into his personal journals that had helped him in his work in Ohio. The gazetteer was a large-scale project, an alphabetized index or dictionary of places that would appeal to a wide readership. A gazetteer differs in subtle ways from a geography, which is a book about a specific place, the distribution of its physical and cultural features, and their relationships. Increase, reader of both gazetteers and geographies, understood the difference. At the time, he may have been the only person in Wisconsin who did.

Writing about his discoveries and observations soon after arriving at a new place had become Lapham's habit. He had done so in Shippingport, Portsmouth, and Columbus. Setting down facts, lists, and prose descriptions, along with sketches and maps, helped him organize and understand the data he was collecting and how it came together to form a complete picture of his new places. Writing gave structure and precision to his thought, helped him know where he was and settle in there. The larger scale of this project may have reflected the new vision that he had formed of himself and his place, a vision arising from his acceptance of Wisconsin as his permanent home and Milwaukee as the city in which he would establish a family and from where he would consider the entire Wisconsin Territory, including the unsettled, yet-to-be-surveyed parts that were still to be absorbed, rationalized, and described by his scientific mind.

Much of the substance of his book, of course, would be based on Lapham's original observations, certainly, the lists of plants and shells, of which he was the area's master, as well as the lakes and streams that he explored while surveying for the Milwaukee and Rock River Canal or when he was off on one of his excursions to the mining district or taking measurements of the Indian mounds of Wisconsin. He filled gaps when he could with information gleaned from articles in magazines and territorial newspapers and with writings of earlier explorers, notably Henry Rowe Schoolcraft, who explored areas of Wisconsin and Minnesota in 1820 and 1832.

In his spare time, especially in winter when short days and inclement Wisconsin weather limited fieldwork, Lapham pieced together his book on Wisconsin, sitting in flickering lamplight at his table, writing from notes in ink in longhand on precious paper, clipping newspapers from Green Bay, Milwaukee, and other places that he gathered during the week, and carefully tipping folded articles onto the pages of his growing manuscript.

Outside, if a former Milwaukee newspaperman-turned-historian is to be believed, much of the rest of Milwaukee spent the dark, cold winter nights raising hell in hovels and on the streets. "There were one hundred and thirty-eight rum holes in Milwaukee in December, 1843, Ho ye thirsty!" exclaimed James S. Buck in his 1881 *Pioneer History of Milwaukee*.[3] With Milwaukee having achieved a population of about 6,000 at the time, simple arithmetic reveals one "rum hole" for every 44 residents. However, Increase, the Quaker-raised, gentle, and studious scientist, was having none of it. Instead, he stayed home, dined with Ann and the children, then excused himself to study, think, and write while Milwaukee's nightlife boiled riotously in the streets outside.

Increase's days were full in the years he was writing. He was elected one of five trustees of the west ward, Kilbourntown, in 1841, and was re-elected in 1842. Among the ordinances passed about this time were acts against hogs and cattle running at large, against shooting in the city limits, against gambling, and for grading streets. Lapham was elected one of three school commissioners that year and was appointed to a three-member committee to raise money for harbor improvements. He was elected as one of three assessors in 1844. Throughout, he was working on the canal.[4]

Had Increase been able to write full time, he might have finished his book before Milwaukee acquired the competence to publish it. In 1842, three years after Increase had begun the project, Philetus C. Hale arrived from Massachusetts to open Milwaukee's first bookstore in a small frame building at 411 East Water Street. Hale also presided over the first circulating library in Milwaukee. Both actions would have attracted Increase's attention. In the spring of 1843, Hale started the first book bindery in Milwaukee in a small frame building on Second Street and began to produce the first record books for Milwaukee and adjoining counties.[5] Just when it was needed by Wisconsin's first serious resident author, Milwaukee had a publisher.

Increase finished writing the book in the winter of 1843 and 1844. "An apology for not writing sooner is that my head has been so full of topography, geography, etc., etc., that it would not contain the material for a letter besides. I have now pretty much emptied it out for the printer, having sent him the last sheets recently, and feel quite relieved," he wrote Darius on March 7. Later in the same letter, Increase shared a thought with Darius that would have confounded his less-informed Milwaukee peers, but one that said much about Increase's insatiable hunger for knowledge: "By comparing the date of this with an old record in the bible you will discover that I am 33 years old!! Can it be possible I have lived so long in this world and know so little about it? This is a humiliating fact is it not, one calculated to humble the wisest and best of us."[6]

Increase handed over to Hale a handwritten, closely edited manuscript with these words on its title page: "*A Geographical and Topographical Description of Wisconsin, with Brief Sketches of its History, Geology, Mineralogy, Natural History, Population, Soil, Productions, Government, Antiquities, &c. &c.* By I. A. Lapham. Milwaukee, Wisconsin." Altogether 1,000 copies

(*right*) Title page to the first edition of *A Geographical and Topographical Description of Wisconsin*. A second edition came out in 1846. These books established Increase Lapham as a writer.

A

GEOGRAPHICAL

AND

TOPOGRAPHICAL

DESCRIPTION OF

WISCONSIN;

WITH

BRIEF SKETCHES OF ITS HISTORY, GEOLOGY, MIN-
ERALOGY, NATURAL HISTORY, POPULATION,
SOIL, PRODUCTIONS, GOVERNMENT,
ANTIQUITIES, &c. &c.

BY I. A. LAPHAM.

MILWAUKEE, WISCONSIN:
PUBLISHED BY P. C. HALE.
1844.

were printed, bound in cloth-covered boards and containing 257 pages.[7] It included a hand-colored map entitled "Wisconsin, Southern Part" drawn by Sidney E. Morse of New York that showed the part of Wisconsin Territory that had been surveyed, almost all of it lying south of the Wisconsin and Fox Rivers.[8]

Despite Increase's having referred twice to his manuscript as a "gazetteer," the book that appeared in 1844 as a result of his labors was not a gazetteer but a geography. Its organization roughly paralleled that of the first true geography of the United States by an American, *The American Geography, or a View of the Present Situation of the United States of America*, published in 1789 by Jedidiah Morse, earning him the title of first American geographer.[9] Morse began his 535-page book with a general description of the nation that had been formed by the Constitutional Convention, followed by separate chapters for each of its thirteen states. This is followed by sixty-six pages devoted to the rest of the known world. Of the 257 pages in Lapham's *Wisconsin*, pages 5 through 96 are devoted to a description of Wisconsin Territory, followed by chapters about each county so far organized, plus chapters about Lake Michigan and the Rock River.

Did Increase misspeak when he told Darius that he was writing a gazetteer? He did not. In fact, Increase wrote *both* a geography and a true gazetteer, almost equal in size (the gazetteer is 215 manuscript pages, the geography manuscript 246), but he published only the geography. Increase preserved each manuscript. Bound together, they reside today at the Wisconsin Historical Society Library in Madison.[10] It is plausible that Increase began writing the geography sometime after 1840, probably as an introduction to the gazetteer, which he had already worked on for more than a year. But the introduction kept growing. "Having made the foregoing general observations we now propose to give Descriptions of Counties, Towns, Rivers, Lakes &c in Wisconsin, alphabetically arranged," Increase wrote on the first page of the gazetteer. He had crossed out "Gazetteer of Wisconsin" that originally appeared between "Wisconsin," and "alphabetically arranged." Then he began directly:

Aissippi—See Shell River
Alkeck Shebe, or Kettle River, See Kettle River.

> Albion a town laid out in Iowa County on the east side of the
> Pekatonica in Town Three range five east.[11]

As the geography grew, it likely became evident to the author that he was
duplicating information he had already written, albeit in gazetteer format.
At some point, perhaps only days before he delivered the manuscript to
Hale, perhaps mindful of the cost and size of publishing both parts as a
single book, Increase decided to print only the geography. Curiously, a
vertical line is drawn completely through the middle of most pages and
parts of other pages of the gazetteer. In that case, he might have utilized
his gazetteer as an alphabetized list of notes. Poring over it, he would draw
a vertical line through all entries that he thought were already sufficiently
covered in the geography: simple, practical bookkeeping.

What did early Wisconsinites learn from Increase's "popular" geography? "The Territory of Wisconsin, as established at present, is bounded as
follows. . . . " This sentence of 203 words walks the reader around the territory from Lake Michigan on the southeast, then north, west, and south to
the middle of the Mississippi River on the southwest, "thence due east to
the place of beginning."[12] Readers who survive this gargantuan sentence,
written by Increase the meticulous surveyor, shortly encounter Increase
the joyous observer, now settled into a conversational tone. "There are no
mountains, properly speaking, in Wisconsin; the whole being one vast
plain, varied only by the river hills, and the gentle swells or undulations
of country usually denominated 'rolling.'"[13] Then Increase talks about
lakes—"Wisconsin abounds in those of smaller size"—then rivers and antiquities— "a class of ancient earth-works...not before found in any other
country, being made to represent quadrupeds, birds, reptiles, and even the
human form."[14]

Increase divided Wisconsin into four geological zones: the north underlain with "primitive" rocks such as granite and basalt, a sandstone-
based western district, the lead and zinc "mineral district" of the southwest, and the "limestone district" of eastern Wisconsin.[15] His book's early
insight may have saved Wisconsin entrepreneurs the cost of exploring for
coal. The bedrock limestones of eastern Wisconsin, the youngest rocks
in the state, were laid down by ancient salt seas before the Carboniferous

This manuscript page of Lapham's gazetteer shows a vertical line through several entries, apparently drawn by Increase as he transferred its information to his geography.

period, when tropical forests that were to become coal flourished. The nearest coal beds to Wisconsin were in Iowa and Illinois. "It appears then, from these facts, that we may not hope to add coal to the other sources of mineral wealth with which a kind Providence has so abundantly supplied us," Increase wrote.[16]

He summarizes a history of exploration from Claude Allouez's mission on Lake Superior in 1665, through Nicholas Perrot in 1670, Marquette and Joliet in 1673, LaSalle in 1679, through the French fur trade until 1763 when Wisconsin "was surrendered to Great Britain." Great Britain governed until 1794 when the Northwest Territory was transferred to the American government. The Americans sponsored expeditions, built forts and stockades, continued the fur trade with Indians, and most importantly expanded the lead-mining district that was begun under the French in southwest Wisconsin. Lapham discusses Indian politics and movements, the Black Hawk War, and the organization of territorial government and public land sales, and he promotes the Milwaukee and Rock River Canal. He discusses a university proposed "at or near Madison," the population, the products, the internal improvements, the altitudes, latitudes, and longitudes of various places, the geology and mineralogy. A section on botany includes "a list of plants which have not before been noticed as indigenous to Wisconsin."[17]

Increase finishes his description of Wisconsin Territory by talking about climate and "general health": "It is believed that the facts here stated will be sufficient to satisfy the reader of the truth of the opinion expressed by our most intelligent physicians, that Wisconsin is, and will continue to be, one of the most healthy places in the world."[18]

That spring and summer after publication, Increase began to favor some of his esteemed scientific correspondents with gift copies. Asa Gray wrote Lapham from Cambridge, Massachusetts, on August 9, 1844: "Returning here a few days ago, after an absence of about a month to recruit my strength, I found on my table your little parcel by Mr. Cushing, and your letter of July 3. Many thanks for the book on Wisconsin. Topographical works of this kind are always useful to me."[19] But the book was not only Lapham's gift to acquaintances; it was selling well, especially to easterners considering a move to Wisconsin. "About two hundred books 'Des. of Wis.' [Description of Wisconsin] have been sold for which I receive 50 cts. each or $100. The remainder is going off gradually," Increase told his father in a letter on September 17, 1844. "I send you fifty dollars to be used as you and Darius may think best for 'the greatest good of the greatest number.'"[20] A second edition, "greatly improved" and now titled *Wisconsin, Its Geography and Topography*, was published in 1846 by I. A. Hopkins of Milwaukee and printed in New York.[21] Lapham corrected some entries from the 1844 edition, brought others up to date, and added new information, including segments on three new counties organized since the first edition, Chippewa, St. Croix, and, temporarily, La Pointe.

Lapham's *Wisconsin* constitutes both Wisconsin's first book of history and "the state's first home-made book of any character to be bound in more durable binding than paper," according to Milo M. Quaife, director of the State Historical Society of Wisconsin (now the Wisconsin Historical Society), writing in *Wisconsin Magazine of History* in 1917.[22] It is a measure of the esteem commanded by Lapham that Quaife's article about Lapham is the lead article of the magazine's first issue.

The book's section on geology did much to establish Increase's reputation as the resident expert on Wisconsin's geology. That reputation was reaffirmed and expanded in 1852 with the publication of David Dale Owen's *Report of a Geological Survey of Wisconsin, Iowa and Minnesota* in which fellow geologist Charles Whittlesey, describing his 1849 fieldwork on the south shore of Lake Superior, reported that "the observations of Mr. Lapham were of the highest value in elucidating, by means of the rocks on the south, those I was studying on the north. I take this opportunity to express my obligations to him, and to state that his examinations, made entirely on private account, are extensive and correct."[23] Whittlesey's report was obviously informed by far more detailed accounts of eastern Wisconsin geology than Increase had published in his *Wisconsin*. Whittlesey noted that Increase "liberally contributed his observations to me, and they are, so far as I know, the only authority on the rocks south of Lake Winnebago and east of the Rock River."[24]

Although we may never know the full circumstances, we can be certain of this: when Increase told Darius that he was writing a gazetteer, he was writing a gazetteer. And much more.

—|—

Though the publication of *A Geographical and Topographical Description of Wisconsin* dominated 1844, the following year was back to work as usual. Increase was busy making sidewalks, "gravelling" streets, selling town lots, collecting rents, and also helping settle the estate of Micajah Williams, Byron Kilbourn's partner and Increase's former boss in Ohio, who had recently died.[25]

But Increase and Ann's main concern in the summer of 1845 was the health of their little boy, William Allen Lapham. Increase wrote his father on July 17, "We are all very well except my little boy.... We hope he is not dangerously sick. *Ann* thinks ours is the prettiest and best baby in the United States and *I* think she is about half right."[26] But the little boy they called Allen was seriously ill and on July 31, he died, "Aged 1y. 7 mo. & 12 days."[27] About two months later, Increase wrote Darius that "We have hardly been able to speak with composure of the death of our only son Allen, and consequently have not written to our friends relating to him."[28]

The Laphams' solace and perhaps the expression of their loss seemed to be in their quiet time together and in the beauty of the world around them. "Our sun risings and moon risings are beautiful beyond description, rendered so in a degree by the broad smooth surface of the Milwaukee River lying just before our door towards the east. They set behind some magnificent hills a hundred feet in height adorned by forest trees."[29]

Increase busied himself with an occupation that had, since his early years in the territory, been a source of pleasure and for which he had not a little skill: mapmaking. It was a natural result of his experience on the canals in Ohio and New York and a necessary effect of his drive to be useful to fellow citizens and also to make a little money. In 1838, Increase had drawn his first important Wisconsin map, showing the route of the Milwaukee and Rock River Canal. Mapmaking was an essential tool of Increase the surveyor, draftsman, engineer, and land agent. He was good at it and he knew it. Through the summer and fall of 1845 when little William Allen was sick and after he died, Increase worked on a sectional map of Wisconsin.

A sectional map, one that showed the boundaries of all thirty-six sections, or square miles, of each township in each county, would inform newcomers to Wisconsin. With it, they could select new home sites and see how their locations related to other features in the territory, including its roads, lakes, streams, and settlements. Such a map would show the part of Wisconsin south of the Fox and Wisconsin Rivers. North of the rivers, section lines had just begun to be laid out by government surveyors. Increase began work on his map of Wisconsin perhaps partly because he was unhappy with the relatively crude maps by Sidney E. Morse of New York that had been tipped into both the 1844 and 1846 editions of Lapham's *Wisconsin*.[30] And mapmaking became another source, albeit unsteady, of income.

The 1846 map was published by partners P. C. Hale and I. A. Hopkins of Milwaukee, whom Increase charged a hundred dollars for publishing rights, and lithographed in Boston. By June, this map was advertised in the Milwaukee *Daily Sentinel* for sale at Hale's bookshop as a sheet or folded as a pocket map.

Increase's map relied in part upon some of the thirty or so maps of Wisconsin that had been produced in the decade before 1846. But mostly it relied upon his own observations as a surveyor in the field and especially on government surveys provided by the US General Land Office. He also wrote local officials such as postmasters to find the latest information about location of roads, mills, or other works.[31]

That year Increase extended his mapmaking skills by teaching himself to use a sextant. He wrote Darius that he got hold of a sextant and went to the north side of Milwaukee bay and, while his wife and children hunted agates on the pebbly shore, Increase watched through the sextant for the time when the sun was at its greatest altitude. "Ain't I the astronomer!" he exclaimed to Darius.[32]

After his 1846 map was finished, hardly a year passed in his life when Increase or his publishers did not revise his Wisconsin map. Some twenty-five versions were made, each new edition adding counties or section lines, roads or railroads, new settlements and new governmental boundaries as Wisconsin developed as a territory and a state. In addition, he prepared maps of early Milwaukee and Milwaukee County, which sold thousands of copies.

—|—

The three years between Increase Lapham's completion of the two editions of *Wisconsin* and beginning his next major work were a relatively quiet time—quiet for Increase, though full of accomplishments and events within the city and territory.

Early in 1846, Increase received a letter from Charles C. Rafn in Copenhagen, Denmark, notifying him that he was elected a member of the Royal Society of Northern Antiquaries, a society that had translated and published an article of Increase's on American Indian mounds and artifacts. This membership was an honor, but it was also a sweet bribe to get Lapham to send them more articles and perhaps Indian artifacts. The honor would have meant more to Increase if they hadn't spelled his name "Ingram," a fact that bothered him so much that he wrote several times to get a corrected copy of this fancy "diploma" with seals and the signature of a prince.[33]

In June and July of the same year, the Lapham family—Increase, Ann, and the two girls, along with niece Elizabeth Stone—traveled by steamboat to Sheboygan, Manitou Island, and then Detroit. With many stops to look for rocks and fossils, the family arrived at the Lapham farm in Ohio on July 7 where they were all reunited when Darius and his wife, Nancy, and their two girls, Phebe and Harriet, arrived a few days later. After a week with his family and taking sister Lorana Lapham with them, the Milwaukee Laphams left on the railroad for home, looking at the geology all the way and passing through Ann Arbor, Michigan, where Increase was horrified to see the specimens in the cabinet of the University of Michigan not arranged or named, with many boxed up in a garret. This new university did not impress Increase. The buildings stood bare in a wheat field: "No shade trees!"[34]

Again Ann Lapham had traveled during a pregnancy. Another son was born to the Laphams December 13, 1846, a few days before what would have been little William Allen's third birthday.

In 1847 we begin to get the benefit of Increase's letters to Ann as he travels around the state and then takes his first and long-awaited trip to the cities of the East. Though at the time they may not have had more than a personal utility, letters such as these are invaluable to those of us who want to know him almost 170 years later. In many ways these travel letters to Ann are more revealing than his letters to Darius. He tells Ann not only what he sees and does but also what he feels, and he often feels lonely, not just because he misses his wife and children, but because Increase really does not like to be alone. In January of 1847 he wrote Ann that he was slipping along on Adelaide "as merrily as one could *all alone*."[35] It may seem that a man who spent a great deal of time thinking and reading and working on intricate scientific endeavors would seek out and savor solitude, but Increase came from a big family and was used to working with them milling around him or at least within earshot. He also wanted to share with others, especially with Ann, what he saw and discovered; he didn't enjoy seeing new things, wonders, alone. He wrote Ann about sights he wished she'd seen with him—like the four flocks of quail he startled in the road.[36]

In Increase's letters to Ann we also can hear what they laughed at to-gether, their private language. Their distance from the dominant Wisconsin culture was often a source of their jokes. Increase wrote Ann as he traveled to Madison in the cold January of 1847 that he "Liquored up" at Parker's Tavern. But Increase, soon to be a member of the Sons of Temperance, did not mean that he had an alcoholic drink; he meant he watered his horse, Adelaide. And he wrote Ann that the Farmington Log House where he stopped for the night, "was full of *raw* New Yorkers" (Increase's emphasis), new immigrants who went off to a dance leaving him with the place to himself.[37] Ann and Increase, old settlers from the 1830s, like many immi-grants, could laugh at the ones who had just got off the boat.

WHI IMAGE ID 43832

Ann Lapham, wife of Increase. Ann was a good match for him—a lover of family, the natural world, education, and Increase, to whom she wrote spirited and warm letters.

Increase Lapham's thoughts as he traveled that winter to Madison are of home and his month-old son. He hopes the little boy sleeps through the night so "mother" does not have to stay up with him. When Increase arrived in Madison, there was a let-ter from Ann saying that their "little nameless one," as Ann "very pret-tily" called him, was healthy.[38] Then, writing to Ann, Increase thought about what to name this month-old boy. Ann had apparently said the matter was up to Increase. So the husband wrote the wife that he did not like names that are too common or "outlandish." Perhaps names like Increase? "But there are some names," he wrote, "to which we attach some peculiar feelings of partiality."

> For instance "Mary" and among boys, there is . . . something of the
> same feeling connected with the short and pretty but rather common
> name of "Henry." . . . I now decide in full and without reserve or equiv-

ocation or jesting to call him HENRY LAPHAM! So there is an end to the much talked of matter, unless, indeed, you demur, in that case the subject may be opened again.[39]

The give and take, the humor, and the loving kindness of a good marriage are all in that exchange.

At several points in his life in Kentucky and Ohio, Increase had expressed a strong desire to see the cities of the East. His first chance came when he was thirty-six at the end of May in 1847. He took with him on the steamboat his sister Lorana, who had been staying with them in Milwaukee since June of 1846. He left behind his thirty-three-year-old wife, Ann; nine-year-old Mary, who was learning to read; Julia, seven; and the infant Henry. Ann's mother, Mehitable Alcott, was likely living with them, as well as a local woman named Mary Waters to help with housework.[40]

On the steamboat *Oregon* traveling up Lake Michigan and down Lake Huron, Increase wrote a journal every day to Ann, telling her about the geological features of the shoreline, the white caps, the reflection of Lake Winnebago on the clouds, the poet who reads aloud on the boat, a cotillion he watched, the music of the black waiters. His letters to Ann clarify his thoughts and deepen their relationship. What he sees is interesting to him but, Increase wrote, "How I do want to see you all again. I must hurry back. I am homesick already."[41] At the same time, Ann wrote to her husband saying she found the gold pen he left behind in his study and thinks she will write to him from his desk where it is quiet, but "to tell the truth I feel much more like having a good cry and going down to my babies." Then, she said, the family all had a good look at a miniature portrait of their father and husband and the children agreed that they would be very sorry to do anything to grieve him.[42] Increase took the time to send at least two articles to the *Milwaukee Sentinel and Gazette*, one about his donation of land for the high school and the other about the fine steamship *Oregon*, which had everything any reasonable person could wish except a "small well selected library of useful and interesting books."[43] What Increase Lapham observes and thinks about, what he feels, he writes to Ann. And where he sees there might be improvement, he writes about it and, in many cases, sends it off to be published.

After a visit with his family in West Liberty, carrying lists from Seneca Lapham and Mehitable Alcott of relatives to visit in the East, Increase left Lorana and traveled from Cleveland to Buffalo with one hundred fat hogs in the "bedlam" of the rattling steamship *Bunker Hill*.[44] On his way to Albany, he stopped in Rochester where the Alcotts were from and made so many duty calls on relatives that he reported to Ann he was sick of "cousining."[45] In Albany he visited the Museum of Natural History, dined with geologist and paleontologist James Hall, then traveled down the Hudson to New York City where he stayed at a hotel on Broadway.

Like many Midwesterners before and since, Increase Lapham didn't know right away what to make of New York City. He reported to Ann that he had a throng of ideas he didn't know how to sort out. He was "woefully disappointed in many things"—the Hudson River scenery, Broadway, Trinity Church, the Astor House, and Central Park; "they do not come up to the pictures I have formed in my imagination."[46]

But Increase was Increase. The next day he visited the Lyceum of Natural History, then climbed the 328 steps to the top of the Trinity Church steeple, before he caught a steamboat around Long Island for New Haven. This was, he wrote Ann, his first ride on salt water. He saw a porpoise from the ship and noticed that salt water makes more froth and foam on the paddlewheels than fresh water and a "fine rattling noise" is caused by the falling of the spray.[47] While in New Haven Increase saw Benjamin Silliman and his son and saw the cabinet of minerals and the gallery of paintings at Yale and walked all over the city in the rain. From New Haven, he went up to Boston and called on relatives, but also took in the Mount Auburn garden cemetery and the Bunker Hill monument.

June 10 was the highlight of his trip. He called on Harvard botanist Asa Gray who had arranged a dinner for Increase, himself, and fellow botanists William Carey and William Oakes, and "didn't we four 'Great Botanists' have fine times!...We dined at 6 o'clock in fine style."[48] His stay in Boston seems both more productive and more enjoyable than his stay in New York. Though soliciting money for the Milwaukee high school was ostensibly one of his purposes for the trip east, asking for money was not something Increase was good at or liked to do, so his attempts in Boston with millionaire Abbott Lawrence were halfhearted at best. But educator Lapham did visit a fine school before engineer Lapham moved on to more interesting things, like taking the railroad to Lowell to see the cotton and wool factories and examining the British steamship *Hibernia*. Back in Boston, patriot Lapham was "almost overpowered with emotions of different kinds when walking all alone in the Faneuil Hall, the cradle of Liberty, hung with portraits of the old patriots of revolutionary and antirevolutionary times."[49] The next day, Sunday, Quaker Increase went to one church in the morning, another in the afternoon.

Though Increase had planned to go to a convention of the Sons of Temperance in Philadelphia, his steamboat was wrecked *twice* in the Long Island Sound, and he was so homesick that he just headed home, with a stop in West Liberty to pick up Darius and his daughter Phebe and their sister Amelia, the youngest Lapham sibling, who were all going to Milwaukee for a month's visit. They were back in Milwaukee June 28 or 29 so Increase could kiss his children, as he wrote to Ann, "*in proprice persona, et non per proxy.*"[50]

Increase ordered this massive walnut bookcase, measuring eight and a half feet tall and almost fourteen feet wide, for his new study in 1847.

COURTESY OF THE MILWAUKEE COUNTY HISTORICAL SOCIETY

In Milwaukee that fall, the Laphams sold several acres so that all "air castles" about being a farmer vanished into thin air.[51] This daydream began back when both Lapham brothers and father were itinerant engineers on the canals in Ohio. Increase had by 1847 a much more practical than romantic idea of farming. He also sold their horse Adelaide, whom they had had for more than eight years. He didn't let his feelings for the horse—"almost a member of the family"—get in the way of practicality; they needed a younger and stronger horse: Billy came into their lives.[52]

Perhaps with money from the sale of the farm, that fall Increase had a very large walnut bookcase with glass doors built for his office on the second floor of their brick building. In the center of the room, with his books and specimens before him in the bookcase, was the table where he examined rocks and fossils, mounted plants, and wrote. Increase was wonderfully organized, as one might expect. He had cubbyholes for each subject and could lay his hands quickly on any document or letter—all of which he folded to fit the cubby spaces and labeled with author, subject, and date received.[53] This organization was a large part of his ability to study many subjects. And the quiet of his out-of-the-way study allowed for concentration and more writing.

Though he was not becoming rich, Increase Lapham was providing a comfortable life for his growing family. And he was providing to those beyond the Wisconsin Territory information that would be useful to emigrants, botanists, and geologists. And for the curious citizens of Milwaukee he increased immeasurably their knowledge of the natural world around them.

CHAPTER 15

Civic Life in the New State

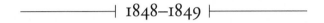

————————| 1848–1849 |————————

IN JULY OF 1848, THE YEAR WISCONSIN BECAME A STATE, Increase and his brother-in-law Samuel Stone of Chicago spent eight days on the road to Blue Mounds west of Madison observing on their way that the limestone bedrock of eastern Wisconsin was mostly hidden by drift, that Waukesha's exposed limestone was compact, hard, and white and its surface scratched and polished by what Increase surmised was "diluvial" action. South of Blue Mounds he collected "fine specimens" of fossil sponges and spiral and bivalve shells.[1] At Portland, in Dodge County, he observed an out-crop of "metamorphic quartz," noting, "It would be a good site for the Penitentiary where the convicts could be employed in quarrying stone."[2]

Increase seldom hesitated to recommend public projects based on his observations. On December 9, 1848, he wrote to Wisconsin's first governor, Nelson Dewey, about his ideas for the penitentiary. There, he explained, the convicts, like their counterparts in Sing Sing Prison, New York, could quarry rock that was "believed to be in every way suited for building purposes, and to possess a degree of firmness and durability which is not found in any of the building materials now used throughout the western states."[3] For reasons that are lost, it did not happen. Dewey decided in 1851 that the state penitentiary would be built in Waupun. Breaking stone does not appear to have been in its curriculum. No doubt Increase would have been part of the efforts had it gone the other way.

Such civic action, making recommendations based on his observations, was merely a continuation of efforts that had started many years earlier. Increase had, from his earliest days in the territory, been an active participant in many of the public undertakings of the city of Milwaukee. He had helped start the state's first public library, served many times as town supervisor and city alderman, served on school committees and as public assessor. In 1848 and 1849, Increase continued this commitment to civic life, though he was compensated for few of his civic activities. Following his family's Quaker beliefs, Increase Lapham took civic duty seriously and expected those around him to do the same.

Perhaps most significant of Increase's civic work was his ongoing commitment to educating the young people of Milwaukee in the high schools and then in a new college. In February of 1848, he gave two lectures, one on Wisconsin and the other on geology to the Milwaukee High School, then meeting in the Unitarian church. In the lecture on geology, Increase expressed his belief that a literal interpretation of the biblical scriptures could give way to scientific evidence, a foundation of education:

> Geology, as I have said in the beginning, attempts to trace the successive changes that have taken place in the structure of the earth, since its first creation. "Changes?" says one. "Was not the earth created just as we find it? Was it not made and finished in six days? and what change can it undergo, since it was finished by an all-wise Creator?"
>
> The inquiring mind of man does not rest satisfied with this indolent and unsatisfactory way of regarding nature. Admitting that all things were created "in the beginning," the question naturally occurs, "Were they originally created precisely as we now find them?"[4]

No, they were not. Lapham explained the phenomena that could change the physical world, including seasonal freezing and thawing that splits rocks, rain that erodes hillsides, and the waves, volcanoes, earthquakes, and floods that remake landscapes. "*Change* appears to be a universal law impressed upon all matter by the Creator. There is nothing permanent, nothing exempted from this great law of nature. All is motion, progress."[5]

Both speeches were dense with details. But in the second one, Increase attempted some levity. In his introduction he noted that geology was becoming more fashionable. And "the ladies" made an appearance: "Even the ladies may now be seen, with their miniature mineral hammers, detaching specimens from the solid rocks of the earth, or gathering agates and carnelians from the lake shore, and gathering health, happiness and buoyancy of spirits."[6] Concluding, he joked: "But I must not weary your patience in listening to this subject.... Rocks are 'hard cases' and we will have nothing more to do with them at present!"[7]

Because his reserve "amounted almost to timidity," as one friend noted, Increase was not an eager platform speaker, and he "did not appear to advantage as a public lecturer." But when called on to speak he would do the work "cheerfully and well." With friends "of congenial tastes and sympathies, he was affable and communicative, as ready to impart knowledge, as to receive it."[8] Increase's lectures would have been long by our standards, but he was speaking to a willing audience of citizens with limited access to books and to ideas.

Science and education were aspects of a single endeavor in Increase Lapham's mind. Why ask scientific questions if the answers were not given to the public? What is the good of science if its benefits can't be shared with all? One medium for sharing would be formal education.

Uncomfortable with his own lack of schooling, especially at the university level, Increase seized every opportunity to advance education to his fellow citizens and to his own children. His contradictory approach reflected his background. He lectured frequently and without pay to local high school students and to fellow members of the Young Man's Association about geology and botany and other fields. Yet he turned down opportunities to teach for a fee at fledgling colleges on grounds that he was academically unqualified, even though he would often lecture on specific subjects to college classes.

In 1846, he had donated thirteen acres of land on the west side of Milwaukee to the city for a high school and then made two trips east without success to raise money for a building. Then on March 28, 1848, he wrote to Darius:

> We are now engaged in another literary enterprise—the establishment of a college. I shall appropriate the high school ground for that pur-

pose, now valued at $3,000. Several persons are to make donations amounting to $500 which entitles them to a 'scholarship,' or the perpetual right to keep a scholar in the college free of expense of tuition. I shall be entitled to six such scholarships.[9]

It seems obvious that in claiming six scholarships, Increase was thinking of his own children, including his two eldest, daughters Mary and Julia. There would be no family debate about whether women should have the opportunity for higher education. In the tradition of the Society of Friends, Quaker women held equal station with men.

When on August 18, 1848, a handbill advertised that the Milwaukee Female Seminary would accept its first students for the fall term starting September 14, there was the name of the "Hon. I.A. Lapham, Milwaukee" leading a list of thirty prominent pastors and scientists from as far away as Rhode Island to be used as references.[10] The school opened with fifty students on the northwest corner of Milwaukee and Oneida (now Wells) Streets, with a faculty of four.

In 1848, the meaning of the word *seminary* was not restricted to a theological or divinity school, as it is today. Then it also meant a private school, especially for young women. The curriculum at the Milwaukee Female Seminary, if not influenced by Increase, the president of the board of trustees, would have pleased him greatly. The new four-year school would teach an array of natural sciences, including botany, physiology, astronomy, chemistry, geography, and geology, subjects not often taught to women, as well as the usual subjects of mathematics, grammar, history, government, and moral science. Everyone would learn reading, spelling, penmanship, composition, and vocal music and would participate in calisthenics. "Young ladies expecting to teach, will receive special instruction," the handbill said.[11]

Two years later, Milwaukee attracted the interest of Catharine E. Beecher of Cincinnati, a member of the famous family headed by Lyman Beecher and including her younger sister Harriet Beecher Stowe. Catharine Beecher was dedicated to the practical and physical education of women. She wrote progressive books on education, including her 1842 classic, *A Treatise on Domestic Economy for the Use of Young Ladies at*

Home and at School. She believed that she would have greater opportunities to test her theories in the developing West than in the traditional East.

After a visit to Milwaukee in 1850, Miss Beecher was impressed with the prospects of Milwaukee and the intelligence of its citizens and proposed to recast the Milwaukee Female Seminary into a new form with a new name, for which she won the approval of Increase Lapham. Essentially, she wanted the school to be a public institution with citizens owning stock and students attending free of charge. She also gave the school "a library and scientific collection worth $1,000."[12] The school was to have insecure times, usually related to money and often rescued by Miss Beecher. Just as often Miss Beecher was frustrated at not being able to manage policy to her satisfaction from hundreds of miles away, and she filed lawsuits or threatened to withhold support.[13] Increase navigated these stormy seas through 1863 as head of the board of trustees.

In March 1851, the Wisconsin legislature incorporated the new school as Milwaukee Normal Institute and High School, and that May, Increase delivered a commencement speech and handed diplomas to the first two women to graduate. Meanwhile, a fund drive was begun to raise money for a new building. Increase and Ann Lapham joined in the efforts. Ann and two other women formed the Ladies Educational Society of Milwaukee, which raised sufficient funds to begin construction of a building on the southeast corner of Milwaukee and Division (now Juneau) Streets.[14] The Normal Institute and High School moved into it in the fall of 1852. "Our building is very beautiful and commodious. We number one hundred and twenty pupils," wrote Mary Mortimer, whose job as teacher in Milwaukee had been secured by Catharine Beecher.

In 1853, the school's name was changed at the suggestion of Increase Lapham to Milwaukee Female College, although "Female" often was dropped from its title in popular usage so that when Mary Lapham, Increase's oldest daughter, enrolled, the school was generally known as Milwaukee College.[15] By this time it offered both a high school and post-secondary education.[16]

Years after she graduated in 1858 from Milwaukee Female College, Mary J. Lapham wrote a reminiscence of her college days:

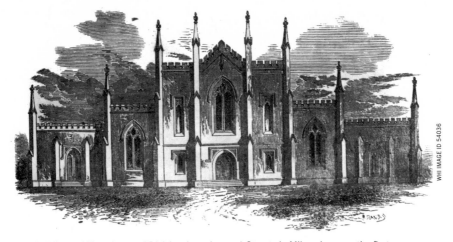

WHI IMAGE ID 54036

This building at Milwaukee and Division (now Juneau) Streets in Milwaukee was the first permanent home of the Milwaukee Female College.

We were at that time in a great quandary as to whether we could rightfully call ourselves the alumni, as we belonged to a new order of beings, women with a college education and diploma! and the question whether we should be called by the feminine form of the word alumnus, or not, remained unsettled for several years....

During the early years of the existence of the Female College a number of excursions were arranged by my father and others, for "nature study" as it would now be called. We went to various points of interest near Milwaukee, the stone quarries of Wauwatosa, the woods and marshes at Brookfield, the Dells at Kilbourn City and to the astronomical observatory at the old Chicago University, with its wonderful telescope, the largest in the country at that time.[17]

She was her father's daughter.

—|⊢

The year 1849 was one of continuations for Increase. He continued his studies and correspondence in botany, sending Asa Gray his "Enumeration of the Plants of Wisconsin." He continued to take meteorological measurements at 6:00 and 9:00 a.m. and at 3:00 and 9:00 p.m., measurements he sent on to Joseph Henry at the Smithsonian. He continued to work "pretty deeply in the map making business."[18] He was re-elected as alderman and served as president of the City Council and occasionally acted as mayor, "in the absence of Simon Pure"—Byron Kilbourn.[19] In that one phrase in a letter to Darius he gives away that he is well aware of the shady dealings of his boss.

Increase took another trip east in September of 1849, picking up Darius on the way to the New York Agricultural Fair in Rochester. Perhaps because Darius was superintendent of the coming Ohio State Fair, the two were admitted before the fair opened. Separated from Darius, Increase rambled around alone all day. Then he sat down somewhere quiet and wrote to Ann, telling her that after the "pop and show and parade," there is "still something wanting to make me feel happy and contented in this world, something to entertain the heart as well as the head." He described for Ann the floral exhibitions and the mechanics' and the manufacturers' pavilions, but he wished that Ann could telegraph herself to him so they might walk the grounds together.[20]

After the fair, Increase traveled on to New York City where he dealt with map proofs. In Philadelphia he walked the city's streets alone. In Washington he went to the Capitol, the patent office, the Smithsonian, the Library of Congress, and the "president's house," but records no detail. He notes that he saw many great men, but he is not sufficiently impressed to mention them by name in his letters or journals.[21] Did he see the president, Zachary Taylor? He doesn't say.

In June and July of 1849, during a cholera scare in Ohio, his parents, Seneca and Rachel, came for a month, visiting the Lapham family at Third and Poplar in what had become the center of the large German immigrant community in Milwaukee. Their mother, Rachel, Increase wrote Darius, stayed in the house and knitted, while Seneca rambled about the city until he had seen most of it.[22] They were happy to finally become acquainted with the city where their son had settled, and to see their latest grandchild. Ann and Increase's third son, Seneca George Lapham, had been born the fall previous, on October 12, 1848.

During August of 1849, Increase Lapham measured and kept meticulous records of the tides on Lake Michigan. He kept a chart that included the daily mean level of the water, the actual daily level of the water, and barometric pressure. These measurements were the first scientific evidence that there were lunar tides on the Great Lakes, a phenomenon that was suspected as early as the mid-seventeenth century by French Jesuit missionaries. He sent his chart to Joseph Henry at the Smithsonian on September 1, and he added that he had taken measurements every three hours on the height of the water of Lake Michigan, "taken with a view of settling, if possible, the long vexed question of 'tides on the lakes.'" Increase told Henry he didn't have the time or expertise to follow up on this, but he added that he had also taken the same sorts of measurements and achieved close to the same results in March of 1846, but the weather was so stormy that his numbers might not be as reliable as those of 1849.[23] Increase's tables were published in the *Milwaukee Sentinel and Gazette* on September 3 of that year. All this documentation would later prove that Increase Lapham, rather than Chicagoan J. D. Graham, had proved that there were tides on the Great Lakes.

It was late in 1849 that Increase expressed some of his most surprising and original ideas about the duty of citizens. "Of the Historical Society of Wisconsin" is Increase's curious memorandum to the officers of the new Historical Society of Wisconsin, an organization formed for the purpose, in Increase's words, of "collecting and preserving materials for the history of Wisconsin." No one, he says, no "intelligent person could doubt the value of such an institution."[24]

On Monday, January 30, 1849, Increase had joined thirty-two other Wisconsin citizens gathered in the Senate chamber in the state capitol in Madison for the purpose of creating the Historical Society of Wisconsin (today known as the Wisconsin Historical Society). He was appointed, along with George Reed and John Y. Smith, to a committee to draft a constitution, which was adopted. He also was on a committee to nominate officers and from this effort Governor Nelson Dewey became president. Among other officers, Increase became the society's corresponding secretary, a post he held through 1853. Also in the first meeting, Increase proposed a resolution calling for Wisconsin's surveyors to furnish the society with sketches and measurements of the "ancient mounds and artificial earth-works in their vicinity."[25]

When Lyman C. Draper became secretary of the society, Increase became a vice president, from 1853 through 1860. In 1862, he became the society president and served as such through 1871. Over the years of his service, Increase was largely responsible for building the society's library to more than fifty thousand volumes. In 1855, Draper arranged for Samuel Marsden Brookes and Thomas H. Stevenson to paint Increase's portrait in oil, to join twenty-nine other portraits of prominent Wisconsin citizens and Indian chiefs in the society's collection. "You have been a pioneer and the literary and scientific man of the state and your portrait should be preserved in our historical gallery," Draper wrote Increase.[26]

At the end of 1849, when Increase wrote "Of the Historical Society of Wisconsin" to the society's officers, his outlook was dark. He seemed overwhelmed by work—perhaps because he realized that with the responsibilities and the idealism of Wisconsin's recent statehood, he himself would still be carrying more than his share of the burden of the state's geology, botany, library organization, work on the schools, as well as the Historical Society. He had written two drafts of an important book for Wisconsin; he had mapped Wisconsin; he had offered an extensive collection to the university. He had up to this point given more to Wisconsin than anyone he knew; most others had not given but sold their time and energies to their state; they were paid. Increase was being asked for more, not just by the Historical Society and by his state, but, more powerfully, by his own sense of citizenship and sense of duty. "[I]t is the duty of each generation to carefully and faithfully preserve and record the history of their own times for the benefit of those who are to come after them," he wrote to his fellow officers.[27]

This is, of course, what Increase Lapham had been doing

This oil portrait of Increase Lapham, commissioned in 1855 by Lyman Draper, was painted by Samuel Marsden Brooks and hangs in the archives reading room of the Wisconsin Historical Society.

WHI IMAGE ID 2758

since he was a boy. He had recorded and measured and charted and written and classified and filed and saved almost every day of his life for the
benefit of those who would come after him. He was right that this saving for posterity is a good thing for citizens to do, but he had lived long
enough to know that most people would not be good citizens in this
respect, or in many other respects. Because of "idleness, intemperance,
and extravagance" and also "apathy, indifference, and indolence," a large
part of the community, he says, will do nothing: "and so it devolves upon
the remainder to perform an undue share of this work."[28] This complaint
was unusual for Lapham. It is an early statement of a feeling that would
surface now and then for the rest of his life—resentment for doing more
than his share of work for less than his share of the reward.

In this document he defines patriotism—perhaps in a self-serving way
and perhaps against some accusation or slur—as the "performance of this
excess of individual duty" for the common good. So we see Increase at this
time as chafing under his workload—much of it self-imposed by his own
idealistic and rigorous vision of the duty of the citizen. In this document
he reveals to us the ideals of citizenship he is trying to live up to.

Lapham reveals also an elitist idea of the "intelligent citizen" and a
harsh and unforgiving attitude toward those who don't do what this intelligent citizen does. Though he makes allowance for the time people
spend "providing the necessary and proper comforts," he seems to not
understand the demands made on most people's lives.[29] He doesn't understand that most people can't do what he can do. He doesn't understand
that, in spite of the fact that he was not formally educated, he has had
rare opportunities. Since he was a boy, he had been surrounded by educated and skilled men and by books and scientific journals. He doesn't
understand how exceptional he is, not only in his background and abilities but more especially in his sense of duty. He also does not understand
that most people simply do not value history the way he and a few others
do. Wisconsin was a state of immigrants who were looking "forward"—a
word that was to become Wisconsin's state motto in 1851. They were not
looking back at where they came from.

Revealed also in this six-page document for the Historical Society of
Wisconsin is Increase's interesting idea that collectors of history should

go behind the facts of their history and find the "hopes and fears," the "reasons, the motives, the latent power that brought about these facts."[30] These are new thoughts for Increase the scientist. Increase's connection with the Historical Society has brought out in his writing for the first time the idea that the "feelings" behind an act are important and powerful.[31] When he was a young Quaker boy working on the canal, his father and older brother brought home to him again and again the idea that what he felt was not as important as what he did, which was to carry out his duty to the family. Seneca and Darius knew that it hurt Increase to be left behind in Shippingport and to stay in Kentucky at a terrible job for more than a year, but they valued the good of the family more highly than Increase's individual feelings. Though these Laphams were apparently a loving and close family, they certainly valued performance of duty over the expression of feelings. The words *feelings* and *emotions* have not appeared in Increase's writing up to this point.

Then, in "Of the Historical Society of Wisconsin," Increase takes this idea of the value of emotion further in a lovely sentence with a beautiful phrase: "We want not only to know that a new town was commenced at a certain time, but why it was commenced; what were the motives, the hopes, the fears, the secret springs that set in motion so strange and important an enterprise." As Increase Lapham the scientist and engineer moves into the realm of history, he thinks of and values the "secret springs" of the settlements of Wisconsin. We see this as well in the questions he suggests that community historians ask of the people around them.[32]

It is with this set of questions that Increase closes his memorandum to the officers of the Historical Society, suggesting that, in a circular to the state's citizens, these questions might elicit the greatest amount of information on the history of each community. We might think he would have these historians begin with the first white settlers of the community—who they were and where they came from—but the white settlers are fourth on Increase's list. Lapham, remarkably, begins with what might be called another "secret spring": the "ancient artificial mounds or earthworks" found in their town. He asks local historians to describe their locality, form, dimensions, and contents if they have been opened. He then instructs them to ask about the Canadian, French, or other traders who have occupied the site. Then, what were the Indian tribes and when did they leave. Only then does he ask about the white settlers.[33]

These questions reveal the deepening of Increase Lapham's abilities. He not only can see a long way back and well into the future, but he is looking deeper into the "heart as well as the head."[34] And he is looking ahead to what will become his greatest work.

CHAPTER 16

The Antiquities of Wisconsin

1846–1855

IN THE FIRST FEW MONTHS AFTER HIS ARRIVAL in Wisconsin Territory, as Increase Lapham leveled streets and mapped Milwaukee County, he measured and drew eleven Indian mounds, some conical, some lizard-shaped—all just blocks from where he lived at Third and Poplar. Within weeks of his arrival in Milwaukee, Increase had not only recorded in his notes these important effigy mounds, but his surveying had assured their destruction. What must he have felt as he watched the heads of these figures destroyed by the shovel and the grader? To what extent did this early destruction affect Lapham's drive to document the remaining Indian mounds? We can't know this, but it's impossible to think he wasn't bothered. In November of 1836 he wrote a letter to the editor in the *Milwaukee Advertiser* decrying the destruction of a turtle mound in present-day Waukesha. Increase was outraged that this 250-foot-long effigy and burial mound had been desecrated by its present proprietor, who had "removed the human bones to make room for his stock of potatoes!"[1]

In his first years in Milwaukee, Increase had his face, Janus-like, on both the future and the past. He was there to promote building, settlement, and development for the white men who lived there or would soon live there. But Increase could also see signs around him of an ancient Indian past that was being obliterated by all this improvement. A decade after he came into the territory, Increase had seen plenty of the effects

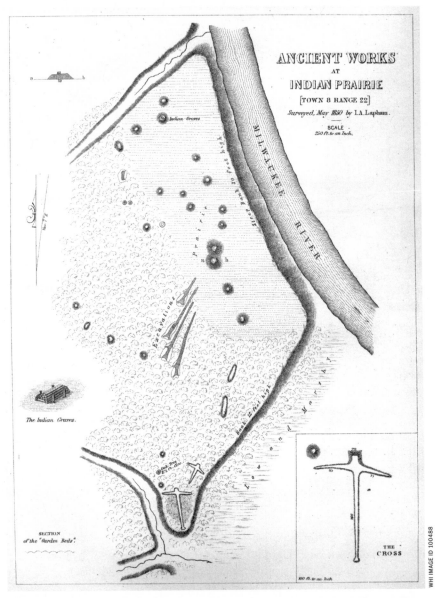

In 1850, Increase surveyed these mounds in Northern Milwaukee County. His sketch appeared in *Antiquities of Wisconsin*.

of the settlement by white men on the land that made up Wisconsin Territory. And the future, hell bent, was tempered only a little by his influence—the schools, the Lyceum, the libraries, the Historical Society. He saw that development could and would go on without him, but that the efforts of *his* mind, *his* observations and values were needed to record not only the natural world he saw around him but also the remnants of the valuable and mysterious past.

In 1855 the American Antiquarian Society and the Smithsonian Institution jointly published what would be Increase Lapham's greatest written work, *The Antiquities of Wisconsin as Surveyed and Described*. Notable in its time, it remains one of the few books on Wisconsin Indian mounds written for a popular audience.[2] In the foreword to a beautiful facsimile edition published by the University of Wisconsin Press, Wisconsin state archaeologist Robert A. Birmingham writes that Lapham's *Antiquities* "stands today as both a remarkable source of detailed maps of the cultural landscape of Wisconsin prior to major settlements and as an important historical document on the evolution of American science and history."[3]

Though Increase Lapham in his earlier notebooks and journals had expressed little interest in the not-yet-named field of archaeology, his connection with it dated back at least to his early days in Kentucky and Ohio. Not only were the surveyors and engineers on the canal an encyclopedia of the ancient curiosities found along the routes of the canals, but Increase would have read of Indian mounds of Ohio and Illinois in *Silliman's Journal of Science* as well. Increase most likely first read of Indian and effigy mounds in Wisconsin when he read Major Stephen H. Long's 1817 *Travels* at the age of sixteen.[4] And all around Ohio were ancient burial and mound sites, including several sites in and around Portsmouth and many along the Scioto River, including one in Circleville that Increase and Darius explored in 1828.[5]

The mounds of Wisconsin were already disappearing when Increase and most white settlers arrived, but Increase had, from the first, done what he could to preserve their memory.[6] The 1844 edition of *Wisconsin: Its Geography and Topography*, includes a passionate section on the territory's antiquities, with three pages devoted to "Antiquities." He wrote that Aztalan, fifty miles west of Milwaukee, was the most visited and most known site and that there existed in the territory "a class of ancient earthworks not before found in any other country, mounds representing quadrupeds, birds, reptiles, and even the human form." These earthworks were rather crude, he wrote, in shapes often difficult to decipher because they were so altered by time. Some resembled the buffalo, or the eagle, crane, turtle, or lizard. A human form near Blue Mounds had been described by Richard Taylor as 120 feet long, with arms and legs extended, and a trunk 30 feet broad. The turtle mound Increase had seen in 1836 was now nearly destroyed. And he wrote that he had heard of an elephant-shaped mound near Cassville. "Should this prove true," he speculated, "it will show that the people who made these animal earthworks were contemporaries with that huge monster whose bones are still occasionally found; or that they had then but recently emigrated from Asia, and had not lost their knowledge of the elephant."[7]

— | —

During 1846, many of Increase's scientific projects had been completed or had fallen through. He was secretary of a new Milwaukee Public Library Association. His major work, *Wisconsin: Its Geography and Topography*, had been revised and enlarged since the 1844 edition. He had abandoned a planned tour for the Lake Superior and Silver Creek Mining Company in which he owned one hundred shares and had great hopes of a fortune.[8] That same September he wrote Darius that he also was not going to Madison as a state legislator, "the people having made the choice of another person!"[9] So in October of the same year, Increase wrote the Royal Society of Northern Antiquaries—the group that had erroneously called him "Ingram"—in Denmark proposing that the society support his methodical study of Wisconsin's antiquities.

> I have made careful measurement and triangulation of such [mounds] as have fallen in my way and hope to be able in future to delineate many more.... Many of these earthworks are but slightly elevated above the surrounding surface, so that their existence would not be suspected in many cases even by careful observers; and hence also they are very soon leveled, when the ground is occupied for agricultural purposes. Many of them have already from this cause been injured or destroyed. To visit the places where these interesting relics of antiquity are found and take a careful measurement of them so as to delineate accurately their form and proportions and occasionally make an exca-

vation to ascertain their contents, before they are "destroyed from the face of the earth" is a task which should be performed immediately, or it will in many cases be forever too late![10]

Increase's letter was accompanied by four drawings. But, there was apparently only silence from the Danes until 1852, when it was too late.

However, three years later, Increase again made essentially the same proposal. Over the course of those years he had gained encouragement from his association with the major scientists of the country, not only in his correspondence, but also in his yearly attendance at conventions of the American Association for the Advancement of Science (AAAS). Though his income may have been uneven, his home life was happy and secure and Ann was very supportive of his work.[11] His was a mind, Increase must have realized, that relished the focus of a big project. And he would have known that he was the only one in the state with the combination of talent, skill, knowledge, and drive needed to carry off such a complex project on the Indian mounds.

In 1848 he had very likely bought the inspiring and comprehensive *Ancient Monuments of the Mississippi Valley* by Ephraim Squier and Edwin Davis, published that year as the Smithsonian Institution's first volume in its Contributions to Knowledge series. This important volume contained many mound surveys—some made as early as 1836—represented in sixty lithographic plates and about two hundred woodcuts. Increase must have been familiar with some of these sites from his years of travel along the Ohio and Scioto Rivers.

So, in late November of 1849, Increase Lapham, saver and filer that he was, brought out the proposal he'd sent to the Danes in 1846 and rewrote it to Samuel F. Haven, secretary of the American Antiquarian Society in Worcester, Massachusetts, suggesting this society support his study and mapping of the antiquities of Wisconsin.[12] Haven must have written right back asking how much all this would cost. On Christmas Eve, 1849, Increase wrote Haven that the project would cost about five hundred dollars for a horse and light wagon, a boy for an assistant, and tavern charges—about four dollars a day.[13] Occasionally it might cost more if he had to hire help to open a mound or buy some artifacts. Haven wrote in

February of 1850 to say that both he and his society liked his proposal and agreed with Increase's terms—five hundred dollars for the wagon, rooms, and a boy. Increase was asking nothing for his own labor.

Increase immediately wrote Haven in Worcester in March that he was grateful for his kindness and for the liberality of the society as to his project. He would begin when the roads were passable in May or June.[14] But Increase's preparations began long before he took to the road. That same day he wrote Joseph Henry at the Smithsonian telling him of his new project and asking if the Smithsonian could give or loan him a pair of barometers to help him in his work: "It would add very much to the value of my observations if I could have a barometer for measuring the elevations of these 'high places' above the neighboring valleys....Can I not do something for you," he adds, "that would entitle me to these instruments as a donation?"[15] That same day Increase wrote his friend, Wisconsin Congressman James Doty in Washington, to press the Smithsonian to send him the barometers.[16] In a week he wrote Charles Whittlesey at the United States Geological Survey and asked him what he knew about the mounds. Might the USGS in Cleveland have some spare barometers in case the Smithsonian doesn't come through? And could he ask prominent geologists David Dale Owen and Joseph Norwood for lists of mounds in the north and west of Wisconsin?[17]

His horse Billy—"a fine looking nag, very gentle, yet sprightly enough, a good 'family horse'" would haul his helper, John H., and Increase and all their gear in the wagon as they traveled the state. At the end of March, Edward Foreman, a secretary at the Smithsonian, wrote Increase saying that Joseph Henry was "desirous of promoting researches among the ancient monuments of the country," and was having two barometers sent. But because of the state of the mail, it would be more than a year before the barometers arrived whole, rather than smashed.[18]

—| |—

At the end of April, the entire Lapham family except for Increase boarded the steamship *Pacific* for Chicago, where Increase telegraphed friends to meet them. The traveling party was Ann, age thirty-six; perhaps Increase's sister Amelia, twenty-one; daughters Mary and Julia, twelve and ten; sons

Henry and Seneca, four and two; and Ann's mother, Mehitable Alcott. No doubt assuming that Increase would be away measuring mounds for months, they planned an extended visit at the Alcott family home in Marshall, Michigan.

A month later, on May 26, after being delayed by problems surveying what would later be called Forest Home Cemetery, Increase and John H. headed south in the wagon. They stopped first in Oak Creek about fifteen miles south of Milwaukee where Increase examined a mound that turned out to be "a work of nature not of art."[19] As often happened during this trip, Increase's experience as a geologist served him well. His knowledge of the action of streams on their banks, their change of position and level, convinced Increase that this was not an Indian mound. He wrote in his notes:

> The ancient channel can be readily traced along the base of the ridge and mound, showing conclusively that they were formed by the action of water wearing away at the ridge so as nearly to form a passage but going entirely around the mound.... Besides, the "mound builders" seldom commenced their work in the "bottom land" when high ground was at hand.[20]

This clear and scientific thinking shows that Increase was the man for the job. Not only could he describe and document real Indian mounds, but he could also debunk common knowledge that called many odd geological formations Indian mounds. The next day he found that what were again supposed to be Indian works were logs of an ancient white cedar swamp buried in sand at the lake shore. Traveling south along the Lake Michigan shore toward Kenosha, Increase discovered more and more sandy ridges formed by the actions of water and wind.

In Racine he met up with his friend Philo R. Hoy, a physician who shared Increase's interest in plants, fossils, and Indian mounds. Hoy gave Increase some good fossils and they got more at Cooley's quarry.[21] Though Increase the antiquarian was off on an expedition to collect data on Indian mounds, Increase the geologist could leave no stone unturned. He collected many geological specimens and notes on these trips for the American Antiquarian Society, just as he had earlier collected data on Indian mounds when he was on trips to look at mines and minerals. Increase and Hoy traced the ancient lake beach south from Racine and then on May 30 through Kenosha to the state line and four miles south, finding ancient mounds as they went. On the prairies west of Kenosha in Salem Increase and Hoy came upon some gravel knolls, on one of which was an ancient mound.

At their lodgings that evening Increase and Hoy were told of the practice of Indians making horseblocks so women and children could more easily mount their horses. Along the trail the Indians cut saplings in two places and then bent down the still-living plants so that in a few years they would be trees large enough to stand on to mount a horse. This sort of ingenuity fascinated Increase. He was also fascinated by the skeleton of a young person that the men unearthed in a mound near Burlington. Increase noted that the young person had a copper ring around its neck. "How it could have got on the neck is a mystery! It was too small to go over the head and the ring was entire."[22]

During the next week, probably leaving Hoy behind, Increase and John went to limestone quarries west of Burlington to Geneva, and then

Increase's *carte de visite* of fellow naturalist Philo R. Hoy of Racine, ca. 1874

WHI IMAGE ID 44414

spent June 4 to 6 near the Pishtaka (Fox) River. In a well he examined what the farmers thought was silver, but Increase was able to identify it as iron pyrite. On top of a hill there he found ancient mounds from which he had a long view toward the south and southwest, "bounded by hills that may have been the ancient lake shore."[23] Someone with no geological knowledge would not have been able to correctly interpret all that he saw and would have missed the complexity and the beauty. This long view ended the first of Increase's explorations of the Indian mounds. But as we can see from his notes, he was studying not just the mounds, but Wisconsin.

By late June, Ann and the family, including Increase's sister Lorana, were back in Milwaukee. Increase's next surveys of the Indian mounds began in earnest. On June 28 Increase and John left about 1:00 p.m. in the light wagon pulled by Billy, heading west on the Watertown Plank Road, arriving at Summit at 7:30 p.m. That night, as on most nights during his travels, he wrote Ann. He told her that he would write his "pencillings by the way" in the form of letters to her. And he asked her to save the letters for his future use. He complained to her of the bad construction of the plank road, made of pine instead of the preferred oak. The soft pine wore away leaving the opposite of potholes—knots.[24] At the Fox River they drove north along the south shore of Pewaukee Lake, Increase noting a limestone quarry and numerous copious springs. Among "Hills and Potash Kettles," they found men building the Milwaukee, Waukesha, and Jefferson Plank Road.[25]

Increase Lapham traveled extensively across southern Wisconsin as he explored the new state.

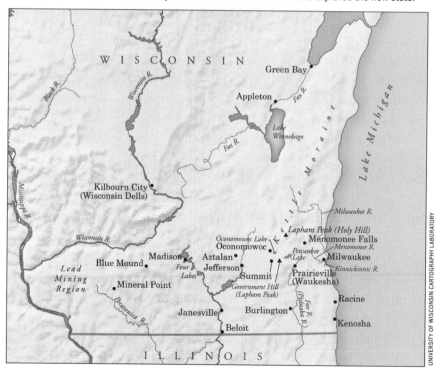

After passing the Nashotah Mission and then spending the night at the "beautiful level plain of Summit," Lapham wrote Ann on June 28 about the curious series of hills and hollows they had passed the day before.[26] He was fascinated by this puzzling landscape of southward ridges interspersed with wetlands. This may be his first use of the term "potash kettle," a pioneer term for the cast-iron pots used to leach ashes for the lye that rural people used to make soap. In this case the phrase described the curious, usually round holes in the hills. The term was to evolve into "Kettle Moraine," which describes the landscape today. He wrote Ann that

> This remarkable series of hills and hollows extends entirely across the state from Green Bay to Illinois. It is apparently composed almost entirely of limestone, gravel and boulders of primitive rocks. The idea, suggested by a writer in the Wisconsin Farmer, that they may contain iron, copper and coal is entirely gratuitous and not supported by any indication of such minerals. Some of the hills are very high and steep with a narrow ridge at the summit. On each side will usually be found deep holes, as if the earth had been thrown up by some giant power from these holes to form the ridges. The holes in many cases are undoubtedly formed by washing away lime or fine sand by subterranean currents, thus allowing the superincumbent earth to fall down.[27]

Increase was right about the metals and coal. Neither commercial-grade metallic ore nor coal was present in the Kettle Moraine. But he struggled to explain how potash kettles formed. Yes, they were excavated by subterranean action, but not by *running* water. What, then, might have formed them? He wouldn't have an explanation until a decade later, when he finally accepted the theory, popularized by Louis Agassiz, that glaciers had formed the landscape of the Great Lakes region. We know today that kettles were formed as huge ice blocks fell from the face of the receding glacier, became buried, and gradually melted to form a slumping surface. Increase was familiar with Agassiz's theory by 1848, when he referred to it in his geology speech to Milwaukee High School, but he rejected it then and was still resisting it in 1850.[28]

Only a few years before, in 1846, Agassiz, a famous Swiss scientist, came to the United States and in 1847 was appointed professor at Harvard

College. Agassiz was primarily a zoologist but in the 1830s his attention was drawn to the glaciers of the Swiss Alps. He noted their movement and especially their impact on surface geology: their moraines; the etched surface rock; the hills, ridges, and potholes they left behind as they receded. Recognizing similar but much larger forms on the landscapes of other places where no glaciers existed, including the mainland of France and Scotland, he visualized a great ice age, a time of continental glaciation. He published his findings in French and German in 1839 and his theory soon made its way into English.

Thus began a twenty-five-year debate that was to settle the mystery of the origin of drift and the land forms now associated with glacial advance and retreat. At first Increase along with most others mocked the ice age theory, deeming it inconceivable that an ice sheet from the north could be vast enough to advance over whole continents, moving with such inexorable force that it could pick up and transport millions of tons of material, from dust to elephant-sized boulders, and deposit it thousands of miles south, leaving behind when it thawed a landscape profoundly altered and strange.

In 1849, Agassiz, an associate by the name of Jules Marcou, and a party of students explored the Lake Superior region, publishing their results in 1850. Agassiz argued that the drift of North America, which he also observed in New England, was laid down by glaciers at the same time as that of Europe. The fact that drift extended southward to the north bank of the Ohio River but not south of it was evidence that it was carried by a glacier and not rushing water. How could rushing water sufficient to move boulders weighing hundreds of pounds come to a halt abruptly at the river?[29] How indeed? Yet for Increase, even if the theory could explain such oddities as erratic boulders and potash kettles, it was not enough to convince him for now.

As other geologists began to interpret the landscape from Agassiz's point of view, and after Increase and Agassiz met and began a correspondence as fellow naturalists, Increase went through stages of skepticism, then cautious reluctance, and finally in the late 1860s, full acceptance. Ironically, once Increase began to promote the theory he would become an important agent in elevating Wisconsin to the forefront of global glacial science.

—| |—

In Summit on his tour of "antiquities" younger than the Kettle Moraine, Lapham and John H. were shown around by a Mr. Spencer and his budding-naturalist son, whom Increase invited for a visit in Milwaukee. The son was the first of many people he met on this trip whom he either asked to come see him or offered the loan of a book or made the gift of an instrument like his "old microscope."[30] Poor Ann often got letters in which she was told to entertain a certain visitor or pack up something and send it. And though Increase found a useful limestone quarry and traces of a mound shaped like an otter, the effigy mounds at Summit were "so much injured that [Increase] could survey nothing."[31] The mounds were disappearing even in Lapham's time, but today more than 90 percent of the thousands of mounds he described have been destroyed by the plunderer's spade, the farmer's plow, and the developer's bulldozer. In many cases Lapham's maps are the only record of what was once there.

The next morning Lapham and John left Summit early, impatient to get to Aztalan. In the Rock River woods, they crossed in nine miles "18 remarkable ridges" west of the Kettle Moraine. The travelers had entered a landscape of teardrop-shaped hills trending southwest, their heads pointing north, and covering most of Jefferson County and eastern Dane County. The hills, we now know, are drumlins, sometimes miles in length and a hundred feet high, a distinctive moraine type that is evidence of the colossal power of glaciers. They were formed under the ice sheet as it advanced, piling up and sculpting gravel, sand, boulders, and clay into parallel ridges. Often streams, marshes, and ancient lake beds lay between them. By contrast, the Kettle Moraine formed between two lobes of the Wisconsin glacier as they thawed centuries after the drumlins were formed. Drumlins were not understood by Lapham and most others of his time. Wisconsin's drumlin fields are famous among geologists for helping to unlock the mystery of continental glaciation.

Increase and John arrived at Aztalan about noon, ate dinner, and settled into a "nice little room in the 'Aztalan House,'" though Increase told Ann that the house was not kept in the best style in the world—not up to Ann's standards. After looking around a little, finding that the south part of the site was in a cultivated field, the north part in woods, he saw that the modern city, as he called the village of Aztalan, lay over part of the ancient city, and many of the mounds had been partially opened "by persons curious in such matters, or by the *money* diggers!"[32]

That afternoon he also went with local resident J. C. Brayton to see the mounds on the east side of the river. The site of Aztalan had been brought to the world's attention in the 1830s and several men had incompletely and inaccurately drawn and described it. Because the mounds looked like Aztec pyramids, one of these men named the site Aztalan and suggested the Aztec people had built the city, which included sites of worship, sacrifice, many dwellings, storage pits, burials, walls of clay and earth, and great palisade walls of tamarack poles interwoven with small branches and plastered with clay. But until that week in 1850 no one had taken the trouble or had the skills to survey, map, and describe nearly the whole site.

That same afternoon Lapham clambered around in the brush with Brayton looking at the mounds, but they must have got to talking about geology and Brayton's new well, because at some point Lapham found himself and a lantern lowered forty feet down into it. In his letter to Ann the next morning he includes a detailed sketch of the inside of the well— four feet of earth at the top, then fifteen feet of soft yellow limestone, twenty feet of hard bluish limestone, then yellow sandstone below that. The man was fearless. And hard on his clothes.

The ancient site seems to have affected him profoundly. On Sunday, June 30, there being no church in Aztalan, and since Increase didn't usually work on Sunday, he wrote letters and then took a walk a mile below Aztalan to see some stone quarries. There he stood for an hour—in his torn coat, with rocks in his pockets—atop a high, ancient mound and contemplated "the folly & vanity of human affairs." The people who lived here had "utterly failed to hand down any certain knowledge of the people who erected them. They are lost—gone—returned to dust—& are forgotten," he wrote that evening.[33] Today, archaeologists know that the Mississippian people related to the culture that originated at Cahokia in Illinois built the great mounds and palisades atop an earlier Woodland peoples' site. Modern archaeologists have built on Lapham just as the Mississippians built on the Woodland people.

From a letter Increase wrote Ann from Aztalan on June 29, 1850, describing its mounds, eighteen glacial ridges he crossed, and the geology he saw in Brayton's well

The next two days at Aztalan, Lapham and John must have worked from sunrise to sunset. They opened mounds and examined the walls of the great enclosure. They pulled up sod, dug several shafts, picked up pieces of pottery—all the while measuring, mapping, making notes. In two old excavations, water stood and the wild iris grew. In remains of the ancient clay walls, Lapham found impressions of grasses so detailed that he could identify the variety of sedge.

The observations, surveying, and mapping he accomplished at Aztalan were just in time. In 1850 Aztalan was already being obscured by farming and development. Today, with remote sensing devices, archaeologists are finding traces of these long-absent people and their civilization—mapped by Lapham—though now completely invisible on the surface. Robert Jeske, an anthropologist at the University of Wisconsin–Milwaukee, says that researchers who have gone to the places Lapham mapped have found his work "very solid." One reason, says Jeske, is that Lapham didn't over-elaborate. "He stuck to a conservative idea of what he was seeing."[34] This is what we see throughout all the works of his life—tirelessness, combined with a clear-eyed trust of evidence and reason. Though he did not embellish, he could extrapolate.

Later Increase would draw a final draft to scale, but the site map of Aztalan published in 1855 is essentially the same as these 1850 notes. Because Aztalan in 1850 was partly obscured by the little village, by forest and brush, and by a field of wheat, it wasn't as easy for him to see what

Increase Lapham's rough drawing of Aztalan, 1850

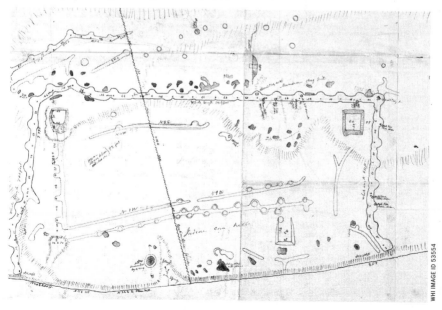

WHI IMAGE ID 53554

we can see now—the two reconstructed flat-topped mounds, the reconstructed palisade walls, the great open space under the sky.

On the morning of July 3, Lapham was up early as usual and wrote
a newsy and teasing letter to Ann, telling her he missed the Milwaukee
newspapers and he missed his children. "If anything important happens
to you or yours—such as marriages, births, deaths, fires, etc. please telegraph to me at once; and if it is deemed necessary I will abandon my trip
and return to you."[35] He would be finishing up at Aztalan that day.

That evening he was invited to tea with Mr. and Mrs. Ostrander, an
older couple who lived in the village. Lapham and other guests spent the
evening talking over the history of the discoveries at Aztalan and "old
times in Wisconsin," as he wrote his wife. Lapham, this geologist who was
perfectly at home scrambling up a cliff for rocks, this botanist who would
slide down a dune for a new kind of grass, was intrigued by the tea table
set by Mrs. Ostrander. He wrote to Ann that the "cakes & many dishes
were covered with little finger napkins folded like the enclosed paper, so as
to keep off the flies etc. By taking hold of the point, these napkins could
be lifted up and would naturally fold into a triangular shape. They appeared to me exceedingly neat, and in this country of insects, exceedingly
appropriate and proper."[36]

This rough man was also a proper Quaker. When a young woman at the
gathering offered to mend the coat he'd torn in the bushes that afternoon,
Lapham was slightly shocked that "she wanted me to take it off for that
purpose!!"[37] We don't know if he took his coat off or left it on, but it was
indeed mended.

From July 4 to July 20 Increase made a big loop southwest, first to
Clinton Creek, Madison, Prairie du Sac, Adams, Otter Creek, Sauk City,
Mineral Point, up Pipe Creek, then on the fourteenth, he began his "retrograde motion" to Sugar River, Janesville, Beloit, Exeter, Monroe, and then
along the Rock River at Lake Koshkonong.[38] During those almost three
weeks, he was not simply alone with John and Billy the horse; he was still
connected to friends. He wrote letters to George Hyer and David Irwin,
who had previously explored Aztalan, asking if they had found there any
cloth or pottery or implements. He made the rounds in Madison, saying hello to judges of the Supreme Court, the chancellor of the univer-

sity, a US Marshal, and the secretary of state, all of whom treated him "very respectably" even without the new coat, which he teasingly said Ann seemed to think constituted his claim to distinction.[39] With a party of eighteen ladies and gentlemen from Dodgeville, he saw lead shot made in a shot tower in Mineral Point. He arranged with William H. Canfield that he would survey works below Adams and south of Baraboo. From his notes, we see that he surveyed Indian mounds on July 3, 8, and 9 after Aztalan, but he made geological observations and notes on at least six of those seventeen days. There was more geology to see than Indian mounds.

What is most notable about that trip in June and July of 1850 is his mood of happiness in his letters to Ann, and his longing for her. He wrote Ann at least eight letters. And these are sprightly and vivid letters in which he teased Ann and his sister Lorana, yet also worried about all of his family in Milwaukee. At Mineral Point on July 12, he wrote in hair-raising detail of his descent on a rope into a mine, no doubt scaring the heck out of Ann. But he assured her he was safe. "I wish you had been there to go down too! It was *so* romantic!"[40]

The language of love was not out of keeping with his writings to Ann. In several letters he asked Ann to meet him in Beloit so they could drive back alone together as a sort of honeymoon. They had been married more than twelve years and really hadn't had a moment to themselves. Their marriage began with two nieces living with them, and the birth of their first child quickly followed. He was thirty-eight years old and that summer of 1850 seemed to be a sort of flowering in his marriage. Ann wrote that she had a dream of seeing him, which she attributed to some "nice Ice cream & strawberries" eaten the evening before.[41] On July 10 from Adams in Sauk County he wrote asking her to come by stagecoach to meet him on the evening of the twenty-second at Janesville. Then they would go together to Beloit to look the place over and think of sending the boys to college there. On July 13 he wrote her again, saying, "We have been separated so much of late I begin to feel that we are quite strangers—I want to renew the acquaintance!"[42] The older children can take care of the younger ones, he says, and after all her housework it will be a relief to get into the country. They can stay at a first-rate hotel in Janesville or Beloit. "Come on your own or on my account—or on account of the Beloit College!"[43]

He said he wouldn't have any surveying of mounds on the way home so that "we shall have the pleasantest kind of a tour." Ann could learn how to drive Billy. But it wasn't to be.

Increase left Beloit and Janesville on his own with Billy and John, and then on July 20 near Fort Atkinson, Increase got a telegram that changed everything. This is what he wrote in his journal: "My brother Darius died of Cholera at Cincinnati 20th of July, 1850, and was buried at the Home Farm, Mt. Tabor, in Champaign County Ohio."[44]

Darius Lapham died on the morning of July 20, exactly two months after his second marriage to Lavinia Seig. He also left two daughters by his first marriage to Nancy, who died young—Phebe, eighteen, and Harriet, about sixteen.[45] Darius, forty-two, was at the time of his death the chief toll collector on the Miami Canal and was chairman of the committee to arrange the Ohio agricultural fair to take place in the fall—the fair that had been postponed from the previous year because of cholera. Darius also had a farm in Champaign County near his father's farm.

Increase is silent in his grief. There are no journal entries for the rest of the year, and only a few letters. We know that he left Milwaukee August 26, this time with his sister Amelia, to be with his family in Champaign County for a month or so.

Darius was his beloved brother. Increase was, of course, very close to Ann, but it was largely the influence of Darius that brought him up, that made him the scientist he became. He was a better scientist than Darius, but it was in his early letters to Darius that the good mind of Increase Lapham tried and forged his ideas. This death of his brother was a great loss, a great grief for him, perhaps as wordless then as it is to us now.

The only thing Increase seemed able to manage for the rest of the year, other than taking care of family matters, was a report to Samuel Haven in Worcester, begun on September 26 and finished on October 2. He tells Haven that since May he has gone over the southern part of the state, except for the immediate valley of the Mississippi River, accumulating a mass of notes, drawings, and topographical sketches. This material, he says, "will require some labor to arrange in the form of an essay. In doing this shall I make the drawings of suitable size for engraving? And if so what is the size most convenient for your 'Transactions'?"[46] And he tells

Haven he won't finish his surveys this season because it's much better to work in the spring and fall when there are fewer obscuring leaves. Then he proposes that if Haven's museum has some articles relating to the ancient works of Wisconsin, "Would it not be well to have them embraced in my proposed essay—so that the 'Antiquities of Wisconsin' may appear complete?" This sentence displays a singular lack of ego, a strong sense of the requirements of science, a feeling of being overwhelmed with his project—or all three.[47]

—| |—

By January 16, 1851, Increase had summoned the energy and concentration to summarize his notes on the antiquities of Wisconsin and deliver them in the form of a lecture to the Young Men's Association at the Free Congregational Church. He began, "The time was...when Wisconsin was a mere wilderness." Then he told of the Indians who roamed there, how their lives were different before the white men came, and of the Indians before them, the mound builders, who left no written history, so that it was up to "the antiquary" to figure out their origin, customs, migrations, and manners. He described the types of mounds, where they are found, what was written about them. He described Aztalan, making clear that his geology and engineering background was helping him avoid common errors in interpretation—telling an Indian mound from a beaver dam, an ice ridge, or a potato hill. He described the animal mounds in Milwaukee and where they are found, noting that nearby on Chestnut Street Hill was a bird-shaped mound that his little daughters helped him survey. And he made sketches, which became the large drawings that he showed the audience.[48]

In May of 1851, after again visiting family at the West Liberty farm in Champaign County, Ohio, he traveled on to Cincinnati to attend the annual convention of the AAAS. On the train, Increase had run into Charles Whittlesey, the geologist; Jared P. Kirtland, an Ohio physician and like Increase an all-around scientist; and Harvard professor Louis Agassiz, all on their way to the meeting. This likely was his first meeting with Agassiz, and the two were to become congenial correspondents. At the meeting he met another of his correspondents. Joseph Henry of the Smithsonian

Institution spoke to him and "seemed quite anxious to obtain the privilege of publishing" his work on the antiquities of Wisconsin.[49]

On June 12 Increase received a letter from Samuel Haven reporting that he had been in communication with Joseph Henry on the matter of both their institutions publishing Increase's work on antiquities. In a letter to Haven later that month, Increase wrote that "[a]s to publication I have no preferences except that I should not like to have it done in a style inferior to the Smithsonian publications." He told Haven that his notes were now 150 pages of drawings, maps, and sketches of fifty localities in Wisconsin. And he concluded this letter to Haven by reminding him that if such a collaboration is arranged, "not to forget that a number of copies are to be furnished me as a sort of compensation for my labor. You only pay for my actual expenses, not for my time and labor."[50] This request for copies would later be the only cause of disagreement, and a major one, between Increase and his publishers.

While all these negotiations were going on, Increase was often in the field. June 10 he left home with Ann and five-year-old son Henry and traveled by plank road to Hartland where he saw damage from a recent storm and an interesting phlox. Then on to Summit Corners to look for a mound shaped like a bird with curved wings reported to be near there. On this trip, and others in the summer of 1851, Increase wrote of his travels as letters to the editor of the *Milwaukee Sentinel*. This trip included Menasha, Lake Winnebago, and a conspicuous mound on Butte des Morts that Increase hoped would be "forever preserved to continue its silent and solemn admonitions to a different race of men."[51] They sailed across Lake Winnebago where Increase noticed the water was filled with floating pollen. Increase visited Congressman Doty in Neenah, then went on to Oak Grove, returning to Milwaukee about June 25.

July 1 he was looking at an ancient mound on the south side of Silver Lake in the town of Summit. At the end of July he took a steamboat up Lake Michigan to Two Rivers, noticing the geology of shoreline, then drove down to Manitowoc where he examined several mounds, then further south to Sheboygan Falls where he examined mounds and found some shells. In August, Increase left for his fourth trip of the summer, accompanied by his farmer brother William and, at long last, carrying his new barometer. They spent four days touring Pewaukee, Waukesha, the west end of Eagle Prairie, and La Grange, then on to Beloit where they visited his professor friend Stephen Pearl Lathrop. They came back by Allen's Grove, where fifty Allens had settled a few years before. Then to Lake Elkhorn where there were mounds on the north shore of the lake.

In September, he seems to have stayed home. But on October 7, traveling with his son Henry, he was in the town of Merton. On October 8 he wrote Ann that he was on a cone-shaped hill that towers over the other hills in the area. Made of drift gravel and boulders, he found three Indian mounds on the summit. His exploration did not go unnoticed. The Irish farmers from the town of Erin were fascinated with this visitor who clambered around the slopes. In his travel notes he wrote:

> It was proposed, in consideration of the interest I took in examining this peak and the mounds and in ascertaining its elevation, to call this prominent hill Lapham's Peak. Wisconsin and especially the southern settled portion present but few such displays of fine scenery and this peak may become a place of resort for summer tourists, who wish to exchange for a season the heat of the city for the pure cool breeze that will always be found at its summit.[52]

Increase had no idea to what extent the hill would attract tourists, but it was in fact called Lapham's Peak for many years. About the turn of the century a Catholic Church replaced a small chapel on its summit and its name was changed to Holy Hill. Later a much larger church maintained by the Discalced Carmelite Friars became a world-famous shrine that attracted thousands of visitors annually. But Increase's Kettle Moraine legacy was not to be ignored. Government Hill, about nineteen miles south, the highest hill in Waukesha County and one housing a weather station on its summit, was later named Lapham Peak.[53]

On October 9 and 10, he found evidence of Niagara limestone but didn't mention Indian mounds in his journal. Spending several days at Horicon taking notes on geology, he returned home about October 16. A few days later he received the letter from Haven, dated October 17, informing him that both the Antiquarian Society and the Smithsonian would publish his *Antiquities*. This must have been a happy day in the

Lapham household. On November 1 he wrote letters to both Haven and Henry responding to this good news and making his season-end report. He told each of them that his work was delayed by the rain of the spring and early summer, which made the roads impassable, and by sickness in his family. Yet he had made five trips that season, totaling about thirty-eight days. Adding this to what he had done the year before, he reported that he had so far on this project worked for seventy-three and a half days at $4 a day for expenses; he had paid $11 for an assistant, $6 for labor, and $6.50 for ancient articles for a total expenditure of $317.50. He had not yet used his allotted five hundred dollars.

In both letters he mentions how important to him as a model was Squier and Davis's *Ancient Monuments of the Mississippi Valley*. He wanted his book to have the same high-quality plates and text. To Henry he says that he knows his book will go through a "severe ordeal" if it is to meet the Smithsonian standards shown in Squier and Davis and asks Henry if he has any important suggestions to help Increase prepare his manuscript for publication. He tells Henry that he has a copy of Squier and Davis, which he will use as "a good model which I hope not to fall behind."[54] Henry does have some suggestions for Increase, including them in a fine statement of the principles of observational research:

> The value of observations of the kind on which you are engaged de-pends almost entirely on their minute accuracy. The facts should be stated just as they are without regard to preconceived opinions or proposed hypotheses. All the measurements ought to be executed with care and every appearance however trivial ought to be noticed. Sketches and notes ought to be taken on the spot and nothing left to after memory.
>
> I think it's important that the relative position of the mounds should be taken and it may be well to accompany your memoir with a map of the country on which the position of each mound is indicated. I think it important that the distribution of the works of the mound builders should be accurately determined & for this purpose it will be advisable to collect all the reliable information as to the locations of mounds which you can obtain though you may not be able to visit them yourself. . . . Do not be afraid of putting down too much—what

is irrelevant can be omitted by those who use the materials in the way of deducing general principles.

Your memoir should be principally a statement of facts and though you may give your hypotheses they should be subordinate to the facts. We are as yet only collecting the bricks of the temple of American Antiquities which are hereafter to be arranged and fashioned into a durable edifice; it is therefore of the first importance that our materials should be of the proper kind.[55]

Increase must have worked up in his study all winter and into the spring of 1852 redrawing and describing the antiquities with these words of Henry in his head and his copy of Squier and Davis at his side. He also made tables of barometric measurements from around the state, corresponded with the chancellor of the University of Wisconsin about the role of the university in the study of sciences, compiled meteorological tables, kept up his correspondence with Asa Gray, worked on his *Grasses of Wisconsin*, gave a lecture on temperance, and somehow made a living to support his large family.

Over the winter, Increase prepared much of the manuscript for publication, including forty-nine plates of the proper size for the Smithsonian volume. On May 24, 1852, Increase wrote Samuel Haven that he was ready to start his final tour of the ancient works of this state and would like the remaining two hundred dollars of the five hundred dollars promised him. He planned to go to the upper part of the Neenah (Fox) River, then across the Wisconsin River up to the Lemonweir River, and "strike the Mississippi at La Crosse, mostly on new ground."[56] Increase left on this "tour of exploration" on May 26 and returned sometime before the end of June.[57]

On this tour, Increase wrote many of his notes as "Glances at the Interior" to be published in the *Milwaukee Sentinel*. He wrote about the state of the new Lisbon plank road and other plank roads, the "beautiful white Waukesha limestone cropping out on the Lisbon Road," how the spring season seemed further advanced away from Lake Michigan, and the wildflowers.[58] He described the neatly built white houses in Oak Grove, Beaver Dam, and Waushara, surrounded by forest trees "which many of the citizens have the taste and good sense to preserve."[59] He wrote of the Indians he saw near Moundsville whose time on their land had expired, but he said they "are probably destined to remain with us and dwindle away and become extinct."[60] At the Wisconsin Dells Increase was shown the many wonders by a Mr. Parker who led him along the narrow shelved rocks with the "boiling cauldron" at a "fearful depth" under his feet.[61] From La Crosse on June 11, Increase wrote to the *Sentinel* and to Ann, telling that he had had an attack of "fever and ague" that was successfully treated by a Dr. S. C. Johnson. Because of his illness, he wrote, his journey "up the Lemonweir across the great dividing ridge down the beautiful valley of the La Crosse, the steamboat ride to St. Paul, the delightful drive of ten miles over the handsomest and best country I ever saw to the far-famed Falls of St. Anthony" and his return to La Crosse, must be passed over with a single sentence.[62] And that was the sentence. La Crosse, he said, reminded him of Milwaukee in 1836—"the same activity, the same array of new unpainted houses, . . . the public houses overflowing, shanties springing up in every direction, movers with their families encamped in the vicinity."[63]

There is much in these letters to the public about everything to be seen in these new and wild countries, but he saved the description of the bird- and man-shaped mounds along the Lemonweir River for *Antiquities*. And he missed Ann. "How I wish you could be with me in my rambles about the country," he wrote to her. "I know you would enjoy it, the very roughness of fare would give you an additional zest for home comforts on our return."[64]

Back in Milwaukee, on July 30, 1852, Increase Lapham wrote Haven that he had completed his survey of the ancient works. The manuscript, he said, was about 390 pages, with fifty-two plates of sizes adapted for the Smithsonian's format, plus one of double that size for Aztalan, one four times that size of the state showing the localities of the known ancient works, and ninety-seven drawings intended to be woodcut illustrations. Increase said he could take the manuscript with him to the AAAS convention in Cleveland in August and deliver it in person to Professor Henry. He also packed up a box of artifacts he would send to Haven. Haven wrote back on August 6 saying that the arrangement was that the

American Antiquarian Society would edit the manuscript first and then send it on to Joseph Henry at the Smithsonian, and he asked Increase to accept his sincere congratulations on the completion of his labors. On August 14 Increase wrote Joseph Henry describing the manuscript he had just sent Haven in Worcester and saying, as many writers do to editors, that the manuscript was not so clean as he wished.[65] And Increase was still feeling the lack of a formal education: he wrote to both Haven and Henry that he had to rely on their proofreaders to "correct any literary defects or awkward expressions" that may arise because of his "[n]ot having had the advantage of a classical education."[66]

Haven's changes in Increase's manuscript were apparently minor, and on March 26, 1853, Joseph Henry wrote Lapham that he had received the manuscript from Haven. He suggested Increase could improve the drawings by taking out distracting township lines (Increase declined for good reasons) and requested a bibliography on the subject, perhaps as a separate pamphlet.[67] Henry told Increase that he was making arrangements immediately for the woodcuts.[68] So the book was being made.

But not very quickly. While the woodcuts and engravings were being made, and while Increase was slowly correcting proofs, a controversy arose among Lapham, Haven, and Henry as to how many copies of the finished book Increase was entitled to. Increase Lapham, not being paid for all the labor of this work, assumed that he would be given enough copies of the finished book so that he could sell some and make some money. Henry was surprised to hear from Increase that he was owed "a number of copies" of the book as a "sort of return for my services in making the survey &c." Increase thought that one hundred copies "would be about right."[69]

Henry wrote Haven right away asking what was agreed on between them. It turned out that no number of copies was written down, but Haven had twenty-five in mind. Henry split the difference and wrote Increase a long letter explaining the mix-up and then agreed to send him fifty copies.[70] More than a year went by with the engravings being made, text set, the proofs read—too slowly, according to Spencer F. Baird, who sent and received the proofs at the Smithsonian. But finally, on November 8, 1855, Baird wrote that forty-eight copies of the book would be on their way to him in a few days—two copies having been sent to friends at Increase's request.[71]

Three years after he finished writing and drawing the book, Increase opened a parcel containing the fruit of his work of more than three years. *The Antiquities of Wisconsin, as Surveyed and Described* was about ten by thirteen inches, containing ninety-two pages of text, fifty-three full-sized illustration plates, and two larger tipped-in plates of the location of the mounds in the state and of Aztalan. Within the text are ninety-one wood-cut illustrations. By any standards, it is a beautiful book, made especially so by the detailed and artistic qualities of Increase Lapham's illustrations and the gracefulness and clarity of his prose.[72]

From the contents we see that Lapham has organized his subject in a very different manner from Squier and Davis, who used at least three systems of classification of the mounds they wrote about: by their purpose, location, and contents.[73] These nineteen chapters were written for a more scientific reader than Increase's. In contrast Increase organized his eight chapters into a narrative of his exploration of the watersheds of Wisconsin, one written for scientists but also for the people of Wisconsin. Increase's topographical knowledge of the state organized his narrative by river valleys rather than by surveyed squares or groups of mounds.

Why watersheds and not the location of the mounds? In his introduction he first delineates the geographic location of the state. About three-fifths of the state, he tells us, lies in the basin of the Mississippi River, while the rest drains into either Lake Michigan or Lake Superior, or the basin of Green Bay. The Mississippi Basin "is naturally divided into five great valleys" of the St. Croix, Chippewa, Black, Wisconsin, and Rock Rivers.[74] Because these valleys or basins and their dividing ridges always remain the same, and because the ancient works lie mostly along streams or on the borders of the small lakes with which the state abounds, Increase used these natural features to organize his descriptions of the ancient works. This description is made clearer by his foldout map of Wisconsin with the location of the mounds clearly marked in relation to all its main watersheds, lakes, and rivers.

Before he describes any of the mounds, in the introduction he gives us an eagle-eye view of Wisconsin. Through three paragraphs we feel as if we fly over the state with this man who never saw an airplane but through his surveying and mapmaking experiences truly had an overview of Wisconsin. We see as if from above the gently rolling land divided into watersheds, which Increase tells us often differed by such slight elevations that a lake or marsh may drain in two different directions. His description then flies north to the high divide southwest of the Porcupine Mountains where the Montreal River flows north into Lake Superior. And he shows us from above that the region around the source of the Wisconsin River is "a grand summit, from which the rivers flow in every direction like the radii of a circle," flowing into the Mississippi River, Lake Superior, and Green Bay. We also see the two southwest running ridges, one of what he calls "broken land" of potash kettles running from near Door County down through Rock County. The other, more understandable to him, is what we now call the Niagara Escarpment, a limestone layer Increase was familiar with from his early days on the Erie Canal in Lockport.[75]

Once we see the whole state, Increase describes the mounds by taking us with him on his tours of the state. So we begin where he began in May of 1850, traveling south along Lake Michigan with Philo Hoy to just south of the Illinois line. He incorporates information he had found out since, information from Professor Lathrop of Beloit College and from Hoy. He tells us a great deal of detailed information in that most accessible of rhetorical forms—narrative. Lapham conflates his tours of exploration in the book to tours of watersheds, so that after he writes about the mounds along the western shore of Lake Michigan, we tour with him the basins of Pishtaka (Fox) River; then the Rock River; the Neenah or Fox River of Green Bay; the Wisconsin River; then the ancient works near Lake Superior. His last chapter is on the contents of the mounds and the remains of ancient workmanship.[76]

And Increase doesn't just describe the mounds as he was charged to do. He also argues against previous conclusions, stating, for example, that Aztalan was not a military site as others had suggested.[77] He also argues that because the mounds were being destroyed so quickly by settlement and agriculture, preservation is an absolute necessity: "Would it not be well to select some of the more important monuments, and, by purchase of the ground, or other means, secure their permanent preservation?

Unless something of this kind is done, and speedily, all knowledge of them will be confined to the scanty records of those who have attempted to describe them."[78]

What Increase predicted did come to pass. Most of the Indian mounds of Wisconsin have disappeared. Yet because of the work of Increase Lapham we can still imagine, and in some cases go to see, the cone-shaped mounds, the turtles and lizards, the panthers and serpents, and the great winged birds flying on the rolling lands near Wisconsin's rivers, lakes, and springs. Because of Increase Lapham, we can see renderings of the mounds in detail, and visit those that have survived. Without Increase Lapham, even the memory of a centuries-old, mystical, mythical bestiary of mounds would have disappeared forever.

CHAPTER 17

Disappointments and False Starts

| 1852–1860 |

By LATE 1852, INCREASE HAD FINISHED his *Antiquities* manuscript and sent it to Samuel Haven for editing. He must have sensed that the book, to be published by the Smithsonian and circulated to scholars nationally and abroad, would elevate his reputation from that of regional scientist to one of national stature. So buoyed, Increase aimed his next two endeavors at national audiences, one a guide to all North American fossils in collaboration with the nation's preeminent paleontologist, the other a comprehensive description of the grasses of the United States, which he would do on his own. On November 12, he wrote John W. Van Cleve, an amateur botanist and geologist at Dayton, Ohio, saying:

> I have matured a plan for getting out a complete work on American Paleontology which shall be to fossils what [Amos] Eaton's and [Asa] Gray's works are to Botany.
>
> It will embrace full generic and specific characters of all North American fossils, followed by the geological position and localities, with reference to authors. It will be illustrated with figures of one species in each of the genera. I have already commenced upon it and shall soon go over the books I have, embracing nearly all the American works, and shall then go to Boston to complete it by use of the Harvard Library.[1]

On January 18, 1853, he set out by rail for the East, stopping first at West Liberty, Ohio, to visit his family. "Mother made so much ado about my coming home that the girls were alarmed.... All the family are to be here tomorrow to help eat the fatted turkey."[2]

On to New York City and then Albany where Increase visited James Hall, state geologist of New York. Hall was a national giant in geology, a hard-driving genius who bullied anyone who dared to question his knowledge, and he seemed to care little if he made enemies.[3] Even so, Increase had visited Hall in 1847 and even dined with him and his wife in their Albany home. And he had invited Hall to call on him in Milwaukee when Hall was visiting eastern Wisconsin in the late summer of 1850.[4] "By means of the specimens I have collected you can form a pretty correct idea of the rocks here. Another year I shall be prepared to report to the world the geology of that portion of the state not embraced in the surveys of Dr. Owen," he wrote then.[5] While it is not clear that Hall accepted the invitation, it is clear that the two had formed a congenial relationship.

When Increase arrived at the Hall home in New York, he reported to his "dear wife" in a February 13, 1853, letter that his host "was in the midst of fossils, preparing the third volume of *Paleontology of New York*. Mrs. Hall came in and occupied herself in preparing the illustrations."[6] He then said that Hall agreed to collaborate with him on his book on the fossils of America. "He has already done something in a work like mine and very readily consented to join me in the proposed work upon equal terms, the details to be arranged on my return. This is what I have wanted and hoped for, but feared I could not obtain."[7]

He then left Albany for Cambridge and Boston, where he spent a day with Haven and where he was impressed upon learning that publishing *Antiquities* would cost five thousand dollars. At Cambridge, he called on Asa Gray, who opened Harvard's library to him. "I am very pleasantly situated here, only one mile from the largest library in the United States, to which I have free access and the privilege of taking out any book I wish!!" he wrote Ann.[8] At Boston, he arranged with a Dr. Warren to see a skeleton of a great mastodon, and he spent much of the time in his room "driving the pen among the technicalities of fossil tertiary shells.... My tin box is now so full that it will hold no more and the remainder must be kept in a parcel by themselves."[9]

Then it was back to Albany, where on March 1 he found an hour to spare to write Ann yet another letter before he was "cabbed" out to Professor Hall's to close their agreement.

> I sent home from Boston by express a box containing "lots of fine things" for you and the children (not omitting Grandma) and my trunk is ready to burst with others.... If my tin box of palaeontology suggests the "essence peddler" my trunk would convince anyone that I was a Dry Goods peddler! Fortunately, I do not have to cross the line into Canada.[10]

That day, Hall and Lapham signed an agreement under which Hall would prepare a work called *American Paleontology* based upon information and fossils supplied by Lapham including descriptions of about two thousand species. They would share equally in the expense and profit of publishing. Yet the work that Increase hoped to pursue after he so successfully completed *Antiquities* had a very different and sadder life story than his book on the Indian mounds of Wisconsin.

Increase saw Hall again that July in Cleveland, Ohio, at a meeting of the AAAS. He wrote Ann from Cleveland on July 31, "Mr. Hall says he has made but little progress in our joint work on North American Paleontology, as he has been absent much of the summer. I presume he will take up the work upon his return and complete it in the course of the fall and winter."[11] But Hall, chronically overloaded with work, let the study of American fossils gather dust. On November 27, 1857, more than four years later, Hall wrote Lapham, evidently in response to a Lapham inquiry that expressed impatience and frustration.

James Hall, the nation's premier paleontologist and collaborator with Increase Lapham on the failed guide to American fossils, ca. 1866

WHI IMAGE ID 45398

In reference to your manuscript, I shall send it if you require it. I had never given up the idea of publishing it in connection with you, but the Brachiopoda required such a thorough overhauling that it seemed to me not wise to do it as the subject then was. . . . I would be glad to pay for the manuscript, particularly that part containing the Paleozoic fossils. I shall, of course, do as you desire.[12]

Increase must have been encouraged by this response, but to no avail. On May 18, 1860, eight years after the initial agreement, Hall finally returned Lapham's manuscript on fossils, accompanied with an apology. The matter was closed.

—| |—

The failed paleontology project was not the only disappointment in the years following Increase Lapham's completion of the *Antiquities* manuscript. In 1853, Increase finished an important article about the grasses of Wisconsin and neighboring states; it was published the following year in volume 2 of *Transactions of the Wisconsin State Agricultural Society*. Ninety-one pages long, it offered formal descriptions of 149 species of grasses in the manner of Increase's friend Asa Gray, to whom he gave ample credit, followed by twelve pages of drawings of grass species, all neatly done by Lapham himself. Increase wrote that he expected that many more species would be discovered but at the time, a "paucity of observers" had not yet been able to cover all of Wisconsin in their search for grass species.[13]

Grasses are vital to human life. They comprise the bulk of food ingested each day, directly as corn, wheat, rice, oats, barley, sugar cane, or rye products or indirectly in the form of animal flesh fed by grazing and foraging grasses. In this knowledge lay the importance of Increase's article and the passion behind it. Now, he hoped to do more credit to Asa Gray's "pets, the grasses," by producing a larger, more ambitious study.

The Wisconsin article had been enthusiastically received. Albert C. Ingham, secretary of the Agricultural Society and an editor of *Transactions*, wrote Increase on January 6, 1854, upon receiving the manuscript: "From

(*right*) Increase's drawings of grasses, such as this millet, were accurate and exquisite.

the limited examination I have given it, I think it is the best and greatest work ever brought out by any Agricultural Society in the Union and I might say in the world, for neither the Royal Society of England nor the Highland Society of Scotland have ever attempted anything like it."[14] Encouraged by such comments and basking in the success of his *Antiquities* project, Increase sent Charles Mason, US commissioner of patents, a copy of the article on June 1, 1855, and proposed that he do a major study of grasses not just of Wisconsin but of the United States as a whole.[15] At the time, the Division of Agriculture was a subsidiary function of the Patent Office.

Unlike the *Antiquities* project, he wanted to be paid for his efforts: two thousand dollars a year for an undefined period of time. Increase also sent his proposal to Rep. Daniel Wells Jr., Democrat from Wisconsin, who remained in contact with Washington agriculture officials and even backed a bill that appropriated some money. The project was discussed off and on for months, moving along positively until July 1856, when Increase went to Washington to directly negotiate with Mason.

On July 14, 1856, in a letter written in Washington, Increase told Mason that he would gather all information regarding the growth, natural history, agricultural, or other value of each species "in the different sections of the Union." In addition he would collect one hundred examples of each plant and arrange them in books to be distributed among state agricultural societies and colleges. He would prepare a drawing of each "ready for the engraver," report from time to time on the progress of the work, and make a final report of all grasses, their common and scientific names, their characteristics, geographical distribution, life cycle, suitable soils for each, and facts regarding the economic value of each species.[16]

He made it clear in a note to Ann written on the back of his copy of the July 14 letter to Mason that he thought the project was approved.[17] Then on August 22, 1856, back in Milwaukee, Increase received a telegram from S. T. Shugert, Mason's chief clerk at the US Patent Office: "Remain at Milwaukee until you hear from Judge Mason."[18] On August 27, five days later, he received a letter from Shugert: "Agreeably to the understanding between you and the Commissioner of Patents, your services are required at this office forthwith, and you can report yourself accordingly."[19]

Evidently Increase failed to notice the date that Shugert had written the letter: August 20, two days before he sent the telegram. Thus, the telegram should have taken precedence. Instead of staying put, Increase left Milwaukee two days after receiving the letter, arriving in Washington on September 3.

There is no description of the meeting Increase had with Mason, if indeed such a meeting took place. What was becoming clear at the time was that the project was shelved indefinitely, if not abandoned. But Lapham had begun work on the grasses project, following the outline he had submitted, and in later letters to Mason he asked to be paid for the work already done, amounting in his estimation to a thousand dollars. Responses from Washington were cautious, tentative, and mostly negative. He wrote letter after letter, sometimes angry, sometimes pleading, plunging into an obsessive fury that stayed with him for years. He enlisted the help of Representative Wells. He was treated politely at times, curtly at other times, but the bureaucracy prevailed.

Years later, it was made clear that politics lay at the heart of his problem. Increase was a Whig in transition to Republican. The president, Franklin Pierce, was a Democrat. In a diary now lodged with the Library of Congress, Commissioner Mason wrote on Wednesday, September 10, 1856:

> I called last night to see the President but he being engaged I was invited to wait on him this morning which I have done. He seems to disapprove of many things I had done and contemplated doing. The employment of . . . Mr. Lapham in particular he objected to and although he did not give his final decision he advised that they should be temporarily discharged and the matter allowed to rest till the Secretary [of the Interior] returned which would be in about five weeks.[20]

It is not clear whether Increase was made aware of these developments, but if he was they did not deter him.

Increase's demands that he be paid for his work continued through a change in administrations. Another Democrat, James Buchanan, was inaugurated president in 1857; Mason left, and Joseph Holt was appointed commissioner of patents. If anything, Holt was even more set against paying Increase than was Mason. "On carefully re-examining the records... I cannot conceive that you have rendered any service to this branch of the Government for which you are entitled to receive pay," Holt wrote Increase on January 20, 1858.[21] This reply wounded Increase deeply. He sporadically and futilely reargued his case through the term of President Buchanan and even into the administration of President Lincoln, when he was told that the Civil War preempted any such expenditure. It wasn't the only time that petty partisanship had thwarted Increase. It happened first in Ohio, when a change of administrations cast him out of state government. Nor would it be the last.

Increase finished his study of grasses as best he could, never receiving a cent in compensation. It exists today as an unpublished manuscript among his papers at the Wisconsin Historical Society—a very different fate from his recently republished *Antiquities of Wisconsin*.

—| |—

The years leading up to the Civil War weren't devoid of small but positive steps. The Wisconsin legislature's approval of three geological surveys of the state in the 1850s represented minor victories for Increase Lapham, even if the surveys were underfunded and sometimes inappropriately staffed. The legislature, eager for information that would lead to increased mining, pressed the geologists to concentrate on quick results at the sacrifice of sound science. Finally, the 1850s surveys set the stage for one of the keenest disappointments of Increase's life.

By the time Wisconsin was granted statehood in 1848, geological surveys had been approved or finished in eighteen states, including Ohio, whose survey Increase had helped organize.[22] Soon after Wisconsin's government was formed, Increase was promoting a statewide geological survey. There were good reasons for it: Lead mining in southwestern Wisconsin had developed on a hit-or-miss rather than a scientific basis. There was evidence from the north that other parts of Wisconsin were rich in metallic ores. Geological understanding was important for agriculture and the physical development of transportation and communities.

But Wisconsin was to have poor luck in its geological surveys, due often, as with other of Increase's projects, to partisan politics. On March 25, 1853, the Wisconsin legislature approved an act creating the first State Geological Survey, budgeting $2,500 a year for four years. Governor Leonard J. Farwell, a Whig, appointed Edward A. Daniels, professor at Carroll College, Waukesha, as first state geologist. Daniels served a year before he was dismissed by newly elected governor William A. Barstow, a Democrat, who in August 1854 appointed James G. Percival, an eccentric, sentimental Yankee poet, lexicographer, medical doctor, chemist, and, curiously, a geologist.

On December 6, 1854, Increase received a letter from Lyman C. Draper, secretary of the Historical Society of Wisconsin: "Dr. Percival who is here expresses a desire to make your acquaintance. I think he must know something of geology but what a timid retired old bachelor he is!"[23] Percival contacted Increase for background on the state's geology before he went into the field. Increase was happy to oblige. Soon Percival, who became known as "Old Stonebreaker," "became a familiar figure in the fields and woods of Wisconsin."[24] Yet his shy and hesitant manner preceded him. As one historian noted, "Some of the boys made sport of him but little children all over the state knew and loved him. He was always poorly clad and suffered greatly from exposure in winter."[25] Percival finished one annual report and had started another when he died unexpectedly on May 2, 1856. After his death, yet another new governor, Coles Bashford, the state's first Republican governor, offered to pay Lapham to finish Percival's second report.[26]

A few months later, Increase delivered a report of III pages that was obviously based on Lapham's understanding of Wisconsin geology as much as Percival's. Even so, Increase began the

James G. Percival was already well known as a poet and linguist when he came to Wisconsin in 1854 to lead its first geological survey.

WHI IMAGE ID 45760

report with a generous tribute to the late poet-geologist. "Died, on the 2d of May, 1856, at Hazel Green, Wis. in the 61st year of his age, Dr. James Gates Percival, eminent as a poet, scholar, and philosopher," it began. It concluded, "Dr. Percival possessed intellectual faculties of a very high order, and few men have exceeded him in variety and exactness of learning."[27] Self-effacing as usual, Increase signed neither the report nor his eulogy of Percival.

And despite his work on the Percival report, on January 5, 1857, Increase told E. Desor, an aide of Louis Agassiz, who had returned to Switzerland: "I have done very little in Geology since the appointment of Mr. E. Daniels by the state except to publish a very small geological map of Wisconsin. My leisure hours have been devoted to botanical pursuits, especially the Gramineae [grasses] of the US."[28]

On March 3, 1857, the legislature approved yet another geological survey, appropriating six thousand dollars a year and appointing James Hall, Dr. Ezra S. Carr, and the same Edward Daniels to share equally in its administration. Carr was a physician who months earlier became professor of chemistry and natural history at the University of Wisconsin in Madison. Daniels was the geologist dismissed from the first survey. Hall already was serving as state geologist for both New York and Iowa and he spent little time in Wisconsin. With Daniels in the field, Carr was left to handle day-to-day administration.[29] The three chief geologists did not get along well. In the first year of the survey, Daniels was the only one to submit a report, sixty-two pages mostly devoted to the iron ores of Dodge and Jackson Counties.

Daniels's report mentioned neither of his colleagues. It also reflected dissatisfaction among state officials. Daniels's edgy defense of the survey appended to his report pointed out that Iowa, Illinois, and Missouri were spending much more than Wisconsin on surveys. "It is but just to allow the Geologist time to mature his results, before finding fault that he has not done more," he told legislators. "In this State we have thus far had constant interruptions and no one has been allowed to carry out to completion any branch of the survey."[30]

The survey stumbled along through 1860, when Hall was appointed principal geologist, under whom a 455-page report was printed. Notwithstanding, in 1862, with the Civil War under way, the Committee on State Affairs suspended the survey, complaining that thirty thousand dollars had been spent without any return whatever except the one-volume report. Carr quit. Daniels joined the Union army as an officer. But Hall, believing himself to be under legal contract—an irony given his lack of progress on the paleontology project with Lapham—labored on. It was said he finished a second volume. If so, it was never published and he received no further pay.[31]

In each of the three surveys, Increase was an interested observer and at times an unpaid contributor. He tried without success to get hired as a botanist on the Carr-Hall-Daniels survey.[32] Other than being paid for finishing the Percival report, his only reward from the surveys came from Hall. In his 1861 report on new species of fossils found in Wisconsin, Hall lavished credit on Lapham for collecting some species and he named a fossil snail found in eastern Wisconsin limestone in his honor, *Murchisonia laphami*.[33] Hall's generosity may have been rooted in conscience. He had, after all, neglected to carry out his part of the bargain with Increase to jointly produce a book about American fossils.

Formal surveys or not, Lapham acted as his own independent geological surveyor throughout this time. Engaged by Edwin Palmer, president of the Wisconsin and Lake Superior Mining and Smelting Co., he went by steamboat to Lake Superior in September 1858, then walked inland to explore iron regions in northern Wisconsin, especially the Penokee range. The Penokee is a range of ancient mountains rising twelve hundred feet above Lake Superior and trending southwest for twenty-two miles in what are now Ashland and Iron Counties. There Increase found a roadless wilderness.

> At Ashland . . . we will doff our coats, substituting red flannel shirts, well supplied with pockets; draw on our "overhauls"; provide ourselves with blankets, provisions and cooking utensils, to be carried by a packer in a large bundle on his back, and we are ready for a tramp through the dense forest, mostly of evergreen trees, to the Penokee Iron Range, whose distant summits we have already seen towards the south-east from La Pointe. I carried with me, as usual on such occasions, a tin box for collecting botanical specimens, and a mineral hammer.[34]

WHI IMAGE ID 43831

Increase Lapham, ca. 1859

He reported enthusiastically about the ore body to Palmer—perhaps too enthusiastically. In fact, Penokee ore turned out to be inferior to ores nearby and, because it was so deep, much more difficult to mine.

A year after Lapham's Penokee trek, the November 11, 1859, edition of the *Milwaukee Free Democrat* headlined an "Important Geological Discovery" made by "Mr. I. A. Lapham." This discovery was closer to home. Lapham found rock of the Devonian period, a bed of limestone on top of the Silurian period limestone that underlies most of eastern Wisconsin's bedrock. The wedge of Devonian rock, the youngest bedrock in the state, lies on the Lake Michigan shore from northeastern Milwaukee County through Ozaukee County almost to Sheboygan. Increase identified it from its characteristic fossils, which included fish, and he wrote about it for Silliman's *American Journal of Science and Arts*.

If he had "done very little in Geology" as he reported to Agassiz's assistant in 1857, he was now proving his worth. Yet despite this small triumph, there would be more partisan politics to contend with in the coming years of the impending Civil War. In fact, the march toward war interrupted not only Increase's geological pursuits but much else as well. It interfered in a profound way with the manner in which he had supported his family and carried on his correspondence. And as a Quaker opposed to war, it confronted Increase with a wrenching decision regarding a son.

CHAPTER 18

A Quaker in Wartime

────────── ┤ 1860–1865 ├ ──────────

POLITICALLY, INCREASE LAPHAM WAS A MEMBER OF THE WHIGS, a fiscally conservative party that supported public improvements such as canals, railroads, and harbors, as well as economic protectionism, and, especially during the presidency of Democrat Andrew Jackson, the dominance of Congress over the presidency. However, with the deaths in 1852 of its two greatest leaders, Henry Clay and Daniel Webster, the party divided over slavery, especially the Kansas–Nebraska Act of 1854 that allowed new states, even those north of the Mason-Dixon Line, to determine whether or not to legalize slavery.

As the Whig Party declined, southern Whigs gravitated to the Democratic Party and northern Whigs joined the fledgling Republican Party, which got its start in Ripon, Wisconsin, and was spreading in the west.[1] By 1856, the Whig Party was defunct and 95 percent of all American Quakers, whose doctrine had condemned slavery as immoral from the seventeenth century forward, began to vote Republican.[2] In 1860, Increase Lapham supported the election of Abraham Lincoln, the country's first Republican president.

Direct evidence is missing, but it is likely that Increase and Ann Lapham saw and may have even met Lincoln in Milwaukee in 1859. John Wesley Hoyt, secretary of the Wisconsin State Agricultural Society and coeditor of its magazine, *Wisconsin Farmer*, invited Lincoln to address the Wisconsin

State Fair in Milwaukee in September 1859. As usual Increase, a life member of the society, had an active role in the fair. This year, "I. A. Lapham and lady, Milwaukee" were two of the six judges of the Fine Arts division of the fair. As such, they distributed awards of two, three, or five dollars for the winners and diplomas for lesser works of merit.[3]

Lincoln was to speak at 11 a.m. on Friday, September 30, the last day of the five-day fair. He arrived at Milwaukee's Newhall Hotel at Broadway and Michigan about midnight Thursday and was met there Friday morning by Hoyt, who led him on a walking tour of the fair, which was held a mile west on open land at what is now Wisconsin Avenue and Twelfth Street. Lincoln's debates with Democrat Stephen A. Douglas in Illinois in 1858 had attracted national attention, and he was known also for his "House Divided" speech. In 1859 his name was being put forward by newspaper editors and Republican politicians as a promising candidate for the presidency.[4]

Abraham Lincoln in June 1860, almost a year after he spoke in Milwaukee and months before he won the presidency

WHI IMAGE ID 23664

The Wisconsin State Fair was held in 1859 at what is now Wisconsin Avenue and Twelfth Street, Milwaukee.

WHI IMAGE ID 33371

Even so, his speech at Milwaukee was mostly nonpolitical and non-controversial. He talked about the usefulness of agricultural fairs like the one he was addressing; he said he was going to talk about agriculture even though he was a politician, not a farmer. He spoke about farm technology and farm economics and he envisioned that educated farmers of the future would succeed in the "most valuable of all arts,... the art of deriving a comfortable subsistence from the smallest area of soil."[5]

He also attacked the "mudsill" theory of labor in which southern politicians held that there was a mudsill, or lowest foundation of labor and slavery, on which the entire economy rested, and that this dictated a society permanently divided between a lower class of laborers and slaves and an upper class of owners, capitalists, and professionals. Theoretically, there could be no movement between them. Lincoln held that the mudsill theory was false, pointing out that many Americans evolved from hired laborers to self-sufficient farmers who depended on neither hired labor nor owners.

Directly after Lincoln's speech, the fair's judges announced the winners of the various fair contests and distributed prizes, which likely required Increase and Ann Lapham's presence. If Increase noted the event in his journal or letters or notes, it has been lost. However, Increase was so enamored of Lincoln that on May 18, 1860, he went to Chicago on the 5:30 a.m. train to attend "the great National Convention at which Abraham Lincoln was nominated for president."[6]

The Chicago convention was held in a huge wooden building thrown up for the occasion at what is now Wacker and Lake Drives. Called the "Wigwam," it could hold ten thousand people but even that proved to be too small. Wisconsin's delegation was led by Carl Schurz of Watertown, an eloquent liberal who fled prosecution for revolutionary activity in Germany in 1848 and arrived in Wisconsin in the late 1850s. An early Republican and abolitionist, he convinced thousands of fellow German immigrants to switch from the Democratic to the Republican Party.

The scene in Chicago was chaotic. Hotels were overbooked and crowds milled around all night in the streets, which were illuminated by bonfires. It was hard to get seating in the Wigwam, where all business had to be done by shouting. Increase attended only the last of three days, the day of balloting that nominated Lincoln. The day began with thirteen candidates, of which five garnered all but a few votes on the first ballot. Although William Seward of New York was favored on the second ballot, Lincoln won on the third ballot. The nominee did not attend; he was at home in Springfield where he was kept informed by telegraph.

Six weeks later, on July 2, Increase and Erastus B. Wolcott left Milwaukee for St. Louis to attend the fifth annual St. Louis Agricultural and Mechanical Fair. After the fair they left St. Louis and stopped for the night at Springfield. Increase's journal for the next day, July 6, states: "Remained at Springfield until 11 a.m. Visited Abraham Lincoln, the Agricultural Society, State Geologists' rooms and other interesting places."[7] It may be that Increase tagged along with Wolcott, an eminent Milwaukee medical doctor with extensive military experience, in his visit to Lincoln. Wolcott would serve as surgeon general of Wisconsin during the Civil War.

The entry is one of Increase's tantalizing records of memorable events that cries out for more information, for more details of the circumstances of the meeting with the future president—where, what was said, who else was present? Yet unlike the younger man, Increase now rarely discussed politics in his correspondence or notes. Abetted posthumously by daughter Julia, the family archivist, Increase tightly controlled what was left behind in order to present the mature Increase as scientist, educator, and patron of good works, but little else.

Several southern states had seceded and Civil War was in the wind when Lincoln was inaugurated on March 4, 1861. Just as the Religious Society of Friends, or Quakers, held slavery as immoral, they famously regarded war with contempt. That Increase was among them was revealed in a letter to Darius September 1, 1847, in which he complained about a book called *Annals of the West*.[8]

Such, however, is history in general, it passes over periods of peace and prosperity and lingers upon battles, wars, murders, etc. I should like to see a history of the world in which war was considered as it should be, an evil, a pull-back upon progress rather than an evidence of greatness. Let the periods of war be passed over instead of the periods of peace.[9]

Increase remained largely silent about the war as it progressed, although other Wisconsin citizens were fully involved. Despite Schurz's efforts, most Wisconsin German immigrants opposed the war. By contrast, Yankee settlers, largely abolitionists like Wisconsin Governor Alexander Randall of Waukesha, were zealous patriots.

—| |—

For Increase personally, the Civil War indeed proved to be "a pull-back upon progress." Among other things, it interrupted his efforts to help establish a national storm warning system, which he had been moving toward since the 1840s when he became aware of the dreadful toll that sudden storms took on ships and men sailing the Great Lakes. In 1848, fifty-five lives were lost on the Great Lakes, seventy-two vessels driven ashore, nineteen totally wrecked, five capsized, and six sunk, Increase reported to the Wisconsin legislature in 1850.[10] Many lives and much property could be saved with storm warnings.

A national storm warning system required three essential components: First, an understanding of how storms form and in which direction and how fast they move. Second, a technology that could move information ahead of a storm. And third, a network of observers across the country who gathered and sent weather information in standardized language and symbols that could be understood by both senders and receivers.

Several American scientists were working on these problems, starting with James Pollard Espy, who had developed a convection theory of storms by the mid-1830s and who discerned that North American storms usually came from the west, and young Joseph Henry, an instructor at Albany Academy in New York in the late 1820s, who developed electromagnets powerful enough to send pulses at some distance over wires. Henry was not the inventor of the telegraph—Samuel F. B. Morse did that in 1837—but Morse's telegraph absolutely depended upon Henry's electromagnetics. Henry went on to become the first secretary of the Smithsonian Institution, a job he held from 1846 until his death in 1878. There he led the effort to publish Lapham's *Antiquities*, but his main interest lay in meteorology. Both Henry and Espy were Lapham correspondents.

Increase began gathering weather information upon arriving in Milwaukee, calculating the length and severity of Wisconsin winters by recording the dates each year when the Milwaukee River froze over and thawed, revealing a five-month-long winter on average. He or, in his absence, Ann or the boys recorded temperatures and water levels on a daily basis. It was convenient. The instruments were on the west shore of the river at the base of Poplar (now McKinley) Street, little more than a block east of the Lapham home.

Some weather events on Lake Michigan could take observers by surprise. Such a one occurred on April 8, 1858, which Increase described in his journal that day:

> A remarkable swell struck the shore of Lake Michigan at Milwaukee quite similar to the "earthquake waves" on the ocean. At the foot of Huron street the water rose, it is said, five feet above its usual level and was depressed as much below, making a total change of ten feet. A vessel entering the harbor at the mouth of the river at a good rate of speed was struck by the returning wave and sent back at a rapid rate. In its progress up the river the wave suddenly upset a small ferry scow on which were nine passengers. So quickly was it done that those in the water did not know what caused it until they saw their hats going off upstream with the current at the rate of ten miles an hour. This was about noon.[11]

Increase consulted his barometric readings, which were recorded for 7 a.m., 2 p.m., and 9 p.m. each day. He found that a deep low had passed over Milwaukee on April 8. Increase had described a "seiche," a phenomenon that occurs only on enclosed bodies of water such as lakes. A high-pressure ridge or a sharp wind passing quickly over the surface will push water in front of it until it reaches the other side. At this point, the high-pressure ridge continues onward, followed by the low, and the wave or surge that had been pushed in front of it reverses itself and moves toward the other side, thus setting up a sloshing effect. Seiches have been known to measure sixteen feet high in the Great Lakes. At 9:30 a.m. Saturday, June 26, 1954, a ten-foot seiche hit the North Avenue pier off Chicago, washing fishermen into the lake and drowning ten of them.

Despite Increase's abiding interest in the weather, an incident arose in 1859 that threatened to distract his attention. In November 1858, Increase's discovery of the lunar tides of Lake Michigan, reported almost ten years earlier in the *Milwaukee Sentinel and Gazette*, came under scrutiny when Colonel James D. Graham of the Army Corps of Topographic Engineers in Chicago rediscovered the tide and reported his findings to the Chicago Historical Society and the American Association for the Advancement of Science. Increase got wind of it, dug out the records of his 1849 measurements and the news clipping and sent them to Captain George Meade, head of the US Lake Survey. Meade replied on July 3, 1860: "Have you Col. Graham's report for last year and his water level observations? I was surprised to find he thinks he is the first to have discovered the Lake Michigan tide. It is rather extraordinary that the quantity he gives in 1859, for the rise and fall is so close to yours determined ten years previously."[12]

But Graham was not backing down, as shown by a letter August 26, 1860, from Samuel Stone, Increase's brother-in-law in Chicago:

The colonel made some remarks one evening at one of our Historical Society social gatherings about the lunar tides. Your daughter Mary was present. After an introduction the same evening Mary told him that you have for years kept a table and had published that fact. I wish to see the Colonel before I write to you again on the subject.[13]

This gnat of a controversy buzzed about for a few months. Increase wrote to Stone on December 17, "It is hardly worth while to trouble Col. Graham on the matters of tides. I only wished you to let me know what he said... in the way of admission of my discovery, but even this is not important as the evidence is conclusive."[14] But he did not hear from Stone directly.

On March 4, 1861, Captain Meade put an end to the matter. In a letter to Increase, Meade said, "I have completed the reduction [*sic*] of the records of the self-registering tide-gauge at Milwaukee and you will be gratified to hear they indicated the existence of a lunar tide almost identical with your previous observations."[15] He included a graph that showed, conclusively, that Increase's and Graham's findings were nearly the same. Nothing more is said of Graham's claim in Increase's letters, and that is just as well; he had other matters to attend to. Graham never conceded the point.

By 1860, Increase had amassed an ever-growing mountain of weather data, which grew more sophisticated with the development of new instruments, some of which he acquired from the Smithsonian through Joseph Henry, others of which came from Captain Meade at the US Lake Survey, created by Congress in 1841 to prepare hydrographical charts and navigational aids for ships and commerce.

On New Year's Day, 1860, Increase wrote in his journal, "Commenced a full series of meteorological observations barometer, thermometer attached, psychrometer, self-registering thermometers (maximum and minimum), anemograph, snow gauge, rain gauge and tide gauge."[16] He was delighted with his new array of gadgets that measured barometric pressure, temperature, humidity, wind, snow, rain, and tide. Now he was operating an up-to-date weather station.

He was not alone. Starting in 1849, Joseph Henry had been organizing a network of observers that eventually numbered five hundred.[17] By 1856, information was being transmitted by telegraph, which Henry posted on a map at the Smithsonian Institution in Washington, presenting a snapshot of the day's weather across the United States, a precursor to today's weather maps on television and in newspapers.[18]

Working in Dubuque, Iowa, 170 miles southwest of Milwaukee, Asa Horr, a medical doctor and, like Increase, a patron of the sciences, also was collecting weather data for the Smithsonian. They naturally became correspondents and agreed in their exchanges about the need for a storm warning system.[19]

Dr. Horr and Increase corresponded frequently during the blustery spring of 1861 and deduced that storms passing through Dubuque likely would hit the Milwaukee area six or eight hours later. Such information could easily become the basis for posting storm warnings on Lake Michigan, where it could save lives. But, alas, Increase wrote Horr on May 10: "The Sentinel declines publishing my meteorological observations the space being wanted for 'War news.' I fear the lake survey will be interrupted on account of 'the times.' Capt. Meade who has charge of it is ordered to another duty."[20]

"Another duty" indeed. Just weeks after settling the lunar tide dispute, Meade left the Lake Survey in Detroit to become a field officer in the Union army. Two years later, as Major General George G. Meade commanding the Union's Army of the Potomac, he defeated General Robert E. Lee's Confederate army at Gettysburg, Pennsylvania, thereby becoming one of the great Union heroes of the Civil War. Increase, who held war in contempt, seldom if ever spoke of him again.

The war also interrupted Joseph Henry's efforts in Washington. The rapid march toward a national storm warning system was halted in its tracks for the duration of the Civil War.

⊣ | ⊢

By 1860, Milwaukee was a nationally important commercial center and the rail transportation center for Wisconsin. Trains gathered the agricultural and mining products from inland Wisconsin and moved them to Milwaukee, which had an improved harbor from which car ferries could move goods across Lake Michigan to railroads and to the East Coast. Or trains could chug south to Chicago, then east. Or schooners could move products through Lakes Michigan, Huron, and Erie to the Erie Canal.

The city now held forty-five thousand persons, more than double its population of ten years earlier, making Milwaukee the twentieth largest city in the nation. Beyond its borders, German, Yankee, Irish, and Norwegian farmers grew more wheat than in any state in the Union. The city became the largest wheat-milling city in the nation. The growing German population not only settled comfortably into the city, but was also the impetus for many enterprises. German breweries, tanneries, and machine shops flourished.[21] A prospering commercial colossus was growing rapidly around the two-story house on Poplar between Third and Fourth Streets in old Kilbourntown where the Laphams still lived.

As wars do, the Civil War brought boom times to the North. Prices and demand rose for Wisconsin products, especially wheat and other foodstuffs, lead, iron, lumber, and manufactures. Salaries rose for laborers and profits for owners. The cost of living rose for everyone.[22] But Increase was neither a salaried worker nor a business owner. He made his living selling land, charging fees for scientific work, and peddling his maps. He seems not to have shared in the wartime prosperity. In fact, much evidence points to the opposite, to genuine hard times for the Laphams.

With the coming of the war, Edward Daniels, a zealous abolitionist, resigned from the Wisconsin Geological Survey to enlist in the Union army as a cavalry officer, creating an opening at the survey. On June 27, 1861, Increase asked Governor A. W. Randall to hire him to prepare a report on how climate affects agriculture, which he would carry out for a thousand dollars a year for as long as the survey continued. The effort came to nothing. The wartime legislature withdrew its support of the survey.

Increase was still agitating to restart his project on the grasses of the United States, a dispute now in its sixth year. On January 3, 1862, he wrote Rep. John F. Potter, a Wisconsin Republican, to enlist his help, arguing that the war added urgency for such a study because farm products were essential "in the vast military preparations rendered necessary by this wicked rebellion." Potter was sympathetic but Increase's argument failed to divert money from the war effort.[23]

By January 1862, Increase's need for money was such that he allowed his subscription to Silliman's *American Journal of Science and Arts* to lapse. This was the magazine that published his first scientific article in 1828 and we can assume that until recently he had an unbroken series in his library. His sons Henry, fifteen, and Seneca, thirteen, somehow scraped together the five dollars needed to restart delivery. In return they received a kind letter from the editors, Benjamin Silliman and James D. Dana.

> Dear Sirs,
> Your esteemed favor came duly to hand covering $5.00 for the renewal of your father's subscription to this journal.
>   Doubtless before this time No. 97 has reached him and although it was rather late for a Christmas present, we hope it is not the less agreeable.
>   We have taken the liberty to send also Nos. 94, 95, 96 which will make his series good, these we shall feel obliged if he will accept from us in token of a long and agreeable acquaintance by letters.
>   Yours respectfully,
>   Silliman and Dana[24]

The entire transaction—the boys' sweet present to their father, the editors' empathy, the pleasure felt by Increase—must have brought a measure of warmth to the Lapham family that cold winter.

Another good thing happened to Increase that Christmas season, though it didn't help his economic situation. Lyman C. Draper, corresponding secretary of the Historical Society of Wisconsin, reported that Increase had been unanimously elected president of the society. Increase had been a vice president since the founding of the society in 1854. He would serve as president through 1871. But the position was unpaid, and the need for money continued. Increase wrote to William D. Wilson, editor of an agricultural magazine in Iowa, on April 28, 1862, offering to write an article about Iowa's forest trees if he was paid two hundred dollars and given free passes over Iowa's railroads. It did not happen.

On April 5, Increase was surprised to receive a note and fifteen dollars from George Ripley, coeditor of the *New American Cyclopaedia,* for a long article about Wisconsin.[25] Increase had written it at the request of Governor Louis P. Harvey and expected to be paid a much greater sum from the state's contingency fund. Ripley's note prompted Increase on May 2 to press the governor's secretary, William H. Watson, for the payment promised by Harvey. His payment probably was overlooked in the turmoil caused by the death of Governor Harvey, who had drowned a month earlier in the Tennessee River on a tour of the Shiloh battlefield in Tennessee. He was succeeded by Lieutenant Governor Edward Salomon.

Increase wrote Ripley that he would return the fifteen dollars "whenever the governor does his duty." He added, "The inadequacy of this sum, as a compensation for the article will be apparent when I say that it embodies results of observations and investigations in geology, botany, meteorology, &c made by me within the past 26 years at a cost of time and money amounting to thousands of dollars, and for which I have received but very little return."[26]

On June 6, an impatient Lapham told Watson: "Please let me know whether Gov. Salomon will have my account for the article on Wisconsin ...properly brought before the Legislature, or whether it will be necessary for me to go to Madison next week and 'lobby' the matter through." Three days later Watson sent Increase a check for one hundred dollars. A day later Increase returned the fifteen dollars to Ripley in New York City.[27]

With summer, Increase was able to do some botanizing with Charles Leo Lesquereux, a guest at the Lapham home. Lesquereux had followed his friend Louis Agassiz from Switzerland to the United States in 1848, settled in Ohio, and was to become the nation's leading expert in mosses. He was profoundly deaf but apparently could read lips in English, German, and French. He and Increase waded into some local marshes in search of rare specimens of moss.

During Lesquereux's visit at the Lapham home, Increase asked his guest to appraise his library and natural history collections. On July 14, 1862, Lesquereux wrote from Cleveland, Ohio, that Lapham's cabinet "contains almost everything that can illustrate and make known the natural history of the state, excepting, perhaps, zoological specimens." It was composed of minerals, fossils, shells, antiquities, and more than eight thousand species of plants. His library of one thousand volumes held complete series of scientific journals that, Lesquereux noted, "I could not obtain elsewhere."[28]

He estimated the entire collection to be worth more than three thousand dollars, not including cases and other appurtenances. "I would not in any way advise you to sell your cabinet and library," Lesquereux noted. "Those inanimate objects that we have put together with care, with much trouble and expense, become the dearest friends of our old age." He hoped that the collection would be kept together in Wisconsin. "No school in the West would have a better arranged and more valuable cabinet than yours."[29] For the moment, the matter was settled. The Laphams would not sell the cabinet.

In 1863 tragedy struck the Lapham home. On Wednesday, February 25, Increase's wife for twenty-five years, Ann Marie Lapham, not yet forty-seven, died unexpectedly. A brief *Milwaukee Sentinel* notice said she helped the poor and was a manager at the Milwaukee Protestant Orphans Asylum. No illness or cause of death was disclosed and as far as we know her husband did not mention her death in his surviving journals or other notes.

Increase's brother-in-law Samuel Stone wrote to his "Dear affected brother" from Chicago on Friday, the day he received a telegram from his daughter Elizabeth announcing that Ann's funeral would be Saturday afternoon. "My dear sir: I regret very much to say that my old pains are just now too severe for me to venture far from my house." Stone remembered Ann "for her continued kind and true parental [*sic*] motherly attention to my infant children" after their own mother, Ann's sister, died.[30]

The Lapham household received many condolences over the next two months, including from John and Charles Hubbard, brothers who, like their cousins Elizabeth and Caroline Stone before them, had become wards of the Lapham family when their mother, Mary Jane, another of Ann's sisters, died in 1851. John and Charles, both soldiers in the Union army, wrote from the front.

Weeks after Ann's death, on April 7, 1863, Charles wrote to his cousin Julia from a camp near Murfreesboro, Tennessee, to say he had read a long letter from his Aunt Ann just days earlier, three sheets of note paper mailed January 23 that had been delayed for weeks at Louisville. It was, he said, "filled with anxiety at my welfare personal safety in the battle before Murfreesboro."[31] It stunned Charles to read of his Aunt Ann's concern for him after she herself had died.

Increase evidently mentioned Ann's death to Lesquereux in a letter in which he acknowledged receipt of a package of specimens. Lesquereux responded on March 21, 1863:

> I read your kind letter just received, with deep and sincere regret and offer you in this great affliction my true affectionate sympathy.... I was scarcely acquainted with your lady ... but I was at once struck with the kind and benevolent expression of her figure and read from it the evident character of a true Christian love. You could not but live happy with such a wife.[32]

As with the death of his brother Darius, his father Seneca, and his son William, Increase seems to have steeled himself for such events through either faith or fatalism. We can only guess how grief affected him in private. In public, he simply carried on. He was busy with regular correspondence less than three weeks later, his household now in the hands of two capable daughters, Mary, age twenty-four, and Julia, twenty-one. His surviving sons, Henry, Seneca, and Charles, were sixteen, fourteen, and six years old respectively.

Throughout these months of continued war and personal loss, Increase's money problems did not abate. While we cannot know if the death of Ann made finances more difficult, we do know that in the spring of 1863, Increase proposed that Byron Kilbourn buy his "library comprising more than one thousand volumes of useful [mostly scientific] books, together with maps, charts, newspaper files, manuscripts &c.—provided the same be made a portion of a public library to be founded in the city of Milwaukee—for the sum of $2,500."[33] Also he would donate his two-story brick house containing his library and office, which he valued at two thousand dollars, so long as he could retain the possession and use of the library and property during the rest of his life.

Kilbourn replied on April 8 that he was having troubles of his own and was not in a position "to incur any expense beyond my actual necessities for the present."[34] Along with his note of rejection to Increase, Kilbourn returned Lesquereux's appraisal.

Increase was to make another proposal two years later, in May 1865, to Milwaukee railroad magnate Alexander Mitchell. One hundred thousand dollars would endow an institution patterned after the Smithsonian, he said, with Increase's collections forming its nucleus and Increase himself taking the lead in "building it up."[35] Nothing came of it.

In June 1864 Increase explored the area around Council Hills, Illinois, near Galena, for lead ore for a client, the Illinois Lead Mine and Smelting Company of Chicago. His report prompted the company to dig deeper in for lead with success. Increase was paid ninety-five dollars for his work and he secured another commission from the company to hunt for coal in Livingston County, Illinois. Again he had success and was paid another fifty dollars.

As 1864 came to an end, Increase finished a small map of Wisconsin showing its counties, cities, major towns, and railroads. In January 1865, the State of Wisconsin ordered one thousand of the maps at twenty-five cents each to be folded and tipped into the annual Wisconsin Legislative Manual, precursor of the Wisconsin Bluebook. The map, which was revised annually as counties were added and as railroads and roads expanded, appeared in the Legislative Manual from 1865 through 1880, providing a stable source of income. In 1869, Governor Lucius Fairchild and his State

Board of Immigration included the map with a thirty-two-page pamphlet written by Lapham extolling Wisconsin's virtues. Some ninety thousand copies were printed, some of them translated into French, Norwegian, Dutch, Swedish, and Welsh and distributed in those countries. It was the most-viewed Lapham map and often formed the first perception of Wisconsin for thousands of immigrants.[36]

But in 1865, Increase had another use for the map, which he employed as the base for one of his most important contributions to understanding Wisconsin's weather. On his early excursions out of Milwaukee, he would

Drawn in 1865 with his most popular map of Wisconsin as a base, Increase's isothermal map showed that Lake Michigan greatly affected inland temperatures in both winter and summer.

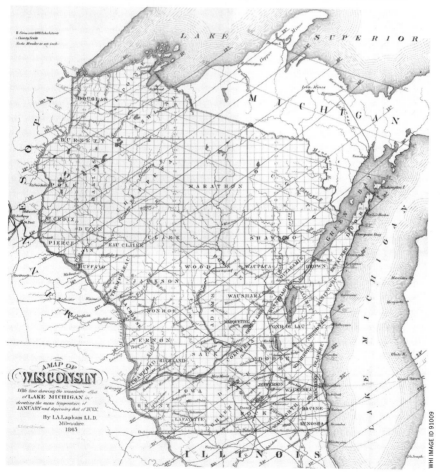

WHI IMAGE ID 91009

notice signs that the weather in Milwaukee differed measurably from the weather a few miles inland, signs such as the timing of fall coloration of hardwoods or the blooming of wildflowers.

Increase and others had recorded temperatures in various places around Wisconsin from the 1830s. When he transcribed the mean temperatures in January and July as isothermal lines on the base map, it was obvious that breezes from Lake Michigan moderated eastern Wisconsin's temperature, cooling it in summer when the lake's water was cooler than the air and warming it in winter when the lake water was warmer than the air. He called his work "A Map of Wisconsin with lines showing the remarkable effect of Lake Michigan in elevating the mean temperature of January and depressing that of July." It was published with an article that Increase wrote for the first issue of the *Transactions of the Chicago Academy of Sciences*.[37]

—┤ ├—

The war years were difficult not only because of strained finances and personal tragedy. For the Quaker Increase, war struck a heavy note. In August 1864 Increase confronted the "wicked rebellion" head on when he desperately argued with his son Henry, who at seventeen wanted to enlist in the Union army. In a letter beginning "My Dear Son" he wrote:

> It is with mingled feelings both of pride and pain, that your letter of the 31st is received; pride in the thought that you are willing to sacrifice so much for the good of our country and pain to think of the necessity there is for a call of men to defend and preserve that country from ruin.
>
> There are many reasons why it would not be wise for you to enlist as a soldier just now; and which make it your duty not to do so.[38]

Increase then listed several. First, Henry should finish his education "as a year of service of an educated man is worth five years of one who is not." Next, he worried that "Hardship and exposure to the weather...will bring on your old scrofulous trouble, and thus very soon render you unfit for service." And he was concerned about his son's youth, arguing that he was "not of age to have the strength necessary to perform the hard service required." But his last and strongest argument had to do with duty and responsibility:

> There are various other ways in which you can serve your country with an equal amount of benefit. If you could train yourself to become a good speaker, and fill your mind with knowledge of history and the principles of government, you might do more good in one public address by persuading men to do right, than you could in a year in the army![39]

It was not Increase A. Lapham the detached scientist, but Increase the distressed father and good Quaker who wrote this letter. He carried the day. Henry did not enlist.

The Lapham family's distaste for war did not lessen their humane concern for the well-being of veterans and their families. According to the *Milwaukee Sentinel* of September 12, 1863, six months after her mother's death and three days before her own twenty-fourth birthday, Mary attended a meeting of the West Side Soldiers Aid Society and brought with her as donations twenty-eight rolled bandages and some old linen to be sent to military hospitals.

She was to become a regular member, seldom missing a meeting and never attending without bringing some donation for Wisconsin's sick and wounded soldiers. On December 16, 1863, the *Sentinel* reported that she brought twenty-four bandages, one pound of sage, and "apples dried by Charles Lapham only six years old." This was followed in subsequent meetings by more rolled bandages, arm slings, artichokes, and blackberry syrup.[40]

By mid-1864, months before the war ended, there were signs that the Lapham financial situation was improving and that life without Ann was settling into a tolerable place. With advances in photography, it became a widespread pastime to collect small mounted photographs, or *cartes de visite*, of celebrities or friends. The fad became known as "cardomania."

Increase began to ask his scientist correspondents for portraits. The little hobby became a family affair when Mary and Julia assumed the task of writing to Increase's far-flung correspondents, enclosing a portrait of their father and asking for one in return. Many correspondents seemed flattered and responded with a portrait and a letter of greetings and good humor; others wrote that they would schedule a sitting with a photographer to provide an up-to-date image. Eventually the Laphams collected almost one hundred photographs of eminent nineteenth-century scientists from the United States and a few other countries. All were put into an album that was given to the State Historical Society of Wisconsin after Lapham's death.

Despite the welcome distraction of the *cartes de visite* for the Laphams, the war still monopolized the country's attention and Increase, like his daughter Mary, was solicited to help the shattered veterans in small ways. He prepared albums containing mosses mounted on boards that were auctioned at fairs held to benefit soldiers and their families in Chicago and Boston. On January 30, 1864, he wrote Miss Ann E. Stone in Chicago:

> The expressman will bring to your door the parcel of mosses that Mrs. Atwater and you requested for a contribution to the funds for the relief of sick and wounded soldiers, and I hope you will not be disappointed with them. Should any one object to paying twenty dollars for this collection, ask them how many cedar swamps and cranberry marshes they would be willing to wade through for that sum—or how many rocky crags they would climb and clamber over at the risk often of a broken neck or limb for that small sum.[41]

He learned later from Mrs. Elizabeth Atwater that the album of mosses sold for thirty dollars.

When the Civil War finally ended in April 1865, prominent women of Milwaukee led by the West Side Soldiers Aid Society planned a "Soldiers Home State Fair" to raise money to attract one of three federally approved National Soldiers Homes to Milwaukee. We don't know whether he was drafted or enlisted but Increase played an important role in the effort as a member of the fair's executive committee and as head of the "Geology, Mineralogy and Natural History" department.[42]

Increase wrote to many of his scientist friends around the nation and Wisconsin, asking them to contribute specimens from their cabinets to be exhibited at the fair. A number of them responded generously. Fair patrons were asked to pay to "vote" on whether the collection later would be contributed to an educational institution such as the State University in Madison, Beloit College, Lawrence University, Racine College, the German-American Academy at Milwaukee, the Milwaukee Female Seminary, or the Milwaukee Public Schools.[43]

The fair opened on June 28 in a large, temporary frame building designed by Milwaukee architect Edward Townsend Mix and erected on Broadway Street at Huron Street (now Clybourn). It ran for two weeks, featuring many exhibits, music, lectures and entertainments, raffles, food, and parades. It was a smashing success, raising more than $100,000. Increase's department raised $1,877.30, which he called a "handsome sum."[44]

Soon after, a Milwaukee delegation went to Washington and persuaded the federal government to establish a soldier's home in Milwaukee. Key to their success was the money raised at the Soldiers Home Fair. Edward Townsend Mix designed the main domiciliary that still dominates a hill that was then some few miles west of the western edge of Milwaukee but that has since been surrounded by urban development.

Over the decades the National Soldiers Home sheltered thousands of veterans not only of the Civil War but of all American wars through the end of the twentieth century. A large national cemetery grew on ground west of the home.

CHAPTER 19

Prophetic Thinking and Forecasts

———————| 1865–1873 |———————

I N OCTOBER 1865, INCREASE ACCEPTED AN INVITATION from John C. Hilton, the president of the Illinois Lead Mine and Smelting Co. whom he had advised on mining questions, for an all-expenses-paid trip with two hundred others to the Pennsylvania region where the world's first successful oil well had been completed in 1859. Among other places, the party visited Pithole City, which had grown from farm fields to fifteen thousand persons in a few months after an oil strike in 1865. The boomtown would decline into a ghost town by 1867, when the oil played out.

At its peak, Pithole had a daily newspaper, a waterworks, a fire company, churches, banks, telegraph offices, saloons, boardinghouses, groceries, machine shops, dance halls, and brothels.[1] It was a brawling place of boomtown excess and low morality, much like pioneer Milwaukee in the 1830s. Though Increase saw Pithole City in its heyday, he wrote little if anything about it. But his visit to the oil fields may have reinforced some prophetic thoughts about fossil fuels, thoughts that were generations ahead of his time and that he refined later in a report about forests.

The Civil War boosted the lumber industry in Wisconsin, and peace did nothing to slow it as the United States entered a gilded age. The cutting of the state's primeval pine forest was to change northern Wisconsin's natural environment as rapidly and completely as agriculture had done in the southern parts. The Mississippi, Wisconsin, and Chippewa Rivers became primary highways for moving white pine logs out of the north to mills. Finished lumber went west, south, and east by rail to become houses for homesteaders on the treeless plains and to expand growing cities throughout the country.

But wood to pioneer Wisconsin residents was not only building material; it was the principal fuel. Increase the geologist was the first to recognize that Wisconsin had no coal; his 1844 book on Wisconsin said that coal prospecting would be a waste of time. Without wood, pioneer Wisconsin would be uninhabitable in winter. "Without the fuel, the buildings, the fences, furniture, and thousand utensils, and machines of every kind, the principal materials for which are taken directly from the forests, we should be reduced to a condition of destitution and barbarism," he wrote in the 1855 *Transactions of the Wisconsin State Agricultural Society.*[2] "The Forest Trees of Wisconsin" was a sophisticated article that spoke of the beneficial effects of forests on climate, soils, and waterways and that emphasized the beauty of trees, "thus increasing our love of home, and improving our hearts."[3] Increase encouraged everyone to plant trees and his article included a guide to Wisconsin's trees, their attributes and uses.

At the time, even though lumbering had emerged as one of Wisconsin's major industries, the pine forest was so vast that few could conceive of its depletion. But after the war the frenetic forest cutting was attracting urgent concern. On March 23, 1867, the Wisconsin Legislature approved "an act relating to the growth of forest trees," creating a three-member commission to report at the legislature's next session "certain facts and opinions relating to the injurious effects of clearing the land of forests."[4]

I. A. Lapham of Milwaukee, Judge J. G. Knapp of Dane County, and former newspaper editor and mayor of Milwaukee Hans Crocker were appointed commissioners. Before the year was out, the commission produced a 101-page report, plus an index, called *Report of the Disastrous Effects of the Destruction of Forest Trees Now Going on so Rapidly in the State of Wisconsin.* It was obvious from its style and organization that Lapham did the writing. Though not voluminous or comprehensive, it was the most far-seeing of any published work that Increase ever produced.

In it Increase relied heavily on George Perkins Marsh, a Vermonter whom Lincoln had appointed as US minister to Italy. In 1864 while in

COURTESY OF THE WISCONSIN STATE HERBARIUM (WIS), UW-MADISON, ACCESSION NUMBER V0081911

These oak leaves, collected in a Wisconsin cemetery in June 1867, were labeled by Lapham as *Quercus coccinea*, or scarlet oak, but were later correctly relabeled *Quercus ellipsoidalis*, or northern pin oak.

Rome, Marsh wrote *Man and Nature*, based on his observations that centuries of deforestation in Greece and Italy had converted both countries into deserts. In concluding that human activity could profoundly alter the natural environment, Marsh had written an early book on ecology. Increase immediately understood its importance and incorporated his work into the Forest Commission report. Marsh was liberally cited in the report and a Marsh quotation appeared on the cover. Echoing Marsh's findings Increase wrote:

On the question of fuel, we are to calculate by ages of the Earth, and not by the life of man. Fuel will be required so long as man shall inherit the Earth, for his comfort and for his existence. Without fuel, humanity would cease to exist. Viewed in this light, the deposits laid up during the uncounted periods of time... in the shape of coal, petroleum and peat, and which man is now drawing out and using for fuel or wasting, must be exhausted.[5]

Here was twentieth-century thinking from a nineteenth-century mind. Two insights stand out: The first is that of an old Earth, an idea Increase had boldly expressed in his early Milwaukee lectures. The second is that some of Earth's resources, including coal and petroleum, could be depleted. By contrast, forests, if properly managed for sustained yield, could be inexhaustible.

The report told how trees and forests tempered winds, supported wildlife, purified the air, enriched the soil, moderated stream flow, and altered climate. It recommended the best trees for cultivation, shelter belts, lumber, and fuel. Yet the commission's report had little influence in slowing the deforestation of northern Wisconsin. The felling of trees continued unchecked into the early twentieth century when the major lumber companies finished the job and moved west. However, ideas similar to Increase's began to flourish thirty-five years after his writing and led to millions of acres being set aside as national forests under the administration of President Theodore Roosevelt. New practices of selective cutting and reforestation, begun in the early twentieth century, mandated by government and endorsed by lumber companies, restored forests to the barren north of Wisconsin.

Other significant milestones shaped Increase Lapham's later scientific work. Chief among these was his relaxed acceptance at last of Louis Agassiz's theory of continental glaciation, revealed in a quiet way in a routine journal entry, after years of resisting it. On June 24, 1868, Increase had gone out of Kilbourn City, today's Wisconsin Dells, to work on a geological puzzle. The town had grown up where the bridge of Byron Kilbourn's La Crosse and Milwaukee Railroad crossed the Wisconsin River, and had taken Kilbourn's name. Increase wanted to check out an area called Dell Prairie, which he knew was underlain with sandstone but where local businessmen were burning limestone to get quicklime.[6] Why would anyone build lime kilns in an area of sandstone bedrock rather than limestone? "I this day visited the locality, and found limestone indeed, but only in the form of boulders connected with the 'Drift,'" he wrote. He then traced the drift northward about twelve miles to Eagle Mound "where this rock is still found in place and from whence doubtless the *glaciers have removed the boulders* now found so abundantly in the drift," he wrote in his journal (emphasis added).[7]

In this maddening way, Increase revealed at last that Wisconsin's odd and variable surface, its lakes, serpentine hills and potholes, the drift, the scratched or polished bedrock surface, the desk-sized erratic granite blocks perched atop hills of gravel, the location of rivers, the layers of yellow clay that made the Cream City brick of Milwaukee, the Kettle Moraine, Lapham Peak, even the shape, depth, and very existence of Lakes Superior and Michigan, could only be understood by envisioning a geologically recent period of continental glaciation.

Now fifty-seven, Increase had struggled with glacial theory for more than twenty years. The nation's leading geologists had been debating the theory since Louis Agassiz published it in Switzerland in 1839, a debate that had intensified when Agassiz applied it to the Great Lakes region in 1850. With Agassiz's long-standing scientific leadership in the United States and with the United States' landscape so obviously a natural laboratory for glacial theory, much of the debate had taken place right in front of Increase, at AAAS meetings, in informal conversations with peers, and in scientific journals, including Silliman's, which Increase read closely. Coming up on Increase's heels were younger men, college-trained geologists, who recognized the logic of the theory and the elegant way it explained so much. At last Increase joined the growing consensus without fanfare, without attracting attention to his subtle abandonment of "diluvial" to "glacial" to explain much of Wisconsin's surface. All it took was one word to reveal that he had come to terms with the Ice Age, one of the nineteenth century's most important earth-science discoveries.

Over the winter of 1868 to 1869, another geological curiosity took hold of Increase's imagination. That winter, Louis Korb, a Washington County

This curious *carte de visite* from Increase's collection, ca. mid-1860s, shows two views of Louis Agassiz.

farmer, came into Milwaukee carrying a sixteen-pound metallic rock that he had unearthed while plowing forty acres on his farm in the town of Trenton, just southeast of West Bend. The rock caught the attention of Carl H. Doerflinger, an active member of the German Natural History Society of Wisconsin. After some rudimentary polishing of the rock revealed characteristic geometric patterns called Widmannstätten markings, Doerflinger recognized it as a fragment of a meteorite.

Doerflinger and Increase were colleagues in collecting. At Doerflinger's urging and noting Increase's "many valuable services to the Natural History Society" the board of directors resolved to present Increase with a piece of the meteorite "not to exceed two pounds in weight."[8] Delighted, Increase immediately sent his specimen to be analyzed by chemist John Lawrence Smith of Louisville, who was assembling what was to become the largest meteorite collection in the country. "Please cut the specimen in two equal parts and return one of them to me, retaining the other for yourself," Increase told Smith on December 22, 1868.[9]

On January 4, 1869, Increase told Smith that he had examined another fragment of the Trenton meteorite that weighed sixty-two pounds. "When you have sufficiently studied the matter please make a 'report' for Silliman's Journal," he wrote. Smith complied with an article in the *American Journal of Science and Arts* on March 18. In his analysis, Smith had noticed, in addition to already-named Widmannstätten markings, smaller, more complex markings that he called "Laphamite Markings."[10]

In a lasting gesture, a portion of the meteorite became the main prop for a stereoscopic portrait of Increase Lapham, in which he is shown peering at its surface through a magnifying glass. He sent the portrait to several other scientists, asking for their *cartes de visite* in return to add to his collection. The meteorite eventually became part of the collection of the Milwaukee Public Museum, where in 1883 Doerflinger was named the museum's first director.

—| |—

The end of the Civil War rekindled opportunities for one of Increase's favored causes, a national storm warning system. The scattered scientists who agitated for a weather tracking network before being interrupted by

In this carefully staged stereoscopic portrait, Increase peers through a magnifying glass at a sizable chunk of the Trenton meteorite. Its odd geometrical patterns came to be called "Laphamite markings."

the war quickly resumed their efforts when the war ended. In a letter dated August 25, 1869, Cleveland Abbe, director of the Cincinnati Observatory, told Increase that his city's Chamber of Commerce had asked for a daily weather bulletin to be compiled from telegraphic dispatches between 10 and 11 a.m. each day. Abbe would "volunteer [his] own services without remuneration" for the first three months, and the telegraph companies would offer "the use of their wires" for the same period. Abbe urged others in the "experiment . . . to furnish us with regular reports."[11]

Increase and Abbe were still in the organizing phase of the Cincinnati experiment when a disastrous storm swept over the Great Lakes. It began when the wind shifted to the northwest on the evening of November 16, 1869. When the tempest passed on November 19, some ninety-seven vessels had been wrecked, thirty-five of them total losses, precious cargo was lost, and scores of sailors drowned. For number of boats lost, it remains the worst Great Lakes storm in history.[12]

Increase quickly made the storm the focus of his efforts to establish a national warning system. On December 3, 1869, Edward D. Holton, a wealthy Milwaukee businessman, addressed the second annual meeting of the National Board of Trade in Richmond, Virginia:

> I hold in my hand a communication from Increase Allen Lapham of my city, a man eminent in science.... This learned gentleman claims that it is within the scope of science to communicate to ship masters and navigators a knowledge of the approach of storms, hours before they reach a certain point. His communication embraces a resolution which I will read,
>
> "RESOLVED, That it is expedient to inaugurate in the United States a system of Meteorological observations by communicating telegraphic information of the occurrence of destructive storms and winds, thus preventing much of the present loss of life and property upon the ocean and lakes."[13]

The executive council of the Board of Trade approved the resolution.

Increase kept busy spreading the word. For the *Bureau*, a Chicago journal, he wrote an article proposing that a national weather service be operated by the Chicago Academy of Sciences, of which he was a member. He wrote a long memorial to Congress describing the rationale for a storm warning system, which he forwarded to Halbert E. Paine, congressman from Milwaukee, who enthusiastically agreed with the premise. On December 14, Paine introduced a bill calling for the creation of a national storm warning system.

After Paine sought advice from Joseph Henry of the Smithsonian and others, he introduced an improved version of the bill on February 2. It passed both the House of Representatives and the Senate and was signed by President Ulysses S. Grant in a week. Less than three months after the disastrous storm of November and after thirty years of effort by the scientific community, the United States had authorized a storm warning service.[14]

Increase took advantage of the months between the bill's passage and the formal beginning of the weather service to travel. The reason for his first trip was professional. He would consult for the Monitor Consolidated Mining Co. in Monitor, California. For Increase, who reveled in seeing new places and landscapes, it was the trip of a lifetime—and the farthest he had been away from home since he settled in Milwaukee almost thirty-five years before.

Increase left Milwaukee April 11, 1870, bound for San Francisco. He crossed the Great Plains aboard the Union Pacific railroad, the trans-continental line that was completed with great fanfare a year earlier with the driving of the golden spike at Promontory Point, Utah. His daughters prepared a big basket of food to sustain him on the train and he seems to have gloried in the journey, writing happily about his travels to his children. He penned long letters to each, starting with the oldest, knowing that each letter would be shared with all his children.[15]

On April 14, he told Mary that he saw several returning Californians, "readily known by their self satisfied air, their large gold chains, and the number of rings upon the ladies' fingers."[16] The next day, he told Julia that his train was detained by a derailed freight train six miles ahead. Indians were suspected of having removed a rail and soldiers were guarding the line. "This will make little difference with the basket, for I am just invited to a supper of speckled Rocky Mountain trout."[17]

He visited silver mines at Virginia City, Carson City, and Genoa, Nevada, and then across the border at Monitor, California, where he was to assess prospects for the Monitor Consolidated Mining Co. Luckily he took a stagecoach. "Had we come by rail from Carson City, as we intended, we should have had a smash up. You will see an account of this accident in the paper I sent from Carson," he told Henry on April 21.[18] Still at Monitor on April 25, he wrote thirteen-year-old Charles, "No railroads here to tell you about except ore trains, which struggle on steep grades."[19]

Increase gathered mineral specimens and wildflowers peculiar to Steamboat Springs, a boiling, sulfur-rich spring between Carson City and Reno. "A long narrow crack far down which we can hear the roar and gurgling of the boiling water and can hear and feel the hot steam arising from it, is a modern example of the way ancient mineral veins have been formed and filled," he told Julia. "I hope this will find you all well—for all is well with me," he closed.[20]

After a brief stopover in Sacramento, Increase went on to San Francisco, where he renewed his acquaintance with Josiah D. Whitney, California's state geologist and a veteran of three geological expeditions in Wisconsin, including one with the Carr-Hall-Daniels survey for which he assessed the lead-mining district. "Some parts of this town," Increase wrote Charles, "are filled with Chinamen, with signs in their language, and stores filled with such goods as they use; queer enough, with their long tails [braids] hanging almost to the ground."[21]

On May 17, Increase wrote Mary from the "Interior of Iowa on the train." "Having a grand time riding through this beautiful state," he said, "seemingly more beautiful on account of the contrast with the wild sage brush of the continental interior."[22]

Arriving home, he finished a report for the mining company by May 25. But he could not stay home. He took field trips in Wisconsin in June, July, and August, twice to Baraboo and once to Eau Claire and Chippewa Falls. He wrote long notes about the origin of the Baraboo Hills quartzite, challenging earlier conclusions of both James Hall and James Percival. He wrote a paper based on the observations of tunnel contractors called "On the Discovery of Fish Remains in Glacial Blue Clay of the New Chicago Tunnel under Lake Michigan," which appeared with a sketch in the *Milwaukee Evening Wisconsin*. "So far as is at present known this is the first indication of the actual existence of animal life of this part of the world during the deposition of the glacial age," he concluded, revealing his growing comfort with glacial science.[23]

Eventually the storm warning service was in place and Increase's months of travel and writing came to an end. The storm warning service was lodged in the army's Signal Corps, headed by Brigadier General Albert J. Myer. It had taken until November 1 for the Signal Corps to set up the observer network stations. On Tuesday, November 8, General Myer met with Increase in Chicago, hiring him for a handsome two thousand dollars a year, and on that very day Increase issued the nation's first weather forecast. It said:

To observers along the lakes (bulletin this at once):

Noon, Chicago, November 8, 1870. — A high wind all day yesterday at Cheyenne and Omaha. A very high wind reported this morning at Omaha. Barometer falling, with high wind at Chicago and Milwaukee today. Barometer falling and thermometer rising at Chicago, Detroit, Toledo, Cleveland, Buffalo and Rochester. High winds probable along the lakes.[24]

Increase's mandate as assistant to the chief signal officer was to compile a daily morning forecast from information telegraphed into Chicago from a network of observers. Then he was to wire it to Myer's office in Washington, DC, as well as to the other observers. Ship captains under way could receive information through signal flags. Almost at once, newspapers began publishing the forecast on a daily basis, a practice that soon became widespread.

At first Increase was pleased with his new job in an office in downtown Chicago. "Things are going on flourishingly here," he wrote Julia on November 27.

I get down town every morning in time to make up a skeleton map of the weather over the whole country from Boston to Cheyenne and from Duluth to Key West. This morning I telegraphed to Gen. Myer at Washington 'a warm current (of air) coming up the Mississippi— cold one coming over the plains' (east of the Rocky Mountains). After noon I have a little time—but scarcely enough to make any visits. Hence I have scarcely entered upon the formidable list you made for me. Look for me at home in the middle or last of this week.[25]

His first monthly paycheck arrived on time and on December 8, he again wrote Julia: "I send inclosed the first considerable sum I have ever received as salary for any scientific work, being a draft for my pay for November."[26]

The work was demanding. The system was unreliable. Poles and wires that carried the warnings could be toppled by the same winds that were the subject of the warnings. So on some days Increase did not receive a complete weather picture. Some data came in from instruments that obviously were not calibrated and Increase spent much of his time writing to observers to adjust their settings. In Chicago, Increase was separated from his library at home, on which he depended heavily for method and theory. In his December 8 letter, he told Julia to ask Mary to send him "the old book of Logarithms, which she will find on the lower shelf at her left hand upon entering the library from the garden door."[27] And the army operated with rigid rules. Increase signed some of his forecasts "Assistant Chief Signal Officer," which resulted in an admonition from a Captain Mallery that his proper title was "Assistant *to the* Chief Signal Officer."[28]

At age fifty-nine, Increase had lived his adult life as a free spirit who set his own schedule, and while he loved to travel, he functioned best in the warmth of his family at home. So from the start, he would take the train up to Milwaukee for a day or two midweek. Soon he simply declared to General Myer that he intended to move his weather office to Milwaukee. Again he was admonished, this time by General Myer himself, who wrote on December 21, 1870, that "[t]he place of the assistant is established at Chicago for especial reason and cannot under any circumstances be removed."[29]

But circumstances had already overruled the general. Five days earlier, on December 16, 1870, Byron Kilbourn died at the age of sixty-nine at his home in Jacksonville, Florida, where he had lived since 1868. When Increase learned the news on December 28, he immediately telegraphed General Myer requesting a leave of absence for the entire month. He followed his telegram with a letter written the same day.

> My business and personal relations with Hon. Byron Kilbourn (dating as far back as 1827) . . . are such that it becomes necessary for me to go to that place without delay. The administration of his estate, which is quite large, both in Wisconsin and in Florida falls upon me by his will and I am advised that his affairs at Jacksonville require immediate attention. . . .
>
> As this is a season when there is but little navigation of the lakes, my presence here seems to be of minimum importance.[30]

Byron Kilbourn had engineered Increase's move from Ohio to the frontier village of Milwaukee in 1836; he had employed Increase as chief engineer of his canal company, and they had been colleagues in several public endeavors as the village grew to a great city. While the men respected each other, they were wholly different, politically, economically, in ambition, and in their visions of what constituted public service. But Kilbourn was always generous to Increase. Having chosen Increase to be executor of his estate, Byron now posthumously treated Increase to a trip to Florida. Judging from his letters home, Increase enjoyed every minute of it.

He left Milwaukee on January 16, 1871, and arrived at Jacksonville, Florida, on January 23, having taken a number of trains through Chicago, Louisville, Nashville, Atlanta, and Savannah. He found that Kilbourn and his mother had been living on the St. Johns River twenty miles north of Jacksonville, where Increase now was "nicely cared for at this beautiful spot," he told Julia in a letter written the day after he arrived.[31]

He wrote Henry the next day: "I have had a delightful time so far; weather like our June!... Please take the river observations regularly. Love to all."[32]

Then a letter to Seneca on January 27:

Day before yesterday I went up Black Creek to its tributary, Peter's Creek. I saw yellow pond lilies and wild violets in flower. I went this morning in a rowboat to Mrs. Mitchell's fine Florida home. It looks rough yet, but will soon be a little Paradise of a place, especially if she succeeds in her attempt to make a grass lawn.... Picked ripe straw-berries and yellow jessamine in blossom today.[33]

From the sixteenth-century town of St. Augustine, he wrote Mary on January 30 that he was at the "most southerly point in my travel....I am at the St. Augustine hotel, room 19, overlooking much of the city and the sea, have rambled through the quaint old crooked streets.... The old Fort, the Governor's house, the city gate and other places of interest I will show you in stereoscope on my return."[34]

On his way home, he took a side trip to Washington, DC, where he visited Joseph Henry, Congressman Halbert Paine, and other acquaintances. On the evening of February 2, he had dinner with Chief Signal Officer General Myer and the next day wrote a jubilant letter to Julia: "I dined last evening with Gen'l Myer at his home on I Street. Have arranged matters satisfactorily—am not to be ordered to Chicago any more."[35]

Increase no longer compiled the daily forecast, but he continued to send regular weather data to Joseph Henry of the Smithsonian and to the Lake Survey in Detroit. He helped devise methods of uniform reporting for the network of weather observers. And he wrote special reports for Myer, who continued to expand the weather reporting system nationwide.

In May, Increase's newfound prosperity enabled him to buy a farm on the south shore of Lake Oconomowoc for Henry and Julia. Meantime, Mary would oversee the Poplar Street household, which included Increase, his widowed mother Rachel, and sons Charles and Seneca. The Oconomowoc property was on the beautiful sheet of water Increase first saw when he was surveying for the Milwaukee and Rock River Canal. A clear spring rose out of the shallow bedrock on the property and flowed into the lake.

Increase referred to the spring by an Indian word, Minnewoc, or "place of waters," a name applied thereafter to the farm itself.[36] Minnewoc remained in the Lapham family until 1888 when Julia sold it to a wealthy Chicago industrialist, George Bullen, who built a lakeshore mansion of limestone quarried on the property. He too called the estate Minnewoc.

And it gave Increase the excuse to create another map, one of his last. A labor of both love and commerce, it was called "Map of the Lakes & Drives Around Oconomowoc, Wisconsin: A Favorite Summer Resort." It showed the major lakes of Waukesha County, connecting streams, locations of roads, limestone outcrops, a cheese factory, important residences, mills, the route of an 1853 tornado, and, just a bit west of center, the location of the new Lapham home on Oconomowoc Lake. When it was finished, he sent bundles of them to daughter Julia and son Henry with instructions to distribute them to retailers.[37]

—|—

"Burnt out," exclaimed Increase's brother-in-law, Samuel Stone, in a letter sent from Chicago on October 12, 1871. "Saved some clothing and furniture. Suffered some. Slept outside of city limits; extreme N. W."[38] After an unusually dry late summer and fall, fire broke out in the city of Chicago on Sunday, October 8. A strong southwest wind pushed it north and east through the very heart of Chicago. Before it burned out on the morning of October 10, it had consumed eighteen thousand mostly wooden buildings in four square miles, killed between two hundred and three hundred people, and rendered one hundred thousand people homeless, including Stone and his daughter Elizabeth, who were sheltered by friends north of the fire zone.

Stone wrote Increase that "you probably know more of present Chicago than we. I have not seen a newspaper since Sunday morning. Thousand reports of the fire, not all to be credited. Many are probably burnt. I was the last person left in the Hist Socy building. Thought I should be burnt."[39] The Chicago Historical Society building also housed the offices of the Chicago Academy of Sciences. Building, collections, library, manuscripts—all were lost.

A day later Stone wrote that he was out of money and asked Increase to send him two hundred dollars in cash for which Stone sent a check to be held by Increase "until you hear when the Bank opens."[40] Increase sent the money. Increase added that he had heard from a friend "regretting death by suffocation of Col. Stone and his daughter. Quite a number of persons have called to know the truth of the report. You must have had a narrow escape."[41]

Unnoticed at the time because of a lack of newspapers and other communications in the far north, the southwest gale simultaneously pushed a larger, more lethal fire through the brittle-dry pine forest of northeastern Wisconsin. A genuine firestorm, it burned 1,875 square miles and killed up to 1,500 people in the little town of Peshtigo and other communities bordering both sides of Green Bay. Before the month was out, General Myer asked Increase to write a report about the weather aspects of the "great fires of the northwest." Increase responded before the end of 1871 with a long essay that was eventually published by the federal government in 1873.[42]

In the essay Increase discussed the drought that preceded the fires and how wind can convert sparks into conflagrations in minutes. He dismissed the theory that the Chicago fire ended with a downpour. The rain that fell as the fire diminished was gentle, amounting to a few hundredths of an inch. The fire ended when it ran out of combustible material in its path, he said. This neither proved nor disproved James P. Espy's theory that great fires produced great rainstorms. They might under conditions of little or no wind but could not with the kind of wind that drove the Chicago fire, Increase concluded.

By now, Increase, sixty years old, was trying to shed some of his regular workload. On December 1, 1871, he wrote to General Cyrus B. Comstock, superintendent of the US Lake Survey in Detroit:

> Having made meteorological observations for half a life time I begin to think that I have done my full share of such work, and must beg that I may after the 1st of Jan'y next be allowed to send you monthly a copy of the very careful observations, made here, under favorable circumstances, with the same kinds of instruments, for the Signal Service.[43]

On January 5, 1872, Increase left Milwaukee's deep snow for Florida to work on Kilbourn's estate, but by January 20, he still had not spent any time on the Kilbourn property. On his way to the Kilbourn house, he spent a night at "Mrs. Mitchell's Cottage," he told Mary in a letter written January 19. It was a cold winter that reached far south, with morning frost even in Jacksonville, "but with a sun so hot as to make my face as red as a toper's."[44] He was home again by January 29.

Again, he was shedding old routines. On February 7, Joseph Henry wrote from the Smithsonian:

> We regret to learn of your withdrawal from the active duties of meteorological observer to this Institution for the city of Milwaukee, but trust that you will not allow your interest in the science you have so long and faithfully labored to advance, to abate in the least. It will always give us pleasure to hear from you as well as to render you any service in our power.[45]

And finally, on May 11, Lieutenant Howgate wrote:

> I am directed to inform you that in view of the limited amount of money at the disposal of this office for the current year and the consequent necessity for the reduction of its working expenses, the ar-

rangement by which your valuable services have been secured to the Chief Signal Officer will terminate at the close of the present month. It is hoped that the Office will be in sufficient funds by that date to liquidate your account in full.[46]

Increase was out of the weather business, except for one small project. On May 31, 1872, he promised General Myer that he would finish "a full list of the disasters upon the lakes from January 1st to this date" in a few days.[47]

On June 24, 1872, Increase, now sixty-one, accepted his last job as consulting geologist. He accompanied Nelson P. Hulst, mine manager for the Milwaukee Iron Company, and William D. Van Dyke, age sixteen, the son of the president of Northwestern Mutual Life Insurance Company, to the Upper Peninsula's iron district. Their train north took them through the devastation caused by the deadly Peshtigo fire months earlier. "Nearly all the way from Green Bay to Marinette we saw along the rail road the dreadful effects of the great tornado of fire in October last. The trees blown down and partially burned, lying in all directions. All the houses, depots and other buildings are new having been built since that time," he wrote to Mary.[48]

In Menominee, Michigan, they met with T. B. Brooks, Michigan state geologist, and two days later, on June 29, entered the woods despite heavy rain and mosquitoes. On July 1, Increase again wrote Mary from "In the Woods, Breen Iron Mine, Upper Menominee River." "Except a wetting Saturday afternoon and some jolting over new rough roads, we have had a good time," he wrote. "We have rough fare and wear rough clothes... We are feasting upon venison just now and hope to have some fine fish from the neighboring lakes and streams. Will [Van Dyke] enjoys the fun very well and eats pork like a man."[49]

Then they had a couple of bad days. Increase's journal entry for July 2, 1872: "Walked to the foot of Little Sturgeon Falls.... Missed the way home. Laid under a log all night in the rain near a creek. Matches wet, good for nothing, hence no fire. Comfortless night."[50] And on July 3: "Spent the day searching for home! 4 of us... rambled northward to a creek which we followed down to Hamilton Lake from whence we knew our way home! Great relief. Got to camp at 3 p.m. hungry and weary."[51]

Increase wrote cheerfully to Mary about his night in the woods, complaining only of "some soreness of feet which will soon be over with." He remained delighted with the adventure now that it was back on track. "We have fresh fish, corned beef (here called 'red horse') with pork, bread, ginger bread, and other good things to eat; good comfortable bunks to sleep in; plenty of fresh air to breathe; and the finest fresh cold spring water (temp. 43 degrees) to drink."[52] Of course, he recorded the temperature of the drinking water.

There were to be twenty more days of exploring both the Wisconsin and Michigan sides of the Menominee River, the explorers transporting themselves in Indian bark canoes on water and walking on land. Finally, on July 23, Increase returned home, but it was just a respite. Just two days later he met with Nelson Hulst at his factory to sort ore and rock specimens from upper Michigan. Based on Increase's favorable report, the Milwaukee Iron Company bought an interest in Breen mine and established a railroad there.

In 1896, Hulst reminisced in a letter to Mary about the party's night in the woods:

> Without shelter from the pitiless rain, and tormented by swarms of gnats, black flies and mosquitoes, there was neither rest nor sleep for us. Our sympathy all went for your father who bore the hardships without a murmur of complaint....
>
> I shall always look back to this trip with your father, except for its hardship to him, with intense pleasure. Companionship with him was an inspiration as well as instruction. His enthusiasm over finding this or that rock exposure or float mineral arouses one's ambition. No wild flower in his path missed his gaze. He was always cheerful, even when there was occasion for despondency, and his gentle lovable manner drew us all to him.[53]

—|—

Though he was slowing down, Increase was not yet done with his scientific pursuits or with offering recommendations. With the purchase of Minnewoc, which was an easy train ride from Milwaukee, Lake Oconomowoc became a new science laboratory for Increase. Using a rowboat on his visits to Julia and Henry, he charted its depth; he analyzed its water; he described its shoreline; he listed its fish species. He was troubled by the practice of overfishing, including commercial market fishing, which in just forty years had already noticeably reduced fish production in Wisconsin's lakes and streams.

On October 29, 1872, he told Wisconsin Governor Cadwallader C. Washburn that state government should consider supporting fish propagation in the lakes, as well as the introduction of new species and "the means necessary to prevent the destruction of our present fish supply. Our small lakes and rivers may be important sources of food supply."[54] In this, Increase was joining a national movement. Congress established a US Commission of Fish and Fisheries in 1871, and Increase's longtime correspondent, Spencer F. Baird, assistant secretary of the Smithsonian, became its first commissioner.

On October 30, Increase wrote to Baird in Washington that he and J. W. Milner, deputy US fish commissioner, had just visited a trout farm in Summit owned by Hercules F. Dousman. "I hope Mr. Milner will succeed in his efforts to introduce new edible fish into our waters and to increase the now rapidly diminishing supply of whitefish, which may have been deemed the fish of the people, both rich and poor."[55] Always one to look for a local solution, he recommended to Baird that stocked ponds use food that is natural to the fish, "*Alburnus Acutus*" (sharp-tailed minnow) that could be taken from under the ice of the Milwaukee River each spring "by thousands."[56]

The matter was taken to the state assembly and passed in March 1873 by an overwhelming majority, with five hundred dollars appropriated for the purpose. Soon Milner arranged a shipment of salmon eggs to Dousman, with Increase receiving the shipment at Milwaukee and taking pains to ensure that it arrived safely at Dousman's fish hatchery at Waterville. Thus began Wisconsin's continuing program to stock fish in public waters.

The 1860s had begun with war and economic hardship for the Laphams; the decade ended with many successes, including a national storm warning system, much enjoyable travel for Increase, and financial stability. Now, with his children having reached adulthood healthy and settled, or nearly so, the scientist seemed to be slowing down. But the new decade brought one last great challenge and an enormous setback that, ironically, would add greatly to his legacy.

CHAPTER 20

A Complete Survey

———┤ 1873–1877 ├———

O N FEBRUARY 13, 1870, INCREASE WROTE a cryptic note in his journal: "Went to Madison and on the 16th assisted in organizing the Wisconsin Academy of Sciences, Arts and Letters."[1] Increase and more than one hundred other Wisconsin men responded to an invitation from John Wesley Hoyt, secretary of the Wisconsin State Agricultural Society, to attend a meeting to organize a state academy of sciences. Included were old friends such as Philo Hoy and upcoming young scientists including Thomas C. Chamberlin, then an instructor in natural history at Whitewater State Normal School, and Roland D. Irving, the first professor of geology at the University of Wisconsin.

Hoy's and Lapham's memberships in the American Association for the Advancement of Science and Lapham's correspondence with Asa Gray, Louis Agassiz, Benjamin Silliman, and others "brought to Wisconsin's Academy a working knowledge of other academies and an acquaintance with the broader horizons of American science."[2] Hoyt became the first Wisconsin Academy president while Lapham was general secretary and editor of the academy's *Transactions*.[3] Hoy was vice president for science, and Chamberlin became a counselor.

The academy's founders wrote a charter that listed six objectives in carrying out its larger mission, which was nothing less than the material, intellectual, and social advancement of the state and the advancement of science, literature, and the arts. The six objectives were included in the first president's report to Governor Cadwallader C. Washburn in 1872, and the second of these objectives had a familiar ring: "A progressive and thorough scientific survey of the State, with a view to determine its mineral, agricultural and other resources."[4] Fully aware of the shortcomings of Wisconsin's earlier surveys, more than one hundred of the state's leading citizens assembled as an academy called for a fifth—and comprehensive— Wisconsin Geological Survey.

The group didn't take long to press its cause. On December 5, 1872, Thomas Chamberlin, the young, ambitious professor of natural sciences at the State Normal School in Whitewater and Increase's colleague in the Wisconsin Academy of Sciences, Arts and Letters, wrote to Increase:

> I have been agitating the subject of a geological survey of the state & so far as I have yet learned the prospect is favorable.... But of course the first thing is to convince the state that the survey if inaugurated will be a creditable one that will justify the necessary expenditure.... Now the important question is would you take charge of the survey if the Act establishing it and the compensation were satisfactory...? Would an act naming you as Director and giving you two assistants ...one to be selected from the University...and the other from the Normal schools...meet your approval?[5]

For the first time, Increase had the opportunity to become head of a survey.

On March 19, 1873, the Wisconsin Legislature, responding to pressure from Chamberlin, Lapham, and others, approved an act calling for "a thorough and complete mineralogical and agricultural survey of the state."[6] It appropriated thirteen thousand dollars a year and said it should be finished in four years.[7] On April 7, Governor Washburn invited Increase to a meeting in Madison that included, among others, the president of the University of Wisconsin and professors Roland D. Irving, geology, and W. W. Daniells, chemistry. They discussed the organization of the fifth state geological survey. "In further conference with the Governor, he intimated that he would commission me chief of the survey with a salary of $2,500," Increase wrote in his journal.[8] This was a handsome sum in a time when the average US household earned well below a thousand dollars a year.

This began Increase Lapham's most important legacy as a geologist. His appointment became effective on April 10. His first act was to recommend three assistant geologists to the governor: Irving, a graduate of Columbia's School of Mines, whose twenty-sixth birthday was days away; Moses M. Strong Jr. of Mineral Point, a Yale graduate and civil engineer, soon to be twenty-seven; and Chamberlin, twenty-nine, professor of natural science at Wisconsin State Normal School, Whitewater, who would soon join the faculty of Beloit College, his alma mater.

No cronies here. With this trio of ambitious, college-trained scientists as his assistants, Lapham confirmed his conviction that the future of science lay in specialization and not in the vanishing breed of all-around generalists of which he was the ideal model. He also showed that he was at once realistic, modest, and wise. This aging man who had tramped every corner of Wisconsin and was the recognized fountain of knowledge about its natural attributes also knew more than anyone then living wherein lay the gaps of that knowledge. Now that he was head of a well-funded Wisconsin Geological and Natural History Survey, he would be able to fill some of those gaps—not personally, but through a cadre of eager young scientists more physically and educationally suited to the task. As it turned

The members of the Geological Survey under Increase Lapham. From left to right, geologist Thomas Chamberlin, mining engineer Moses M. Strong Jr., and geologist Roland Irving.

BELOIT COLLEGE ARCHIVES

WHI IMAGE ID 45363

out, this act of appointing young specialists, his first act as state geologist, became his most important contribution to the survey and the one that assured its enormous success.

Lapham's acceptance of Chamberlin was based on their mutual interest in the survey. As founding members of the academy, both had lobbied for it.[9] Irving had contacted Lapham shortly after coming to Wisconsin in 1870 to join the University of Wisconsin's faculty, asking for his judgment of the age of the Baraboo quartzite. Lapham's replies have been lost but these and later Irving letters reveal the growing congeniality of the correspondence.[10] Strong was the son of Moses M. Strong, a pioneer lawyer from Mineral Point and a force in Democratic politics. After graduating from Yale, the younger Strong studied for two years in Germany before returning to Wisconsin as a mining engineer.

In May, Chief Geologist Lapham mailed formal assignments covering the first season to his three assistants. Each letter was a model of military-like precision and clarity. Obviously, Increase intended to build an impeccable, formal record of the survey. The letter to Roland Irving began:

> The Governor having commissioned you as one of the Assistants in the Geological Survey of Wisconsin you will proceed to organize a party and supply yourselves with the necessary outfit and instruments to explore the iron and copper ranges in Ashland County, commencing the field work as soon after the first of June as possible.
>
> It will be your duty, besides examining the iron and copper ores and their relation to the adjoining rock formations, with a view to discoveries in other districts, to note also all facts throwing any light upon any of the special matters required to be considered in the law authorizing the survey.[11]

WHI IMAGE ID 102203

Leaving much to Irving's personal judgment, Increase nonetheless suggested some investigations of "special importance," including the age of certain rocks of interest to miners, the fertility of soils, and the nature of drift. Conceding that it was impossible to visit every square mile in the county, Irving was to extend his work west into Douglas County. "Two months' time will probably be sufficient for the examination of these two counties, leaving time to visit Black River Falls, Ironton and such other points as may be deemed advisable before the close of the working season," he wrote.[12]

Upon returning to Madison, Irving was to work with Professor W. W. Daniells in analyzing the chemistry of specimens of ores, minerals, and clays. Then he was to write a full report by December 1 and expect it to be published under his name as part of the survey's annual report.

Lapham instructed Chamberlin to explore all of Wisconsin south of Shawano County and west of the Niagara limestone of eastern Wisconsin, excepting the lead-mining region. The eastern part of his range would put Chamberlin in the Kettle Moraine. This proved fortunate. It was in the Kettle Moraine that Chamberlin acquired the insights that were to establish him as the most influential American glacial geologist of his time.

Strong was to revisit the lead-mining district. Lapham suggested that he define his district by drawing a north-south line from the Illinois border to about thirty miles north of the Wisconsin River and an east-west line through Grant, Iowa, and Green Counties, with the lines intersecting at Mineral Point.

As might be expected, complications arose in managing a survey that was much larger in mission, cost, and workload than its predecessors. Irving experienced serious health problems starting in July and Lapham appointed Elias T. Sweet, a recent graduate of the University of Wisconsin, to replace him temporarily in September. Legislators, mine owners, and entrepreneurs, all impatient for results that might return quick profits, demanded immediate and tangible information about mine locations. Perhaps most seriously, the chemist Daniells was overwhelmed with the number of specimens to be analyzed, but Increase believed that the law prevented him from hiring a second chemist.[13]

When the survey's first season ended, Increase finished an article called "On the Relation of the Wisconsin Geological Survey to Agriculture," for *Transactions of the Wisconsin State Agricultural Society*.[14] On Sunday, January 5, 1874, one day before the deadline proscribed by law, Increase delivered the first annual report of the survey to the governor in Madison, but it was not Governor Washburn. That very day, William Robert Taylor, a Reform Democrat, assumed the office, having been elected by an unusual coalition of railroad interests, German anti-temperance voters, and a farm advocacy organization, the Grange.[15]

Lapham's report reviewed the work of all three assistants and their crews, emphasizing in detail Strong's report from the mining district. Strong had found a "wonderful increase of production within the past two or three years" of zinc sulfide ore, or sphalerite, found in association with lead sulfide ore, or galena. The ore had long been thought to be worthless, but with the demand for zinc growing, "owners of abandoned mines find themselves in the possession of unexpected wealth," Lapham wrote.[16] Yet,

> It is to be regretted that the want of a cheap fuel in the lead region prevents the smelting of these ores within our own state. The construction of a railroad from Milwaukee directly to the source of supply of these zinc ores, by cheapening the cost of transportation, might render . . . a business of profit on the shores of Lake Michigan.[17]

Lapham's analysis proved to be shortsighted. What became the nation's largest zinc smelter was built at Mineral Point in the early 1890s. Trains brought in fuel and carried out zinc.

After submitting his report, Increase left Milwaukee with Julia on February 2 for a month-long pleasure trip in the South. At Louisville, Increase led his daughter on a tour of his old "tramping grounds" when he was a boy in Shippingport.[18] They saw Civil War sites and, most enchanting to Increase, climbed Stone Mountain outside Atlanta in a fog. Then the fog lifted and "opened to us a grand panorama of swelling hills, yellow farm enclosures, green pine forests, with here and there the smoke of a village, or the steam of a passing railroad train, creeping leisurely along."[19]

From Atlanta, the pair took a sleeping car to Washington, where Increase casually dropped into his journal one of his most tantalizing phrases. "Went to the Smithsonian, White House, saw President Grant; to the Treasury department and other places of interest."[20] That's two visits to two presidents in fourteen years, described in three words each, neither further elucidated. "Visited Abraham Lincoln" in 1860. "Saw President Grant" in 1874. Father and daughter arrived home on March 4.

On April 16, Increase wrote to Samuel Stone: "I am now perfecting arrangements for the continuance of our geological survey which will be prosecuted by three or four parties in as many different parts of the state."[21] He said he hoped to get Thomas Benton Brooks, the Michigan state geologist whom he'd met on his upper Michigan trip, to return to the iron region of the Menomonee River and report on it for the survey.

Increase delivered the second annual report in January 1875. Again Chamberlin, Irving, and Strong led crews into the field for another season. In addition, "Maj. T. B. Brooks, late of the Michigan Geological Survey," and his crew surveyed the Menomonee River iron district. "I am glad to be able to state that the survey is progressing with reasonable rapidity; the amount of work done being considerably in excess of that of 1873," Increase wrote. Then, somewhat defensively, he acknowledged that his first report lacked details because his three field assistants ran out of time. Now their 1873 accounts were included as "supplementary reports."[22] Other than that, there was not a hint of trouble.

But lightning struck on a February day in 1875. Governor Taylor removed Lapham from his position as chief geologist, replacing him with Orlando Williams Wight of Oconomowoc and later of Milwaukee. Wight was a Yankee fluent since childhood in Greek and Latin, translator of classics, writer of history, philosopher, theologian, esteemed medical doctor, lawyer, and much else.[23] But he was no geologist. His sole qualification for the job was that he had devoted 1873 to campaigning for the Reform Democrats and Governor Taylor.

Increase might have learned about his dismissal first in the newspaper. The reason that it could happen at all was that, through an oversight, Governor Washburn had neglected to send his name to the Senate for confirmation.

The first of a blizzard of angry letters protesting this cruel treatment of Increase came from Horace A. Tenney, Madison newspaper editor, Republican politician, and assistant state geologist in the 1853 survey.

"Outrage," "indignation," "disgust," "abhorrence," "shameful," "misfortune," "regret," "disaster," "legislative wrong," and "mortified" were among the words used by friends, scientific colleagues, and correspondents.[24] Milwaukee and Madison newspapers ran editorials expressing outrage. Letters of protest poured into the governor's office.

Lapham's son Seneca wrote to his uncle Samuel Stone in Chicago:

> Enclosed I send you some newspaper cuttings &c. to show how the "Reform party" affects father. Father being a republican was obnoxious to governor "Farmer Taylor" and he has appointed a great politician in his place.
>
> The only complaints which have been made against the survey, were by men who had interest in iron mines or mineral springs for which father and his assistants could not conscientiously make a report satisfactory to the owners....
>
> Mr. Paul, the R. R. commissioner, went to the governor as soon as he heard of the appointment, and told him it would hurt the party more than anything he could have done and that he had better withdraw the appointment. The governor said he could not, for Dr. W. [Wight] had been after him day and night ever since he had been in office to give him an appointment, and this was the first chance....
>
> Dr. W. called on father and among other things said he expected to be governor next election and would be willing to have the office return to father! He also asked father to lend him his books on geology, especially geology of Wisconsin, as he knew nothing of the subject except from elementary works.[25]

On February 18, Thomas Chamberlin wrote Increase that he, Irving, and Daniells offered to resign. "What course we shall pursue beyond that and beyond an attempt to vindicate you, is yet uncertain. We await developments." Four days later, Daniells told Increase, "We, Irving and I, were strongly advised not to resign and leave the survey in the hands of such men, but both felt we could not do less."[26]

"Our geological survey has gone the fate of its predecessors—or rather a worse one," Irving wrote James Hall in February. "The governor has appointed a disreputable politician to Dr. Lapham's position, leaving the survey still unorganized. . . . One reason of the trouble was my refusal to call the Penokee ores so rich as Col. Whittlesey makes them to be. Wisconsin has certainly had ill luck with her surveys."[27]

Apparently Governor Taylor, perhaps to avoid further political damage, did not accept the resignations. On April 20, Roland Irving told Increase, "I am to work this year just as planned for me by yourself. We follow indeed your scheme throughout. I hope I have done right in continuing on the survey. It was with no little misgiving that I withdrew my resignation from the governor's hands."[28]

In May, the ultimate scientific judgment was cast upon Lapham's treatment when the prestigious *American Journal of Science and Arts* declared, "A great wrong has been done to Dr. Lapham, and a greater to the State. But we cannot believe that the State of Wisconsin will be satisfied to thus stultify itself before the world by sustaining the appointment to a scientific position of one who confessedly knows nothing of its duties."[29]

Meanwhile, Increase cooperated in the transfer of power, moved into the Oconomowoc farm with his children, and turned his thoughts to other subjects. Throughout, he maintained a quiet dignity.

Wight served only a year as Wisconsin state geologist. Republican Harrison Ludington, former Milwaukee mayor and wealthy business owner, defeated Taylor in the next election and assumed office on January 3, 1876. One of his early actions, on January 19, was to receive Wight's annual report and send it to the Senate. Days later he accepted Wight's resignation. In February, Ludington appointed Chamberlin as state geologist.[30]

Chamberlin, who had a prodigious capacity for work, served as state geologist through 1881. Not only did he return to the field, but he wrote the largest fraction of the 3,100 pages of the four-volume *Geology of Wisconsin*, which was hailed as a model for other surveys. Its success led to Chamberlin's becoming one of the leading scientists of his time. He was to head the Glacial Division of the US Geological Survey, serve as president of the University of Wisconsin, and organize the geology department of the University of Chicago where he was professor from 1892 through 1918. He founded *The Journal of Geology*. He was president of the Geological Society of America; the Wisconsin, Chicago, and Illinois Academies of Science; and the American Association for the Advancement of Science.[31]

His work for the *Geology of Wisconsin* volumes included essays on continental glaciation based on his studies of the Kettle Moraine. The nature of drift revealed to him that several glacial advances and recessions had occurred over thousands of years, all ending with the recession of the Wisconsin glacier starting ten thousand years ago. He saw the directions its lobes moved and he saw the relationship between the lobe that excavated Lake Michigan and its smaller neighbor, the Green Bay lobe, which excavated the bay, Lake Winnebago, and Horicon Marsh and sculpted the land south to Janesville.

He realized that the Kettle Moraine—he invented that term—was an "interlobate" moraine, that is, it was formed between the Lake Michigan and Green Bay lobes as they repeatedly advanced and thawed, laying down the ridges and creating the kames and eskers, outwash plains, and kettles that made the landscape so enticing.[32] And he gave the name "driftless area" to southwestern Wisconsin, explaining that other glacial lobes diverged halfway down western Wisconsin, leaving the southwest untouched as the only preglacial landscape in the state today.

All of this demonstrated that once a scientific hypothesis is shown to be generally correct—in this case Louis Agassiz's—skilled scientists can fill in its details. Chamberlin declared that the Kettle Moraine, which he thought beautiful, also was the world's primary outdoor laboratory of its kind for glacial science. Much of the best of it has been preserved by the State of Wisconsin as the Kettle Moraine State Forest. Today the thousand-mile Ice Age Trail traces the southernmost limits of the Wisconsin glacier from the border with Minnesota at Interstate Park eastward and then south to Janesville, where it makes a grand U-turn and rides the ridges of the Kettle Moraine northeast all the way into Door County.

Revitalized, Strong and Irving remained with the survey and attracted other promising young professionals, including Charles R. Van Hise, a recent graduate of the University of Wisconsin and a student of Irving's.[33] Van Hise was to later serve as president of the university for fifteen years. He became a national voice in the young conservation movement championed by Theodore Roosevelt, and he oversaw the development of "the Wisconsin Idea," under which the University of Wisconsin system reached every corner of the state through its extension service.[34]

Strong died tragically on August 18, 1877, while working for the survey. His party of three was heading by skiff and canoe down the Flambeau River to explore the geology of Chippewa River and its tributaries. Navigating a rapids, a companion went overboard and Strong went in to rescue him and did not resurface. In his written eulogy, Chamberlin lamented a "shadow of a deep loss...thrown across the history of the survey."[35]

The four volumes of the survey were published between 1877 and 1883. In 1876, the legislature determined that volume 1 should be a general explanation of geology, specifically as it applied to Wisconsin, and was to be written in a language laypeople could understand to make it appropriate to distribute to Wisconsin schools and libraries. Chamberlin realized that the law basically demanded that the survey be summarized in the first volume. Thus, he published volume 1 last, in 1883. The first volume he finished, volume 2 in 1877, gave a history of the early years of the survey. In it Chamberlin published Lapham's 1873 and 1874 annual reports along with Wight's 1875 report, although he took pains to point out in footnotes where he disagreed with Wight's conclusions. "The fullest opportunity has been offered the author of this report to revise it for this volume, but not having been accepted, the delicate duty has devolved upon very unwilling hands," he said. "I have deemed myself under obligation to publish everything of a geological nature, even where dissenting from the views presented."[36]

In his introduction to volume 1, Chamberlin wrote in conclusion,

> And now, as with this first which has become last, I lay down, with inexpressible relief, the burden of this work, which has, notwithstanding, been largely a labor of love, I have greatly to regret its imperfections, of which no one can be more painfully conscious than myself. Such as it is, it is presented to the magnanimity of a generous people.[37]

The work Increase had done throughout his life informed the entire survey. Chamberlin and the others cited him often as a reliable source of infor-

---

(right) This drawing by Increase of *Bronteus laphami*, a large trilobite, was based on a fossil found in Niagara limestone near Kaukauna, Wisconsin, and named in honor of Increase after his death.

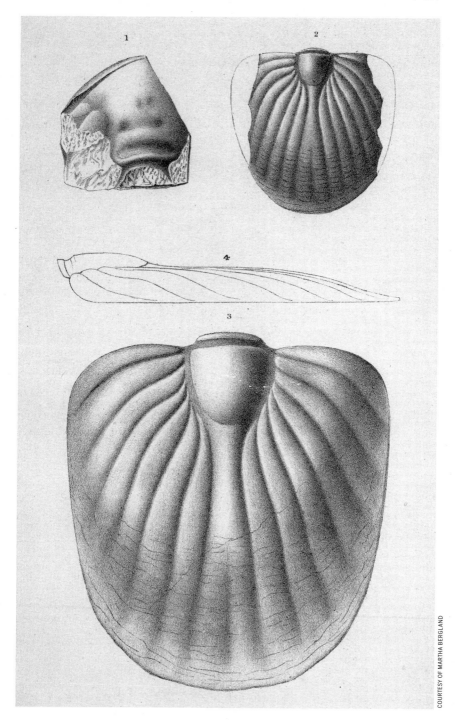

mation, a fossil was named for him, his annual reports are included, and Roland D. Irving reprinted verbatim the original instructions he received from Increase to guide his conduct in the field that first season.

Increase Lapham haunts *Geology of Wisconsin* like a beneficent ghost that moves lightly through its pages, then from time to time comes fully into view to offer a sharp insight before fading again into the background.

CHAPTER 21

So Calm and Peaceful

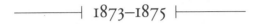

I NCREASE ALLEN LAPHAM DIED ON SEPTEMBER 14, 1875, six months before his sixty-fifth birthday, seven months after he had been dismissed
as Wisconsin state geologist, and more than a year before the first volume
of the Wisconsin Geological Survey was published, the volume that included his first two annual reports to the governor. He did not live to
share in its success, to which he had contributed so much.

While the crass politics that led to his dismissal generated a storm of
public protest that spread far beyond Wisconsin's borders and engulfed
much of the nation's scientific establishment, Increase lived the last
seven months of his life quietly, even serenely, dividing his time between
his old Milwaukee house on Poplar and the Minnewoc farm on Lake
Oconomowoc.

We don't know the source of his calm but his health may have been
paramount. He had collapsed months earlier and his children had been
speaking among themselves about their father's failing heart. And it was
in his character, as we have seen, to set aside personal grief and disappointment and devote his time to useful works, in accordance with the Quaker
imperative that guided him from childhood.

One of his few references to his dismissal came only days after it occurred when he wrote a notably matter-of-fact message to Assistant State
Geologist Moses M. Strong Jr. on February 17, 1875.

I have your letter of the 12th with the manuscript and drawings of pre-historic mounds, which, like all your other work is well done, and highly interesting and valuable.

My successor having been appointed, this will be my last official communication.

Inclosed is my check for your salary for the present month, after which time you are to look elsewhere for such direction as may be needed. Please have everything in readiness for any contingency that may happen.[1]

Strong replied from Mineral Point a day later:

It is with the utmost regret that I learn of the termination of the official relations which have so long and so pleasantly existed between us.

I trust however, that the friendly acquaintance formed during the past two years which I have found both pleasant and instructive may still continue to exist.[2]

As usual, Increase soon turned to other projects. He promised an article on cranberries to F. W. Case of Madison, who wanted to publish it in *Transactions of the Wisconsin Horticultural Society*. He provided information on Milwaukee's Cream City brick to geologist Elias T. Sweet of Madison, who was writing an article for a technical journal. When spring arrived, he recorded outdoor activities in his journal. On April 22 he measured the temperatures of the water of Lake Oconomowoc and Minnewoc Spring, finding that the spring water flowing from the Lapham farm into the lake was 46 degrees Fahrenheit, whereas the lake water was 36 degrees. On May 1 he noted a snowstorm. On May 17 he took soundings and recorded the temperature of nearby Lower Nemahbin Lake.

In May the *American Naturalist* magazine published Increase's article titled "The Law of Embryonic Development the Same in Plants as in Animals." In it, he argued that the embryonic stages of "higher orders of animals" resemble full-grown animals of the "lower orders," and that the same phenomenon could be observed in the world of plants.[3] The paper, only four pages long, was the closest he came to dealing with Charles Darwin's theory of evolution.

That month Increase was asked to contribute to the natural history exhibits at what would become the first American world's fair, the Philadelphia Centennial Celebration in 1876, marking the one hundredth anniversary of the signing in that city of the Declaration of Independence. All states and US territories were invited to help plan and participate in the celebration.

On May 7, in response to a request from his longtime friend Joseph Henry, secretary of the Smithsonian Institution, Increase sent a set of Wisconsin clam and snail shells from his collection to be exhibited at the Centennial celebration and he said more would be shipped later. That very day he received a letter from Elias A. Calkins, a Milwaukee newspaper editor and a member of the Wisconsin Centennial board, asking him to take charge of preparing a display of Wisconsin "soils minerals, &c." for the exhibition.[4]

Increase told Calkins that he could provide the materials and set up a comprehensive geological display for Wisconsin "provided ample funds be supplied to meet all expense... including transportation to and from Philadelphia."[5] Calkins did not follow through and Increase assumed that his role in the Centennial exhibit had been forgotten.

—|—

·On July 26 Increase and Julia joined a party for a visit to Kilbourn City, today's Wisconsin Dells. The "dalles," or rapids, on the Wisconsin River cut through sandstone cliffs on either side, sculpting the stone into magical, striated formations. Increase first explored the place in 1849, before Byron Kilbourn extended the La Crosse and Milwaukee Railroad across a bridge he built over the mighty river and then developed the town that came to be named for him. Kilbourn's railroad later was absorbed into the Milwaukee Road line that carried Increase and Julia to their outing.

Now in 1875, Kilbourn City was the most-visited tourist destination in Wisconsin, due largely to the efforts of H. H. Bennett, a promoter and photographer whose exquisite photographs of cliffs circulated throughout the nation. Bennett and Lapham were acquaintances. Bennett had photographed Increase inspecting the geology of Taylor's Glen in 1869 for a stereoscopic image.

By the time of this visit, which was all for pleasure rather than work, steamboats had replaced rowboats as the vehicles from which to view the river's formations. The group was treated well. "Found some of the citizens had been out before breakfast this morning and built a board walk through 'Witch's Gulch' for the benefit of our party," Increase wrote in his journal. "Also found that the steamer was placed at our disposal for the day! We can go where we please and stay as long as we please!"[6]

The tourists went on to Sparta, visiting Castle Rock, and returned to Devil's Lake at night. On July 29 they "crossed the lake in a row boat, climbed the rocks—saw a rattle snake! Drove to Baraboo."[7] They returned home from Madison on July 31.

On the day he returned to Milwaukee, Increase received a letter from W. W. Field, secretary of the Wisconsin Board of Centennial Managers, apparently inquiring into Increase's progress on the Centennial geological exhibit. Increase replied that he had been informed that the board had taken no action on the matter and he supposed that the project had been abandoned. "It is now too late to get up a creditable show of our rocks and minerals," he said, in a letter dated August 2.[8]

Ten days later Increase and this time Mary were off on another jaunt, first to Madison, then to Spring Green and its environs. "Not having a compass we got lost in the woods among the hills—roads not well defined."[9] They visited Wisconsin's Natural Bridge northeast of Spring Green and the site of a well in which the skull of a musk ox was found, its presence in Wisconsin being associated with the Ice Age.

Upon returning to Milwaukee, Increase received yet another request to participate in the Philadelphia Centennial Celebration. Spencer F. Baird, assistant secretary of the Smithsonian Institution in Washington, asked if Increase could supply him with models of outstanding examples of Wisconsin's effigy mounds. This time Increase seized the opportunity. He told Baird on August 21:

I have consulted with a skilful carver in wood, who will do the work as soon as I can prepare the proper designs. We adopt the size you suggested (4 ft x 6 ft) but find it hardly desirable to divide the table into sections as you will see by the accompanying sketch. I presume it

Increase Lapham, photographed in 1869 by H. H. Bennett in Wisconsin Dells, observing the rock formations in Taylor's Glen. A companion identified as A. Holly is seated on a rock in the background.

is the intention of the commission to pay for this work. I shall want the privilege of exhibiting it here and perhaps making a few copies.[10]

It did not take him long to draw up a plan for models of fourteen of the more interesting mounds. According to an article in the *Evening Wisconsin* for September 11, 1875, the models were exhibited at the State Fair's Department of Fine Arts in Milwaukee from September 6 through September 11.[11] Among the fourteen were renderings of "The Man Mound" and "The Long-Armed Man," both of Sauk County, as well as the bear, buffalo, otter, squirrel, elephant, eagle, and lastly, the turtle of Waukesha County, the first mound surveyed by Lapham in 1836.

In his letter to Baird, Increase mentioned he had been studying the small lakes in the Oconomowoc area "with a view to their capabilities for Fish production." He had toyed with the idea earlier, having told his Racine friend, Dr. Philo R. Hoy, of his interest in May. "If the temperature of Oconomowoc does not rise above 65 degrees in the deep water in the summer, then the speckled trout would flourish there, and what an attraction it would be to have these beauties inhabiting the lake," Hoy had replied on May 7.[12]

On September 13 Increase entered into his notebook the results of soundings he had taken from a rowboat to find the depth of the lake from Cisco Bay east, each sounding separated by "30 strokes of oars," recording the deepest sound as fifty-one feet. Apparently, he had disembarked on the north shore of the lake, for his notebook relates that he "gathered grasses, yellow clover, *Lythrum alatrum* [winged loosestrife] and *Nesaea verticillata* [swamp loosestrife].... Never found the latter growing before."[13] The next day, Increase finished his paper, "Oconomowoc Lake, and Other Small Lakes of Wisconsin, Considered with Reference to Their Capacity for Fish-Production." It was a dry scientific piece in which he discussed the size, shape, temperature, chemistry, seasonal changes, currents, geology, and plant and animal life of the lake. But he allowed himself a few rapturous moments, calling Oconomowoc Lake

> one of those beautiful sheets of clear, cold water that may be taken as a type or representative of hundreds of others within the State of Wisconsin....
>
> The banks of the lake consist mostly of high grounds which are selected as sites of beautiful, often costly residences, which, especially when duplicated by reflection from the smooth surface of the water, form landscapes worthy of the pencil of the painter.[14]

When he finished the paper to his satisfaction he left his table where he had been working in the house on Minnewoc farm, walked to the edge of the lake that had been the subject of his writing, stepped into his rowboat, and pushed off from shore. That evening when he had not returned, his lifeless body was found in the bottom of the boat.

Increase Lapham was buried in the family plot at Forest Home Cemetery, on the outskirts of the south side of Milwaukee, the cemetery he had helped to lay out decades earlier. His grave joined a family group that included the graves of his mother, Rachel, his wife, Ann, and the two children they lost as babies. All were marked by modest headstones, especially when compared with some of the massive monuments that proclaimed to all visitors that Milwaukee now was home to families of power, pride, and wealth.

His death reignited the furor that had begun with his firing as state geologist, with much the same rhetoric from many of the same sources as before. Some implied a causal relationship between his being fired and his death only months later, although there is little evidence for that.

It also brought an outpouring of deeply felt tributes from some old friends and colleagues. Dr. Hoy of Racine, now president of the Wisconsin Academy, told the academy that his friendship with Increase began

> in 1846, when one morning there landed [at Racine] from the steamer *Sultana* a small man with a huge collecting box hanging at his side. He came from Milwaukee and intended returning on foot along the lakeshore in order to collect plants and shells, no easy journey, encumbered, as he soon would be, with a well filled specimen box.... In after years we were often together, studying the mounds, quarries, forest trees, etc., near Racine and my impression of his energy, perseverance, enthusiasm, accuracy and extent of information were all deepened by our subsequent meetings.[15]

In Milwaukee, Samuel S. Sherman, who had been president of the Milwaukee Female College, of which Lapham was a founder, read a long biography outlining Lapham's writings and accomplishments to the Old Settlers' Club of Milwaukee on December 11, 1875. Toward the end he said, "When asked, by a gentleman well known in scientific circles, in what department of science he was laboring, he replied 'I am studying Wisconsin.'"[16] Sherman also read a letter from Asa Gray, the Harvard botanist and longtime Lapham correspondent, who called Lapham "the pioneer botanist of your state.... The impression I have of him is that of a man thoroughly to be respected, and implicitly to be relied upon;—a modest, retiring industrious excellent man;...I have the idea that he had a happy, as well as a useful and honored life. What more could be asked?"[17]

Meanwhile, the State of Wisconsin bought Lapham's library; his cabinet of minerals, rocks, shells, and fossils; and his herbarium for a handsome sum. The minerals, rocks, and fossils became the nucleus of the State of Wisconsin's exhibit at the Philadelphia Centennial Celebration, which was housed in an annex of the Main Building, then the largest building in the world with an interior area of more than twenty-one acres. The celebration was attended by an estimated ten million visitors during its run from May 10 through November 10, 1876.

Elias T. Sweet, a University of Wisconsin–educated geologist and secretary of the Wisconsin Academy, took charge of the exhibit and wrote in a catalog of the Wisconsin State Mineral Exhibit:

> The scientific collection is quite full. In fact, it is much more complete than was at first anticipated. It comprises nearly a thousand specimens of rocks, ores, minerals, and fossils, mainly from the extensive collection of the late Dr. I. A. Lapham of Milwaukee. The cabinet, which Dr. Lapham was many years in gathering was recently purchased by the state, and donated to the University.[18]

Sweet noted that other specimens came from the Wisconsin Academy of Sciences, Arts and Letters and from the State Geological Survey, specifically from Lapham's assistants Roland Irving, Thomas Chamberlin, and Moses M. Strong. The Smithsonian exhibited Lapham's model of Wisconsin effigy mounds.[19]

At the Centennial fairgrounds, the state constructed a two-story Wisconsin Headquarters, one of eleven state buildings on State Avenue. Its centerpiece was Old Abe, the famous, still-living bald eagle that served as the battlefield mascot of the Eighth Wisconsin Infantry during the Civil War. Wisconsin's industries were another important presence at the celebration. Racine's J. I. Case Co. brought threshing machines; Milwaukee's Edward P. Allis & Co. showed off sawing and milling machines;

WHI IMAGE ID 2219

The last known photograph of Increase Lapham

Milwaukee's Valentine Blatz and Joseph Schlitz brewing companies each won awards for their lagers; and no fewer than sixteen Wisconsin companies won for cheese or butter.[20]

It was a true world's fair, the nation's first, and it brought in exhibits and food from many other countries, from Brazil to China. The fair's clear statement was that the United States had arrived as a first-rank player on the international stage in all categories. Gone were the frontier, the Civil War, slavery; major American industries were competing on a global scale; railroads and telegraph had tied the nation together from the Atlantic to the Pacific and from Canada to Mexico. Alexander Graham Bell demonstrated his telephone; the Remington Co. showed off its typewriter, which had been invented in Milwaukee; and the gigantic Corliss steam engine was started up by Presidents Ulysses S. Grant and Dom Pedro of Brazil and supplied power to every machine in Machinery Hall.

—⊢ | ⊢—

On November 3, 1875, Seneca Lapham, living in Milwaukee, wrote to his sister Mary Lapham at Lake Oconomowoc.

Dear Sister,
The advices so far received indicate the election of Mr. Ludington by about 5100 majority and I have had faith all along that the Geological Survey would be put in better hands than it now is.
I think Father would have been pleased if he could have seen today—not at the prospect of being reinstated, but that the people have refused to endorse the actions of Governor Taylor. And now all that can be done is to have some competent geologist at the head in whose

report confidence will be had. Then it seems to me that Father will receive the proper credit and if the reports are ever published his work will be found most invaluable.

It does seem hard that he could not have had the satisfaction of knowing and seeing the fulfillment of his desires, but both he and Mother have taught us by their practice if not so much by words to believe that whatever God does is done for the best. He would not have cherished any hard feelings against anyone and would not wish that we should.

You said you feel almost as though you would go wild when you think of father being out on the lake alone that afternoon, but do you not think it would have been harder for anyone with him. Suppose he had fallen, when one of you were with him the stroke would have been as instantaneous and as fatal—but you would not have known what to do. You would have thought, perhaps I can help him by rubbing—perhaps I should go for assistance. Then think of one's feelings while leaving him alone in the boat while going to the house and back.

To me there seems something beautiful in his death. So calm and peaceful as it must have been, so much in consonance with his life. And we know so far as human knowledge can judge that he was prepared to meet his Creator at any moment. Do you feel that there would have been consolation enough in the care which we know you and Julia would have shown for him to compensate for a lingering illness in one who so much disliked to be confined to the house by anything?

I do not know but I may be wrong in my feelings, but it seems to me as though Father's death under all the circumstances ought not to affect our lives, as far as such a thing is possible, than would a long journey in his work have done.[21]

—|—

The Philadelphia fair showed that we had entered a new age. It was being brought about by scientists and engineers who had become ever more deeply specialized in both education and practice. The era of the all-in-one botanist, zoologist, meteorologist, geologist, archaeologist, and naturalist was gone. The need for the generalist observer with a keen, all-encompassing eye that was essential for a time and place, essential to interpret a virgin country on the frontier, had come and gone within the lifetime of Increase Lapham.

And Increase Lapham was the man for an age and for a place, for the young Wisconsin he studied so passionately and so well.

ABBREVIATIONS

DL: Darius Lapham

IAL: Increase Allen Lapham

JAL typescript: Julia Ann Lapham typescript, Increase Lapham Archives, Wisconsin Historical Society, Madison, Wisconsin

Lapham MSS: Increase Lapham Archives, Wisconsin Historical Society, Madison, Wisconsin

PREFACE

1. Sherman, "Increase Allen Lapham," 51.

CHAPTER 1: *A Habit of Observation*

1. Thomas L. Nichols, *Forty Years of American Life* (London, 1864), quoted in Way, *Common Labor*, 174.
2. Diary of Lyman Spalding, quoted in Riley, *Lockport: Historic Jewel*, 19.
3. Way, *Common Labor*, 139–140.
4. Increase Allen Lapham (hereafter cited as IAL) autobiography (1830), in Thomas and Conner, *The Journals*, 111 note 1.
5. IAL autobiography, 1875, Increase Lapham Archives, Wisconsin Historical Society, Madison, Wisconsin (hereafter cited as Lapham MSS).
6. Howe, *What Hath God Wrought*, 140.
7. Ibid., 216–217.
8. Ibid., 217.
9. Ibid.
10. Shaw, *Canals for a Nation*, 1–2, 39.
11. Ibid., 161.
12. Trevorrow, "The Laphams—A Canal Family," 31.
13. Ibid., 1.
14. IAL to Lyman Draper, May 16, 1859, Lapham MSS.
15. Darius Lapham (hereafter cited as DL) to IAL, December 9, 1829, Julia Alcott Lapham typescript (hereafter cited as JAL typescript), 131.
16. Howard Brinton, "The Quaker Contribution to Higher Education in Colonial America," *Pennsylvania History* 25 (1958): 245–246, quoted in Frost, *The Quaker Family*, 209.

17. Aldridge, *Laphams in America*, 13–14.

18. Larkin, *The Reshaping of Everyday Life*, 73.

19. Ibid.

20. Aldridge, *Laphams in America*, 13–14.

21. IAL autobiography (ca.1875), Lapham MSS.

22. Ibid.

23. Shaw, *Canals for a Nation*, 174.

24. IAL autobiography (1830), in Thomas and Conner, *The Journals*, 112 note 1.

25. Frost, *The Quaker Family*, 218.

26. IAL autobiography (1830), in Thomas and Conner, *The Journals*, 112.

27. Niagara Escarpment Resource Network pamphlet, Lakeshore Natural Resources Partnership, 2007.

28. Spafford, *A Pocket Guide for the Tourist and Traveller*, 45.

29. "History, 1816–1840," Historic Lockport website, http://elockport.com/history-lockport-ny.php.

30. Shaw, *Canals for a Nation*, 41.

31. "History, 1816–1840," Historic Lockport.

32. IAL autobiography (ca. 1875), Lapham MSS.

33. Ibid.

34. Ibid.

35. Ibid.

CHAPTER 2: *The Difficulty of Youth*

1. IAL to Lyman Draper, May 16, 1859, Lapham MSS.

2. Shaw, *Erie Water West*, 88–89.

3. David Thomas, *Travels through the Western Country in the Summer of 1816* (Auburn, NY: David Rumsey, 1819).

4. IAL autobiography (1830), in Thomas and Conner, *The Journals*, 112.

5. Buley, *The Old Northwest Pioneer Period*, 1:4, 530–533.

6. Mansfield and Drake, *Cincinnati in 1826*, 45.

7. IAL journal entry, August 30, 1826, in Thomas and Conner, *The Journals*, 112.

8. Berquist and Bowers, *Byron Kilbourn and the Development of Milwaukee*, 9.

9. Banta, *The Ohio*, 284–285.

10. Rhodes, *John James Audubon*, 54.

11. Ibid., 53, 52.

12. McMurtrie, *Sketches of Louisville and Its Environs*, 119.

13. Trollope, *Domestic Manners of the Americans*, 34. Mrs. Trollope, mother of novelist Anthony Trollope, traveled in America from 1827 to 1830; her book was first published in 1832.

14. Rhodes, *John James Audubon*, 54.

15. IAL journal entry, June 12, 1828, in Thomas and Conner, *The Journals*, 45.

16. IAL autobiography (1830), in Thomas and Conner, *The Journals*, 113.

17. IAL journal entry, February 12, 1826, JAL typescript, 5.

18. IAL to DL, Louisville, February 27, 1827, quoted in JAL typescript, 5–6. We don't have the original of this letter, only Julia Lapham's typescript of her father's letter, which is why the spelling is correct; she made him out to be a better speller than he was.

19. IAL to DL, August 5, 1827, JAL typescript, 7–8.

20. Ibid.

21. Benjamin Silliman to IAL, December 27, 1826, in JAL typescript, 5.

22. *American Journal of Science and Arts* 11, no. 1 (1826): 1–39.

23. Ibid., 39–53.

24. *American Journal of Science and Arts* 11, no. 2 (1826): 238–246, 349–359.

25. No author, "The Divining Rod," *American Journal of Science and Arts* 10 (1826): 211.

26. The articles mentioned in this paragraph appear in the *American Journal of Science and Arts*, volumes 11 and 12.

27. Increase had begun keeping a "Meteorological Table" in his small blue notebook by September 1827. Lacking instruments, he recorded whether the days were "Fair," "Variable," "Cloudy," "Windy," or "Rainy." IAL, blue notebook, 66–71, Lapham MSS.

28. The articles in this paragraph can be found in the *American Journal of Science and Arts*, volume 11, issues 1 and 2.

29. Winchell, "Increase Allen Lapham," 18.

30. Ibid.

31. IAL to Benjamin Silliman, June 24, 1827, JAL typescript, 6.

32. Benjamin Silliman to IAL, July 22, 1827, JAL typescript, 6, and original letter in Lapham MSS.

33. IAL to Benjamin Silliman, November 10, 1827, JAL typescript, 14–21.

34. Ibid. The original is lost. When Increase's daughter, Julia Lapham, was compiling his biography from his journals and letters, she had in front of her either the draft or original of the letter he sent Darius, including the appended article draft. The article draft was not kept for the archive.

35. Ibid., 21.

36. Benjamin Silliman to IAL, December 25, 1827, JAL typescript, 24.

37. Ibid.

CHAPTER 3: *"A Journal of Science & Arts with Miscellaneous Nonsense"*

1. See John Syme's 1826 portrait of Audubon in Rhodes, *John James Audubon*, 275.

2. Quoted in Rhodes, *John James Audubon*, 52.

3. IAL journal entry, October 6, 1827, in Thomas and Conner, *The Journals*, 11.

4. IAL, blue notebook, page 2, Lapham MSS.

5. Wright, *Culture on the Moving Frontier*, 71–72. Not only were books widely available but there were enough booksellers and printers in Kentucky by 1805 that in Lexington, there was a meeting of the Booksellers and Printers of the

Western Country. Advertisements for books were common in the newspapers of the time.

6. Greene, *American Science in the Age of Jefferson*, 220.
7. IAL journal entry, February 17, 1828, in Thomas and Conner, *The Journals*, 24. Hereafter, we will reproduce IAL's idiosyncratic spelling without the intrusive *sic*.
8. See, for example, journal entries December 24, 27, and 29, 1827, and January 1, 2, 4, 5, 10, 11,12, and 14, 1828, in Thomas and Conner, *The Journals*, 18–20.
9. Ibid., December 26, 1827, 18.
10. Ibid., March 20, 1828, 29.
11. Ibid., March 24, 1828, 30.
12. Ibid., March 25, 1828, 30.
13. Ibid., February 21, March 14, March 24, and April 30, 1828, 25, 29, 30, and 39.
14. Ibid., October 24, 1827, 12.
15. Ibid., November 12, 1827, 12.
16. Ibid., December 15, 1827, 16.
17. Sandburg, *Abraham Lincoln*, 66–67.
18. IAL journal entry, April 12, 1828, in Thomas and Conner, *The Journals*, 35.
19. Ibid., June 21, 1828, 47.
20. Ibid., April 22 and May 14, 1828, 37 and 41.
21. Mr. Joyes was a state representative from Jefferson County.
22. IAL journal entries, March 3 and 4, 1828, in Thomas and Conner, *The Journals*, 26–27.
23. Ibid., February 21, 1828, 24–25.

CHAPTER 4: *Natural History and the Lapham Brothers*

1. IAL to DL, December 3, 1829, in JAL typescript, 126.
2. IAL journal entry, May 3, 1828, in Thomas and Conner, *The Journals*, 40.
3. Ibid., April 12, 1828, 35.
4. Ibid., March 31 and April 1, 1828, 33.
5. Ibid., March 16, 1828, 29.
6. The joke begins on March 8, 1828, in an IAL journal entry. Thomas and Conner, *The Journals*, 27.
7. Ibid.
8. IAL journal entry, June 10, 1828, in Thomas and Conner, *The Journals*, 45.
9. Ibid., April 13, 1828, 35.
10. Ibid.
11. Ibid.
12. Ibid.
13. Ibid., March 30, 1828, 31–32.
14. Ibid., January 10, 1828, 19.

15. Ibid., February 4, 1828, 22.
16. Ibid., February 23, 1828, 25.
17. Ibid., March 7, 1828, 27.
18. Ibid., February 14, 1828, 24.
19. IAL to Benjamin Silliman, January 28, 1828, JAL typescript, 30.
20. Ibid., 30–31.
21. IAL journal entry, May 8, 1828, in Thomas and Conner, *The Journals*, 40.
22. Ibid., May 16, 1828, 41.
23. Ibid., May 30, 1828, 42.
24. Ibid.
25. Ibid., June 1, 1828, 44.
26. Ibid.
27. Ibid., June 6, 1828, 45.
28. IAL expense list for 1828, June 11, 1828, blue notebook, 67.
29. IAL journal entry, June 12, 1828, JAL typescript, 49. In JAL's transcription of this diary entry, Julia Lapham leaves out everything that tells of Increase's surprise and anger that he was left behind. She changes his wording, "We expect to be off tomorrow," to "The family expect to be off tomorrow." Increase's mild cursing is left out and she has him calmly say, "Today father with all the family and all the goods belonging to them left this little place for Chillecothe, O." Lapham MSS.

CHAPTER 5: *First Year on His Own*

1. IAL journal entry, June 12, 1828, in Thomas and Conner, *The Journals*, 46.
2. IAL expense list for 1828, June 11, 1828, blue notebook, 67, Lapham MSS.
3. IAL journal entry, July 4, 1828, in Thomas and Conner, *The Journals*, 52.
4. IAL to DL, June 22, 1828, JAL typescript, 53.
5. IAL journal entry, September 28 and October 5, 1828, in Thomas and Conner, *The Journals*, 64, 66.
6. Ibid., October 5, 1828, 66.
7. Stuart, *Lives and Works*, 103–105.
8. "Fifth Annual Report of the President and Directors of the Louisville and Portland Canal Company," 1830, Lapham MSS.
9. IAL, Index to Field Book on Canals, 1828–1832, Lapham MSS.
10. See, for instance, S. P. Hildreth to IAL, September 23, 1828, JAL typescript, 85; IAL to S. P. Hildreth, October 23, 1828, JAL typescript, 87–90. Copies of more letters may be found in the Lapham MSS.
11. IAL to DL, March 9, 1829, JAL typescript, 100.
12. Amos Eaton, *A Geological and Agricultural Survey of the District Adjoining the Erie Canal, in the State of New-York*.
13. Greene, *American Science in the Age of Jefferson*, 249–250.
14. Eaton, *A Geological and Agricultural Survey*, 89. Eaton spelled *greywacke* with an *e* in the British manner; American geologists usually spelled it "graywacke."

15. This term is spelled both ways in writings of the time. Later it was standardized as "corniferous."
16. DL to IAL, June 16, 1828, JAL typescript, 51.
17. Ibid.
18. IAL journal entry, June 21, 1828, in Thomas and Conner, *The Journals*, 47.
19. IAL to DL, June 22, 1828, JAL typescript, 52–54, Lapham MSS.
20. DL to IAL, June 23, 1858, JAL typescript, 58.
21. Around June 27, Increase wrote in his journal that the letter he received on June 21 (and probably the next lengthy letter) contained excerpts from Darius's journal. IAL, journal entry, June 27, 1828, in Thomas and Conner, *The Journals*, 47.
22. DL to IAL, June 23, 1828, JAL typescript, 57.
23. Ibid., 57–59.
24. Ibid., 59.
25. DL to IAL, June 26, 1828, JAL typescript, 60.
26. DL to IAL, July 2, 1828, JAL typescript, 64–65.
27. IAL journal entry, July 6, 1828, in Thomas and Conner, *The Journals*, 52.
28. Ibid., 52–53.
29. IAL to Darius, July 8, 1828, JAL typescript, 62–64.
30. Ibid.
31. Ibid.

CHAPTER 6: *Until a Better Opportunity Presents*

1. IAL journal entry, June 23, 1828, in Thomas and Conner, *The Journals*, 47; DL to IAL, June 23, 1828, JAL typescript, 60.
2. IAL to DL, July 12, 1828, JAL typescript, 66.
3. Ibid.
4. DL to IAL, June 23, 1828, JAL typescript, 60.
5. IAL to DL, July 12, 1828, JAL typescript, 67.
6. IAL to DL, July 27, 1828, JAL typescript, 68–71.
7. Trescott, "The Louisville and Portland Canal Company," 692.
8. DL to IAL, August 4, 1828, JAL typescript, 73–74.
9. Seneca Lapham to IAL, August 8, 1828, JAL typescript, 74.
10. IAL to Seneca Lapham, August 24, 1828, JAL typescript, 76.
11. DL to IAL, September 1, 1828, JAL typescript, 78.
12. Increase began a new journal book on November 1. The one that ends with October had many empty pages in it which were filled later by Darius, reinforcing the idea that Increase's journal was not meant to be a private book but was written for Darius's eyes as well.
13. IAL journal entry, November 3, 1828, in Thomas and Conner, *The Journals*, 72.
14. Ibid., November 4, 1828, 72.
15. Hedeen, *Big Bone Lick*, 114–115.
16. Donald Smalley, introduction to Trollope, *Domestic Manners of the Americans*, xxvi.

17 IAL journal entry, November 4, 1828, in Thomas and Conner, *The Journals*, 72.

18. Ibid., November 7, 1828, 73.

19. Ibid., November 8, 1828, 73.

20. Ibid., November 8–13, 1828, 73.

21. Ibid., November 24, 1828, 74–75.

22. Ibid., November 4–29, 1828, 72–76.

23. Ibid., December 25, 1828, 77.

24. Ibid., 78.

25. Ibid., January 18, 1829, 77–78.

26. Ibid., February 2, 9, and 11, 1829, 80.

27. IAL to Benjamin Silliman, February 11, 1829, JAL typescript, 98–99.

28. IAL to Canvass White, March 5, 1829, Lapham MSS.

29. IAL journal entry, March 9, 1829, in Thomas and Conner, *The Journals*, 80.

30. Ibid., March 10–31, 1829, 80–83.

31. IAL to Seneca Lapham, April 1, 1829, JAL typescript, 104–105.

32. IAL journal entries, April 1–30, 1829, in Thomas and Conner, *The Journals*, 83–87.

33. DL to IAL, April 22, 1829, JAL typescript, 107.

34. IAL journal entry, May 29, 1829, in Thomas and Conner, *The Journals*, 94.

35. Ibid., May 11, 1829, 90.

36. Ibid., May 20, 21, and 26, 1829, 92–93.

37. Ibid., May 21, 1829, 91–92.

38. Ibid., May 25, 1829, 93.

39. Ibid., May 30, 1829, 94.

40. Ibid., June 11, 1829, 97.

41. Ibid., June 12, 1829, 97.

42. Ibid., June 25, 1829, 99.

43. Ibid., January 5, 1830, 124 note 201.

CHAPTER 7: *A Man's Estate*

1. IAL journal entry, July 20, 1829, in Thomas and Conner, *The Journals*, 103.

2. Ibid., July 19–31, 1829, 103–104.

3. Ibid., August 9–10, 1829, 105.

4. IAL to DL, August 5, 1929, JAL typescript, 116–117.

5. Ibid.

6. IAL to William Darlington, September 2, 1829, JAL typescript, 120.

7. IAL to DL, August 30, 1829, JAL typescript, 118.

8. IAL journal entries, September 4, 7, 13, 16, and 29, 1829, in Thomas and Conner, *The Journals*,106; William Darlington to IAL, September 2, 1829, JAL typescript, 120.

9. IAL to DL, August 5, 1829, JAL typescript, 117.

10. DL to IAL, August 31, 1829, JAL typescript, 118–119.

11. Ibid., 119.

12. Ibid., 118–119.

13. IAL to DL, September 13, 1829, JAL typescript, 120–122.

14. Ibid.

15. IAL journal entry, October 7, 1829, in Thomas and Conner, *The Journals*, 106.

16. IAL journal entry, October 14, 1829, blue notebook, 60, Lapham MSS.

17. Ibid., October 17, 1829.

18. Darius Lapham and Increase A. Lapham, "Observations on the Primitive and other Boulders of Ohio," 300–303.

19. Woods, *Ohio's Grand Canal*, 9.

20. Ibid.

21. IAL journal entries, October 27, 1829, and December 1, 1829, blue notebook, 60–61, and JAL typescript, 123–128.

22. IAL to Benjamin Silliman, November 5, 1829, Lapham MSS.

23. IAL to Benjamin Silliman, December 1, 1829, JAL typescript, 125.

24. Tappan, "On the Boulders of Primitive Rocks," 291–297.

25. Ibid.

26. IAL to DL, December 3, 1829, JAL typescript 126

27. DL to IAL, December 9, 1829, JAL typescript, 130.

28. IAL to DL, December 17, 1829, JAL typescript, 132.

29. Benjamin Silliman to IAL, December 24, 1829, JAL typescript, 133.

30. Tappan, "On the Boulders of Primitive Rocks," 296.

31. DL and IAL, "Observations on the Primitive and other Boulders of Ohio," 300.

32. Ibid., 301.

33. Ibid.

34. Ibid., 303.

35. IAL journal entry, January 5, 1830, JAL typescript, 134.

36. IAL to Rachel and Seneca Lapham, January 8, 1830, Lapham MSS.

37. IAL journal entry, January 8, 1830, JAL typescript, 135.

38. IAL to Rachel and Seneca Lapham, January 8, 1830, Lapham MSS.

39. IAL journal, January 9, 1830, JAL typescript, 135.

40. IAL to Seneca and Rachel Lapham, February 8, 1830, JAL typescript, 137–138.

41. IAL journal entry, January 23, 1830, in Thomas and Conner, *The Journals*, 107–108.

42. Ibid., February 18, 1830, 109.

43. IAL to Seneca Lapham, February 21, 1830, JAL typescript, 139–140.

44. IAL journal entry, February 27, 1830, in Thomas and Conner, *The Journals*, 110.

45. Ibid., March 3, 1830, 111.

46. Ibid., March 5, 1830.

47. IAL journal entries, February 24 and March 3, 1830, JAL typescript, 141–142.

48. Ibid., March 13 and March 20, 1830, 142.

49. IAL to DL, June 14, 1830, JAL typescript, 146.

50. Thomas and Conner, *The Journals*, 125 note 223.

51. IAL, "Answers to Questions by Daniel Carroll," pages 4–5, Box 7, Folder 1, Lapham MSS.

52. Ibid., 7

53. Ibid.

54. Ibid., 11, 17.

55. IAL journal entry, June 12, 1830, JAL typescript, 146.

56. Ibid., June 25 and 26, 1830.

57. Ibid., June 27–29, 1830, 146–147.

58. IAL journal entry, March 29, 1830, in Thomas and Conner, *The Journals*, 111. Interestingly, Julia Lapham leaves mention of this conversation with Mr. Henry out of her manuscript biography of her father.

59. IAL journal entry, June 30, 1830, JAL typescript, 147.

60. Ibid., July 2–3, 1830.

CHAPTER 8: *Acquainted with Everybody but Intimate with None*

1. IAL journal entry, July 5, 1830, JAL typescript, 147–148.

2. *History of Lower Scioto Valley, Ohio*, 185.

3. Ibid.

4. IAL to DL, July 25, 1830, JAL typescript, 148.

5. Ibid., 148–149.

6. Ibid., 148.

7. "Waverly, Ohio." http://www.touring-ohio.com/southwest/waverly/waverly .html. Accessed May 11, 2010.

8. "Captain Francis Cleveland, History of Scioto County Ohio," http://www .heritagepursuit.com. Accessed September 21, 2010.

9. Hagerty, McClelland, and Huntington, *History of the Ohio Canals*, 27.

10. IAL to Rachel Lapham, January 1, 1832, JAL typescript, 183.

11. IAL to DL, September 21, 1830, JAL typescript, 153.

12. IAL journal entry, August 28, 1830, JAL typescript, 150.

13. F. Cleveland, drawing, and W. Woodruff, engraver, "Topographical Map of Portsmouth and its Vicinity with the Southern Termination of the Ohio Canal" (Columbus: Ohio Historical Society, 1830).

14. According to the writing on the map, it must have been drawn after the number of steamboats passing through in 1829 was calculated and before the Louisville Canal opened in December 1830. There is no mention on the map of the 1832 flood at Portsmouth, which was much worse than the 1815 flood and which changed plans for the canal. The map is hand dated 1833 in pencil, but that has to be a mistake. The engraver was W. Woodruff, an active and accomplished engraver in Cincinnati.

15. IAL journal entry, March 30, 1832, JAL typescript, 158.

16. IAL, "The Stranger," *Portsmouth Courier*, January 4, 1831.

17. Rachel Lapham to IAL, September 20, 1830, JAL typescript, 151.

18. IAL to DL, July 25, 1830, JAL typescript, 148–149.

19. DL to IAL, August 20, 1830, JAL typescript, 150–151.

20. IAL to DL, November 7, 1830, JAL typescript, 156.

21. IAL journal entry, December 16, 1830, JAL typescript, 158.

22. IAL to DL, December 22, 1830, JAL typescript, 158–159.

23. IAL journal entries, August 21, 1831–September 8, 1831, JAL typescript, 139–141.

24. Woods, *Ohio's Grand Canal*, 13.

25. IAL to Seneca Lapham, September 4, 1831, JAL typescript, 177.

26. IAL journal entry, September 18, 1831, JAL typescript, 140.

27. *History of the Lower Scioto Valley*, 228.

28. IAL journal entry, February 22, 1831, JAL typescript, 165.

29. IAL journal entry, November 18 and 19, 1831, Portsmouth journal, Lapham MSS. From 1830 to 1832 Increase Lapham wrote new journal entries at the back of a journal in which he had recopied and corrected spelling of his first two journals, which cover 1811–1829. Julia Lapham left some of this Portsmouth material out of her typescript.

30. Ibid., November 22, 1831.

31. Ibid., December 1, 1831.

32. DL to IAL, July 3, 1831, JAL typescript, 175.

33. IAL to Seneca and Rachel Lapham, July 22, 1831, JAL typescript, 175.

34. IAL to DL, March 10, 1831, JAL typescript, 167.

35. IAL journal entry, August 6, 1831, JAL typescript, 175–176.

36. IAL to DL, August 21, 1831, JAL typescript, 176–177.

37. Ibid.

38. Ibid.

39. IAL to Seneca Lapham, September 4, 1831, JAL typescript, 178–179.

40. IAL journal entry, December 31, 1831, Portsmouth journal, 169.

41. IAL to Rachel Lapham, January 1, 1832, JAL typescript, 183.

42. IAL journal entry, March 8, 1832, JAL typescript, 208.

43. Darius Lapham and Increase A. Lapham, "Observations on the Primitive and Other Boulders of Ohio," 300–303.

44. IAL journal entry, March n.d., 1832, JAL typescript, 198–201.

45. IAL journal entries, February 5–23, 1832, Portsmouth journal, 153–157.

46. Ibid.

47. IAL to DL, May 11, 1832, JAL typescript, 206.

48. IAL to DL, June 6, 1832, JAL typescript, 208.

49. Ibid.

50. IAL to DL, July 17, 1832, JAL typescript, 208.

51. IAL journal entry, August 1, 1832, JAL typescript, 210.

52. IAL to Seneca Lapham, September 1, 1832, JAL typescript, 210.

53. IAL to DL, October 18, 1832, JAL typescript, 212.

54. IAL to DL, November 8, 1832, JAL typescript, 214.

55. Rachel Lapham to IAL, November 10, 1832, JAL typescript, 215.

56. IAL to DL, November 8, 1832, JAL typescript, 214.
57. IAL journal entries, December 14–18, 1832, Portsmouth journal, 160–161; JAL typescript, 215–216.
58. IAL journal entry, December 19, 1832, JAL typescript, 216.
59. DL to IAL, March 11, 1832, JAL typescript, 193–194.
60. Ibid., 193.
61. Ibid.
62. IAL journal entry, December 19, 1832, Portsmouth journal, 161.
63. IAL journal entry, December 18, 1832, JAL typescript, 216–217.
64. IAL journal entry, December 22–31, 1832, Portsmouth journal, 162.
65. IAL journal entry, December 19, 1832, JAL typescript, 216.
66. IAL to DL, March 21, 1833, JAL typescript, 222.

CHAPTER 9: *In the Capitol*

1. IAL to DL, May 8, 1833, JAL typescript, 223.
2. Ibid., 223–224.
3. Cole, *A Fragile Capital*, 36–38.
4. Cyrus P. Bradley, quoted in Cole, *A Fragile Capital*, 82–83.
5. IAL to DL, May 8, 1833, JAL typescript, 224.
6. Kilbourn, *The Ohio Gazeteer*, 10th ed. (Columbus: John Kilbourn, 1831), 154.
7. Ibid., 152–153.
8. Ibid.
9. IAL to DL, June 25, 1833, JAL typescript, 225.
10. Ibid., 225–226.
11. Forman, "The First Cholera Epidemic," 413.
12. IAL journal entry, July 14, 1833, JAL typescript, 226.
13. Ibid., September 10, 1833, 229.
14. Forman, 410.
15. Ibid., 415.
16. IAL to DL, July 26, 1833, JAL typescript, 226.
17. Forman, 419.
18. Ibid., 413.
19. DL to IAL, September 1, 1833, JAL typescript, 229.
20. Forman, 425.
21. IAL to DL, August 15, 1833, JAL typescript, 230–231.
22. IAL to Naaman Goodsell, August 12, 1833, JAL typescript, 242.
23. Ibid.
24. Ibid., 238–239.
25. Ibid.
26. Ibid.
27. I. A. Lapham, "Agriculture in Ohio."
28. IAL to DL, January 19, 1834, JAL typescript, 245.
29. IAL to DL, February 23, 1824, JAL typescript, 247.

30. I. A. Lapham, "Observations on the Geology of Scioto County, Ohio," address to the Historical and Philosophical Society of Ohio, JAL typescript, 250–257; IAL to DL, May 18, 1834, JAL typescript, 258.
31. S. P. Hildreth to IAL, October 1, 1834, JAL typescript, 269–270.
32. IAL to DL, May 18, 1834, JAL typescript, 258.
33. IAL to DL, January 3, 1835, JAL typescript, 272.
34. DL to IAL, January 6, 1835, JAL typescript, 273–274.
35. IAL to DL, January 30, 1835, JAL typescript, 274.
36. Ibid., 275.
37. Ibid. "Wolf scalp vouchers" refers to the bounty paid by the state to Ohio farmers who sent such proof of wolves they had killed. Increase had recently set his brother Pazzi up in the printing business in Columbus, borrowing money from Mr. Brown for a year. The $504 paid to a printer probably refers to a state job sent Pazzi's way.
38. Ibid.
39. Ibid.
40. Stuart, *Lives and Works*, 289.
41. IAL to DL, January 30, 1835, JAL typescript, 275.
42. "Tappan, Benjamin (1773–1857)," Biographical Directory of the United States Congress, 1774–present, http://bioguide.congress.gov/scripts/biodisplay.pl?index=T000039.
43. IAL to DL, January 12, 1836, JAL typescript, 308.
44. IAL to DL, January 30, 1835, JAL typescript, 275–276.

CHAPTER 10: *Leaving Ohio*

1. IAL to DL, July 7, 1835, JAL typescript, 284.
2. Charles W. Short to IAL, October 25, 1835, JAL typescript, 286.
3. IAL journal, October 15–November 1, 1835, JAL typescript, 286.
4. John L. Riddell to IAL, November 13, 1835, Lapham MSS.
5. Dexter, "The Early Career of John L. Riddell, 184–185; Riddell to IAL, November 13, 1835, Lapham MSS.
6. Robert Buchanan to IAL, December 9, 1835, JAL typescript, 287–288.
7. Samuel P. Hildreth to IAL, December 10, 1835, JAL typescript, 188; copy of the original letter, MSS 216, Lapham Papers, Box 1, Folder 4, Ohio Historical Society, Columbus.
8. George Graham to IAL, December 10, 1835, JAL typescript, 288.
9. IAL to J. M. McCreed, December 10, 1835, JAL typescript, 289–293.
10. Meeting minutes, Historical and Philosophical Society, Columbus, Ohio, December 30, 1835, JAL typescript, 294.
11. IAL to J. M. McCreed, December n.d., 1835, JAL typescript, 295.
12. IAL to DL, January 12, 1836, JAL typescript, 308.
13. S. P. Hildreth to IAL, January 11, 1836, JAL typescript, 305.
14. IAL to DL, March 8, 1835, JAL typescript, 279.

15. IAL to DL, January 12, 1836, JAL typescript, 307.

16. IAL to DL, January 27, 1836, Lapham MSS.

17. C. J. Ward to IAL, January 18, 1836, JAL typescript, 276–277.

18. IAL to DL, January 27, 1836, JAL typescript, 308–309.

19. Rachel Lapham to Hannah and Increase Lapham, February 11, 1836, JAL typescript, 310–311.

20. DL to IAL, February 17, 1836, JAL typescript, 312.

21. Berquist and Bowers, *Byron Kilbourn and the Development of Milwaukee*, 36–37.

22. IAL to DL, January 27, 1836, JAL typescript, 308–309.

23. IAL to DL, March 21, 1836, JAL typescript, 314–315; John Riddell to IAL, March 31, 1836, Lapham MSS; Hildreth to IAL, April 1, 1836, Lapham MSS; IAL to DL, April 7, 1836, JAL typescript, 315; John Locke to IAL, April 14, 1836, JAL typescript, 315–316 and original in Lapham MSS.

24. IAL to DL, April 7, 1836, JAL typescript, 315.

25. Ibid.

26. IAL to DL, February 18, 1836, JAL typescript, 313.

27. IAL to DL, April 7, 1836, JAL typescript, 315.

28. Berquist and Bowers, *Byron Kilbourn and the Development of Milwaukee*, 45–46.

29. IAL to DL, April 22, 1836, JAL typescript, 315.

30. DL to IAL, April 26, 1836, JAL typescript, 316–317.

31. IAL to DL, May 13, 1836, JAL typescript, 318.

32. IAL, "Notes in Regard to Geological Survey," May 19–24, 1836, JAL typescript, 319–320.

33. IAL to DL, May 23, 1836, Lapham MSS.

34. IAL, "Notes in Regard to Geological Survey," May 24, 1836, JAL typescript, 320.

35. DL to IAL, May 26, 1836, Lapham MSS.

36. IAL to DL, June 21, 1836, JAL typescript, 324–327; original letter in Lapham MSS.

37. Ibid., 325.

38. Ibid., 326.

39. Ibid.

40. Ibid.

41. Ibid., 326–327; Berquist and Bowers, *Byron Kilbourn and the Development of Milwaukee*, 67.

42. IAL to DL, June 21, 1836, JAL typescript, 326.

43. Wallin, "Douglass Houghton: Michigan's First State Geologist, 1837–1845."

44. IAL to DL, July 7, 1836, Lapham MSS.

45. Ibid.

46. Ibid.

47. Minton Baker, "Charles Minton Baker's Journal," 398.

48. IAL to DL, July 7, 1836, Lapham MSS.

CHAPTER 11: *Begin We Then at Milwaukee*

1. Wheeler, *The Chronicles of Milwaukee,* 75.
2. IAL to DL, February 25, 1837, JAL typescript, 354.
3. Wheeler, *The Chronicles of Milwaukee,* 74–75.
4. IAL journal entry, July 3, 1836, JAL typescript, 327.
5. Lucius Barber, quoted in Gregory, *Southeastern Wisconsin,* 213.
6. Wheeler, *The Chronicles of Milwaukee,* 40–41.
7. IAL journal entry, July 3, 1836, JAL typescript, 327.
8. IAL to Seneca Lapham, July 27, 1836, JAL typescript, 330.
9. IAL to C. W. Short, August 17, 1836, JAL typescript, 335.
10. Wheeler, *The Chronicles of Milwaukee,* 87–88.
11. IAL to C. W. Short, August 17, 1836, JAL typescript, 336–337.
12. IAL to Seneca Lapham, October 1, 1836, JAL typescript, 338–340.
13. Wheeler, *The Chronicles of Milwaukee,* 60.
14. Berquist and Bowers, *Byron Kilbourn and the Development of Milwaukee,* 1–8.
15. Wheeler, *The Chronicles of Milwaukee,* 57.
16. Darius Lapham and Increase A. Lapham, "Observations on the Primitive and other Boulders of Ohio," 300–303.
17. IAL, "Geological and Mineralogical Notes of Wisconsin" (unpublished notes), 1836, JAL typescript, 348.
18. IAL to C. W. Short, August 17, 1836, JAL typescript, 335.
19. Ibid.
20. Asa Gray to IAL, June 28, 1836, JAL typescript, 331.
21. Quaife, "Increase Allen Lapham," 9.
22. I. A. Lapham, *The Antiquities of Wisconsin.* Plate VI, "Ancient Works, Second Ward, Milwaukee, from a sketch made in 1836," shows the club-shaped mounds truncated by the very streets Increase was surveying.
23. IAL to C. W. Short, August, 17, 1836, JAL typescript, 336; IAL to Seneca Lapham, October 1, 1836, JAL typescript, 339.
24. IAL to Seneca Lapham, October 1, 1836, JAL typescript, 340.
25. Ibid.
26. Ibid., 339–340 and 348.
27. IAL to Seneca Lapham, October 15, 1836, JAL typescript, 340–343.
28. Ibid.
29. Byron Kilbourn to James Kilbourn, January 23, 1836, Ohio History Center, Kilbourn Papers, quoted in Berquist and Bowers, *Byron Kilbourn and the Development of Milwaukee,* 39.
30. IAL to Seneca Lapham, October 1, 1836, JAL typescript, 339–340.
31. Strong, *History of Wisconsin Territory,* 217.
32. Schafer, *Four Wisconsin Counties,* 61–64.
33. Since the formation of the Republic, Congress had debated whether to protect settlers from speculators or speculators from settlers. Depending on whether the Congress at the time favored capitalists or westward expansion, it would

pass "non-intrusion laws" to ban squatters from settling on public lands or "preemption laws" that granted squatters the privilege of buying lands that they had improved. The booming national land market of 1835 and 1836, and its abuses—more than 20 million acres were sold in 1836 alone—led to President Jackson's controls. A lucid history of preemption laws is found in Paul W. Gates, *History of Public Land Law Development* (Washington, DC: Public Land Law Review Commission, Government Printing Office, 1968), 219–247.

34. Strong, *History of Wisconsin Territory*, 241.
35. See Schafer, *Four Wisconsin Counties*, 70.
36. IAL to Seneca Lapham, March 19, 1837, JAL typescript, 356.
37. IAL to DL, May 12, 1837, JAL typescript, 362.
38. IAL to DL, September 27, 1837, JAL typescript, 369.
39. IAL to Seneca Lapham, October 1, 1836, JAL typescript, 340.
40. DL to IAL, October 30, 1836, JAL typescript, 344.
41. IAL to DL, December 3, 1836, JAL typescript, 345.
42. IAL to DL, February 25, 1837, JAL typescript, 353.
43. Darius's letters are lost, but Increase quotes them in IAL to DL, May 12, 1837, JAL typescript, 361.
44. IAL to DL, May 12, 1837, JAL typescript, 362.
45. IAL to DL, August 8, 1837, JAL typescript, 368.
46. IAL to DL, December 9, 1837, JAL typescript, 372–374.
47. DL to IAL, January 14, 1838, JAL typescript, 375; Zachary Taylor, with eight hundred men, attacked four hundred Seminoles in their home territory on Lake Okeechobee and lost, though it was called a victory against the Indians; one man, not "fifteen or twenty," was killed by the Canadian navy on this attack on the steamship *Caroline*, supplying William MacKenzie's rebellion. Mackenzie had recently declared himself the head of a provisional Canadian government.
48. IAL to Rachel Lapham, April 5, 1838, JAL typescript, 382.
49. Ibid., 381, and IAL to DL, April 5, 1838, JAL typescript, 380.
50. IAL to DL, May 18, 1838, JAL typescript, 376.
51. IAL to Rachel Lapham, April 5, 1838, JAL typescript, 382.
52. IAL to Hannah Lapham, September 27, 1837, JAL typescript, 371.
53. Ibid.
54. IAL to DL, April 5, 1838, JAL typescript, 382.
55. IAL to Rachel Lapham, April 5, 1838, JAL typescript, 381.
56. Increase quotes his brother's letter to him in IAL to DL, June 10, 1838, JAL typescript, 387–389.
57. Ibid.
58. Ibid.
59. C. W. Short to IAL, August 7, 1838, JAL typescript, 392.
60. Ibid.
61. IAL to DL, September 2, 1838, JAL typescript, 394–396.

62. Seneca and Rachel Lapham to IAL, August 13, 1838, JAL 396.
63. IAL journal entry, November 28, 1838, JAL typescript, 401.
64. "Mrs. I. A. Lapham, Ann Maria Allcott, aged 16, Composition," 1832, unpublished school essay, Box 7, Folder 2, Lapham MSS.
65. IAL to DL, December 9, 1838, JAL typescript, 400–401.
66. Professor James H. Eaton, 1875, quoted in Scott, "An Appreciation of Increase Allen Lapham," 26.

CHAPTER 12: *The Milwaukee and Rock River Canal*

1. IAL to Seneca Lapham, July 27, 1836, JAL typescript, 331.
2. I. A. Lapham, *A Documentary History*, 5.
3. Ibid.
4. Schafer, *The Wisconsin Lead Mining Region*, 46.
5. Strong, *History of the Territory of Wisconsin*, 596.
6. *Milwaukee Advertiser*, February 18, 1837, as cited by Schafer, *The Wisconsin Lead Region*, 69.
7. I. A. Lapham, *A Documentary History*, 6.
8. Ibid., 10.
9. Ibid., 12–13.
10. Ibid., 13.
11. Ibid.
12. Ibid., 9.
13. IAL to "Father, & mother & brother & sisters," March 1, 1838, JAL typescript, 377–378.
14. IAL to DL, December 6, 1839, JAL typescript, 414–415.
15. Berquist and Bowers, *Byron Kilbourn and the Development of Milwaukee*, 98.
16. IAL to "all the folks at home—," March 17, 1838, JAL typescript, 398.
17. IAL to Ann Lapham, April 5, 1838, JAL typescript, 401–402.
18. Berquist and Bowers, *Byron Kilbourn and the Development of Milwaukee*, 32.
19. Ibid., 107.
20. Ibid.
21. Smith, *The History of Wisconsin*, 451.
22. Byron Kilbourn to Micajah Williams, June 13, 1838, Williams Papers, Ohio Historical Society, quoted in Berquist and Bowers, *Byron Kilbourn and the Development of Milwaukee*, 99.
23. Seville, "Milwaukee's First Railway," 89–90.
24. Berquist and Bowers, *Byron Kilbourn and the Development of Milwaukee*, 47.
25. IAL to William Lapham, November 27, 1837, JAL typescript, 370–371. Lapham likely visited the site of the Battle of Wisconsin Heights between Madison and Blue Mound. Lapham errs on the year. The Black Hawk War began and ended in 1832.
26. *Milwaukee Sentinel*, January 1, 1837, quoted in Gregory, "Museum Origins in Milwaukee," 54.

27. *Milwaukee Sentinel*, January 8, 1837, quoted in Gregory, "Museum Origins in Milwaukee," 55.

28. The history of the Milwaukee Lyceum, including excerpts from the *Sentinel*, are included in Gregory, "Museum Origins in Milwaukee," 54–62.

29. IAL Lecture to Milwaukee Lyceum, January 23, 1840, JAL typescript, 421.

30. Ibid., 422.

31. Jackie Loohauis, "Public Museum Makes a Mammoth Catch," *Milwaukee Journal Sentinel*, July 31, 2007.

32. IAL Lecture to Milwaukee Lyceum, January 23, 1840, JAL typescript. 423.

33. Ibid., 425–426.

34. Ibid., 427–428.

35. Ibid.

36. Ibid., 428–429.

37. Ibid., 430.

38. IAL to DL, February 2, 1840, Lapham MSS.

CHAPTER 13: *The Botanist*

1. Buck, *Pioneer History of Milwaukee*, 81–116.

2. IAL to DL, January 17, 1841, JAL typescript, 459.

3. Ibid.

4. Ann and Increase Lapham to Rachel and Seneca Lapham, February 28, 1841, JAL typescript, 463–464.

5. IAL to Seneca Lapham, April 13, 1841, JAL typescript, 470–471; IAL to DL, May 6, 1841, JAL typescript, 471–472. The nature of Pazzi's and the Hubbards' illnesses are not recorded, if they were known.

6. Garno Family Tree, 1840 United States Federal Census, Milwaukee West Ward, Milwaukee, Wisconsin Territory, roll 580, page 169, image 363. Records of the Bureau of the Census, Record Group 29, National Archives, Washington, DC. Ancestry.com.

7. Nowhere does Increase record that they took their two-year-old daughter with them. But they might not have wanted to leave her with the grieving Mary Jane Hubbard and Mehitable Alcott and the three little boys. The similar daguerreotypes of Darius and Increase with their daughters may also be evidence that Mary was with them.

8. DL to Seneca Lapham, June 22, 1841, JAL typescript, 476.

9. DL to Seneca Lapham, July 9, 1841, JAL typescript, 476–477.

10. IAL to Seneca Lapham, July 24, 1841, JAL typescript, 477.

11. IAL to Seneca Lapham, April 13, 1841, JAL typescript, 471.

12. This may have been true at first, but sections of Gray's and Torrey's *Flora* were being published and he also consistently read articles on botany in *Silliman's Journal*.

13. IAL to DL, May 21, 1843, JAL typescript, 550.

14. IAL to DL, September 10, 1843, JAL typescript, 555.

15. IAL to DL, May 6, 1841, JAL typescript, 472.

16. Amos Eaton, "Botanical Grammar and Dictionary," appended to Eaton, *Manual of Botany*, 2, 35.

17. Eaton, *Manual of Botany*, 380.

18. IAL Lecture to Milwaukee Lyceum, March 10, 1842, Milwaukee Lyceum, JAL typescript, 506–507.

19. Ibid.

20. J. Barrett to IAL, May 17, 1841, JAL typescript, 473.

21. M. A. Curtis to IAL, August 9, 1841, JAL typescript, 479–480.

22. IAL to C. W. Short, April 21, 1834, JAL typescript, 249–250; C. W. Short to IAL, March 10, 1843, JAL typescript, 546.

23. In examining nineteen mounted specimens of *Habernaria hooker's tor* [Hooker's Orchid] at the Milwaukee Public Museum—Increase Lapham's among them and the oldest specimen—one sees that his specimen is the most informative. Not only can you see more of the details of this mature plant, including part of the root structure, on Increase's specimen, but on the page his specimen makes a bold and artistic design.

24. IAL Lecture to Milwaukee Museum, on Botany, March 10, 1842, JAL typescript, 507–508.

25. Asa Gray to IAL, June 28, 1836, JAL typescript, 333–334.

26. Ibid. We don't have Increase's letter, but Gray on June 28 records that he is responding to a letter of June 18, 1836, from Detroit.

27. Ibid.

28. Asa Gray to IAL, September 19, 1842, JAL typescript, 525.

29. Ibid.

30. Asa Gray to IAL, March 11, 1843, JAL typescript, 546.

31. IAL to Asa Gray, December 21, 1844, JAL typescript, 575–576.

32. Ibid.

33. Asa Gray to IAL, May 2, 1846, JAL typescript, 598.

34. "John Torrey," http://etcweb.princeton.edu/Campus WWW/Companion/torrey_john.html, from Alexander Leitch, *A Princeton Companion*, copyright Princeton University Press (1978).

35. John Torrey to IAL, August 26, 1840, JAL typescript, 449–450.

36. John Torrey to IAL, May 7, 1841, JAL typescript, 472–473.

37. IAL to John Torrey, September 29, 1841, JAL typescript, 482.

38. IAL to John Torrey, November 20, 1841, JAL typescript, 482.

39. John Torrey to IAL, July 19, 1845, JAL typescript, 472–473.

40. IAL to C. W. Short, February 17, 1834, JAL typescript, 246; C. W. Short to IAL, October 25, 1835, JAL typescript, 286.

41. IAL Lecture on to Milwaukee Lyceum, March 10, 1842, JAL typescript, 508.

42. Ibid., 511–512, 514.

43. The botanical diaries of Increase Lapham can be found in the Lapham MSS at the Wisconsin Historical Society, Box 13, Folder 6, and Box 20, Folder 17.

44. Clipping from *Milwaukee Advertiser*, July 22, 1837, Box 13, Folder 6, Lapham MSS.

45. Increase was reluctant to call it the Fox River because Wisconsin had a bigger Fox River that drains into Green Bay. He correctly believed that two rivers so named would cause confusion. He preferred both Indian names: Pishtaka and Neenah.

46. Clipping from *Milwaukee Advertiser*, July 22, 1837, Box 13, Folder 6, Lapham MSS.

47. See the 1827–1828, 1831, 1833, and 1835 botanical journals, Box 13, Folder 6, Lapham MSS.

48. See these lists in IAL's journals, Box 13, Folder 6, Lapham MSS.

49. Another plant book titled in pencil "Wisconsin Plants," undated and unpaged, may date to just before this one. In this five-page, four-by-six-inch booklet he lists 395 species of Wisconsin plants. Box 20, Folder 17, Lapham MSS.

50. IAL, "Plants of Wisconsin 1845," Box 20, Folder 17, Lapham MSS.

51. Manuscript of "Plants of Wisconsin," p. 3, Box 20, Folder 17, Lapham MSS.

52. IAL to "Dear Parents, Brothers & Sisters," February 17, 1842, JAL typescript, 497.

53. IAL to DL, February 18, 1842, JAL typescript, 498.

54. IAL to DL, October 31, 1842, JAL typescript, 532.

55. Ibid.

56. Buck, *Pioneer History of Milwaukee*, 132.

57. IAL to DL, December 10, 1842, JAL typescript, 533.

58. IAL, "Journey from Milwaukee, the Greatest, to Green Bay, the Oldest City of Wisconsin," JAL typescript, 535–544.

59. IAL to DL, May 21, 1843, JAL typescript, 550–551.

CHAPTER 14: *The Uses of Writing and Publishing*

1. IAL to Seneca Lapham, September 17, 1844, JAL typescript, 570. His tenants were D. Van Deren and R. N. Messenger.

2. IAL to DL, February 4, 1838, JAL typescript, 375.

3. A "rum hole" was a shack, a backroom, or even a dugout in which alcoholic drinks, often distilled by the proprietor from questionable ingredients, were served to any and all customers. Such hovels often were shelters for the homeless, the transient, and the alcoholic. They were not as uplifting or dignified as taverns and the opening of Milwaukee's first brewery came as a relief to many of the city's respectable drinkers, especially its burgeoning community of Germans.

4. Buck, *Pioneer History of Milwaukee*, 2:103, 104, 138,144, 165, 198.

5. Buck, *Pioneer History of Milwaukee*, 2:121–123.

6. IAL to DL, March 7, 1844, JAL typescript, 564.

7. Hawks, "I. A. Lapham, Wisconsin's First Scientist," 135.

8. Edmonds, "Increase A. Lapham and the Mapping of Wisconsin," 165.

9. "Jedidiah Morse," *The New Columbia Encyclopedia*, 4th ed. (New York and London, Columbia University Press, 1975), 1837.

10. See Lapham MSS, Box 27, Folder 1.

11. Ibid.

12. I. A. Lapham, *A Geographical and Topographical Description of Wisconsin*, 5.

13. Ibid., 9.

14. Ibid., 11, 13, 14.

15. Ibid., 53–66.

16. Ibid., 65.

17. These quotations fall into the first general chapter on Wisconsin Territory, pages 5 through 96. See particularly pp. 77–84 on botany.

18. Ibid., 92.

19. Asa Gray to IAL, August 9, 1844, JAL typescript, 567.

20. IAL to Seneca Lapham, September 17, 1844, JAL typescript, 570.

21. The equally grandiose full title of the second edition was *Wisconsin: Its Geography and Topography, History, Geology, and Mineralogy: Together With Brief Sketches of its Antiquities, Natural History, Soil, Productions Population and Government.*

22. Quaife, "Increase Allen Lapham, First Scholar of Wisconsin," 9.

23. Owens, *Report of a Geological Survey*, 452.

24. Ibid.

25. IAL to DL, September 15, 1845, JAL typescript, 587.

26. IAL to Seneca Lapham, July 17, 1845, JAL typescript, 584.

27. Entry in Lapham Family Bible, Lapham MSS.

28. IAL to DL, September 15, 1845, JAL typescript, 587.

29. Ibid.

30. Edmonds, "Increase A. Lapham and the Mapping of Wisconsin," 165.

31. Ibid., 171.

32. IAL to DL, March 26, 1846, JAL typescript, 593.

33. Charles C. Rafn to IAL, January 29, 1846, JAL typescript, 592; IAL to Charles C. Rafn, October 16, 1846, Lapham MSS.

34. IAL journal entry, July 20, 1846, JAL typescript, 602–604.

35. IAL to Ann Lapham, January 15, 1847, JAL typescript, 609.

36. Ibid., 609–610.

37. Ibid.

38. IAL to Ann Lapham, January 20, 1847, JAL typescript, 610.

39. Ibid., 610–611.

40. Seventh Census of the United States, 1850 United States Federal Census, Milwaukee Ward 2, Milwaukee, Wisconsin, roll M432_1003, page 271B, image 212 (National Archives Microfilm Publication M432, 1009 rolls), Records of the Bureau of the Census, Record Group 29, National Archives, Washington, DC. Ancestry.com. Images reproduced by FamilySearch.

41. IAL to Ann Lapham, May 26–29, 1847, JAL typescript, 614–617.

42. Ann Lapham to IAL, May 30, 1847, JAL typescript, 620.
43. IAL to Ann Lapham, May 29, 1847, JAL typescript, 617.
44. IAL to Ann Lapham, June 3, 1847, JAL typescript, 622.
45. IAL to Ann Lapham, June 7, 1847, JAL typescript, 624.
46. Ibid.
47. IAL to Ann Lapham, June 8, 1847, JAL typescript, 626.
48. IAL to Ann Lapham, June 10, 1847, JAL typescript, 627.
49. IAL to Ann Lapham, June 12, 1847, JAL typescript, 627.
50. IAL to Ann Lapham, June 15, 1847, Lapham MSS.
51. IAL to DL, September 1, 1847, JAL typescript, 633.
52. Ibid.
53. James H. Eaton, quoted in Scott, "An Appreciation of Increase Allen Lapham," 26.

CHAPTER 15: *Civic Life in the New State*

1. IAL, "Geological Notes of a Tour from Milwaukee to Blue Mounds and Back, July 12–20, 1848," JAL typescript, 720.
2. Ibid., 721.
3. IAL to Governor Nelson Dewey, December 9, 1848, JAL typescript, 745.
4. IAL, "Geology: A Lecture for the Milwaukee High School, Delivered March, 1848," JAL typescript, 680–699, 685.
5. Ibid., 691.
6. Ibid., 682.
7. Ibid., 699.
8. S. S. Sherman, "Increase Allen Lapham, LL.D.: Biographical Sketch," pamphlet (Milwaukee: The Old Settlers' Club, 1876), 26. Read before the Old Settlers' Club, December 11, 1875, Milwaukee, Wisconsin.
9. IAL to DL, March 29, 1848, JAL typescript, 704.
10. Milwaukee Female Seminary flyer advertising the school's opening September 14, 1848. Replica in Milwaukee County Historical Society Research Room, Lapham Collection, Box 4, File 5.
11. Ibid.
12. Kellogg, "The Origins of Milwaukee College," 399.
13. Kieckhefer, "Milwaukee-Downer College Rediscovers Its Past,", 213.
14. Article in the *Milwaukee Sentinel and Gazette*, April 14, 1851, quoted in Hawks, "Increase Lapham, Wisconsin's First Scientist," 116.
15. Bascom, Butler, and Kerr, *Centennial Records of the Women of Wisconsin*, 14.
16. Today, after a series of institutional changes, the campus is part of the University of Wisconsin–Milwaukee. In 1895, Milwaukee College merged with Downer College of Fox Lake, Wisconsin, to become Milwaukee-Downer College, which was housed in an elegant campus in the city of Milwaukee. Milwaukee-Downer was absorbed by Lawrence University, Appleton,

Wisconsin, in 1964, and the Downer campus became an integral part of the University of Wisconsin–Milwaukee.

17. Mary J. Lapham, "A Four Minute Fagot," n.d., signed, handwritten manuscript, Lapham MSS.

18. IAL to DL, April 29, 1849, JAL typescript, 758.

19. Ibid.

20. IAL to Ann Lapham, September 12, 1849, JAL typescript, 774.

21. Ibid., 777.

22. IAL to William Lapham, June 27, 1849, JAL typescript, 759.

23. IAL to Joseph Henry, September 1, 1849, JAL typescript, 770.

24. IAL, "Of the Historical Society of Wisconsin," 1849, JAL typescript, 798.

25. Draper, *Collections of the State Historical Society of Wisconsin*, xxx–xxxvi.

26. Lyman C. Draper to IAL, May 11, 1855, JAL typescript, 847.

27. IAL, "Of the Historical Society of Wisconsin," JAL typescript, 798.

28. Ibid., 796, 799, and 798.

29. Ibid., 797.

30. Ibid.

31. Ibid., 800.

32. Ibid., 797.

33. Ibid., 799–801.

34. IAL to Ann Lapham, September 12, 1849, JAL typescript, 774.

CHAPTER 16: *The Antiquities of Wisconsin*

1. IAL, "Letter to the Editor," *Milwaukee Advertiser*, November 24, 1836, quoted in Nurre, introduction to *Antiquities*, xi.

2. Birmingham and Eisenberg, *Indian Mounds of Wisconsin*, 5. Written almost 150 years after Lapham's, Birmingham and Eisenberg's book is only the second work for a general audience on this subject.

3. Birmingham, foreword to *Antiquities*, vii.

4. I. A. Lapham mentions Long's book on December 13, 1827, when he was sixteen years old. Thomas and Conner, *The Journals*, 16 and 115.

5. DL to IAL, July 12, 1828, JAL typescript, 73.

6. In his writing Increase referred often to Richard C. Taylor's 1838 article on effigy mounds near Madison. See Taylor, "Notes Respecting Certain Indian Mounds," 88–101.

7. I. A. Lapham, *Wisconsin: Its Geography and Topography*, 16.

8. IAL to DL, September 9, 1846, JAL typescript, 604.

9. Ibid., 605.

10. IAL to Charles Rafn, the Royal Society of Northern Antiquaries, October 16, 1846, Lapham MSS.

11. We actually know very little about his finances; all those papers were probably destroyed by his daughter Julia, who served as executor of her father's papers.

12. Founded in 1812 as an independent library, with early members including

Thomas Jefferson and John Adams, the American Antiquarian Society still exists. Samuel·F. Haven was secretary from 1838 until his death in 1881, editing their *Transactions* and presiding over the building of their new library, which stands today.

13. IAL to Samuel F. Haven, December 24, 1849, JAL typescript, 785.

14. IAL to Samuel F. Haven, March 12, 1850, JAL typescript, 808.

15. IAL to Joseph Henry, March 12, 1850, JAL typescript, 808–809.

16. IAL to James Doty, March 12, 1850, Lapham MSS.

17. IAL to Charles Whittlesey, March 19, 1850, JAL typescript, 809–810.

18. IAL to Ann Lapham, May 1, 1850, JAL typescript, 814; Edward Foreman to IAL, March 29, 1850, JAL typescript, 811. "John H." is never given a last name in Increase's letters or notes. It may be that he was John Hubbard, Increase's nephew, and that Increase would not want to be accused of nepotism.

19. IAL, "Notes," Racine, May 27, 1850, JAL typescript, 822.

20. Ibid., 823.

21. Ibid.

22. Ibid., May 30, 1850, 826.

23. Ibid., May 31, 1850, 827.

24. IAL to Ann Lapham, June 28, 1850, JAL typescript, 827, and original in Lapham MSS.

25. Ibid, 828.

26. Ibid.

27. IAL to Ann Lapham, June 28, 1850, Lapham MSS.

28. IAL, public lecture, "Geology of Wisconsin," JAL typescript, 666.

29. Later glacial scientists explained that the Ohio and Missouri Rivers formed at the southern edges of the thawing continental glacier and carried away vast amounts of meltwater to the Mississippi River. See Merrill, *The First One Hundred Years of American Geology*, 629.

30. IAL to Ann Lapham, June 29, 1850, Lapham MSS.

31. Ibid.

32. I. A. Lapham, *Antiquities*, 50.

33. IAL to Ann Lapham, July 3, 1850, JAL typescript, 833.

34. Robert Jeske, interview by Martha Bergland, February 7, 2009, Milwaukee, Wisconsin.

35. IAL to Ann Lapham, July 3, 1850, Lapham MSS.

36. Ibid.

37. Ibid.

38. Increase mentions his "retrograde motion" in a letter to his wife. IAL to Ann Lapham, July 13, 1850, JAL typescript, 843.

39. IAL to Ann Lapham, July 10, 1850, Lapham MSS.

40. IAL to Ann Lapham, July 13, 1850, Lapham MSS.

41. Ann Lapham to IAL, July 6, 1850, Lapham MSS.

42. IAL to Ann Lapham, July 13, 1850, Lapham MSS.

43. Ibid.

44. IAL, journal entry, July 20, 1850, JAL typescript, 844.

45. According to notes on Lapham family history in the Lapham MSS, Nancy Lapham died at age thirty-one. The date of her death is not known, nor is the date of her birth given in these pages on family history. See Lapham MSS, Box 1, Folder 7.

46. IAL to Samuel F. Haven, October 2, 1850, JAL typescript, 848.

47. Ibid.

48. IAL, "Lecture delivered January 16, 1851, before the Young Men's Association and Citizens at Free Congregational Church, Milwaukee, by I. A. Lapham," JAL typescript, 854–871.

49. IAL to Samuel F. Haven, June 28, 1851, JAL typescript, 884.

50. Ibid.

51. IAL, "Letter to the Editor," *Milwaukee Sentinel*, June 16, 1851, JAL typescript, 880.

52. IAL, notes, October 8, 1851, JAL typescript, 887–888.

53. Both Holy Hill in Washington County and Lapham Peak (now Lapham Peak State Forest) in Waukesha County are "kames," or cone-shaped hills of unsorted boulders, cobbles, gravel, sand, and clay that formed as glacial debris was carried by rushing water down a vertical shaft in a melting glacier, the drift piling up below like sand in an hour glass.

54. IAL to Samuel F. Haven, November 1, 1851, JAL typescript, 893–894, and original in Lapham MSS; IAL to Joseph Henry, November 1, 1851, JAL typescript, 894–895, and original in Lapham MSS.

55. Joseph Henry to IAL, November 22, 1851, Lapham MSS.

56. IAL to Samuel F. Haven, May 24, 1852, JAL typescript, 929–930.

57. IAL, journal entry, May 26, 1852, JAL typescript, 930.

58. IAL, "Glances at the Interior, Watertown, May 27, 1852," JAL typescript, 931–932.

59. IAL to Messrs. Editors, *Milwaukee Sentinel* (Fox Lake), May 28, 1852, JAL typescript, 932–934.

60. IAL to Messrs. Editors, *Milwaukee Sentinel*, June 1, 1852, JAL typescript, 936–937.

61. IAL to Messrs. Editors, *Milwaukee Sentinel*, June 2, 1852, JAL typescript, 936–937.

62. IAL to Mssrs. Editors, *Milwaukee Sentinel*, June 11, 1852, JAL typescript, 941.

63. Ibid., 942.

64. IAL to Ann Lapham, June 14, 1852, JAL typescript, 949–950.

65. IAL to Samuel F. Haven, July 30, 1852, Lapham MSS; Samuel F. Haven to IAL, August 6, 1852, Lapham MSS; IAL to Joseph Henry, August 14, 1852, Lapham MSS.

66. IAL to Samuel F. Haven, draft, August 14, 1852, Lapham MSS.

67. Increase wrote a bibliography and sent it on to Joseph Henry on May 6,

1853. Samuel F. Haven, seeing it, decided it should be more extensive and completed a bibliography that was published in 1856 under his own name. Samuel F. Haven, "Archeology of the United States, or, Sketches Historical and Bibliographical, of the Progress of Information and Opinion Respecting Vestiges of Antiquity in the United States," Smithsonian Contributions to Knowledge, vol. 8 (Washington, DC: Smithsonian, 1856).

68. Joseph Henry to IAL, March 26, 1853, Lapham MSS.
69. IAL, quoted in Joseph Henry's letter to IAL, July 27, 1854, Lapham MSS.
70. Ibid.
71. Spencer F. Baird to IAL, November 8, 1855, Lapham MSS.
72. The work was also published as one of four articles in the Smithsonian's Contributions to Knowledge series, volume 7.
73. Even more specific are the sections within each category. Under "Purpose" were mounds of defense, sacred enclosures, mounds of sacrifice, and mounds of sepulture. This was followed by several chapters organized simply by location of the mounds—the south or the northwest. Their third system of classification was by the sort of implements or ornaments found in the mounds—metal, stone, bone. See the table of contents in Squier and Davis, *Ancient Monuments of the Mississippi Valley*.
74. I. A. Lapham, introduction to *Antiquities*, 2.
75. Ibid., 2–3.
76. Lapham did not do all the work of the book himself. He refers to many other writers on the subject, including Squier and Davis. And he includes quotations from and plates based on drawings of L. L. Sweet, Logan Crawford, William H. Canfield, Stephen. P. Lathrop, and his friend, Philo Hoy.
77. Though much of Aztalan is thought to have had ceremonial use, parts were later found to be fortifications. Birmingham and Eisenberg, *Indian Mounds of Wisconsin*, 160–161.
78. Lapham, *Antiquities*, 50.

CHAPTER 17: *Disappointments and False Starts*

1. IAL to J. W. Van Cleve, November 12, 1852, JAL typescript, 956–957.
2. IAL to Ann Lapham, January 23, 1853, JAL typescript, 959.
3. Merrill, *The First One Hundred Years of American Geology*, 233–234.
4. IAL to James Hall, August 12, 1850, JAL typescript, 844–846.
5. Ibid., 846.
6. IAL to Ann Lapham, February 13, 1853, JAL typescript, 965.
7. Ibid., 965.
8. Ibid., 966.
9. Ibid., 967.
10. IAL to Ann Lapham, March 1, 1853, JAL typescript, 967.
11. IAL to Ann Lapham, July 31, 1853, JAL typescript, 977–978, and original in Lapham MSS.

12. James Hall to IAL, November 27, 1857, Lapham MSS.

13. I. A. Lapham, "The Grasses of Wisconsin," 397–488; 398.

14. Albert C. Ingham to IAL, January 6, 1854, JAL typescript, 985.

15. IAL to Charles Mason, June 1, 1855, JAL typescript, 1031.

16. IAL to Charles Mason, July 14, 1856, JAL typescript, 1032–1033.

17. IAL to Ann Lapham, July 17, 1856, JAL typescript, 1034.

18. S. T. Shugert to IAL, August 22, 1856, JAL typescript, 1035.

19. S. T. Shugert to IAL, August 20, 1856, JAL typescript, 1035.

20. Charles Mason, diary entry, September 10, 1856, trans. Charles Mason Remey, Charles Mason MSS, Library of Congress Archives, Washington, DC. The "Secretary" mentioned in the letter was Robert McClelland, secretary of the Interior and Mason's superior.

21. Joseph Holt to IAL, January 20, 1858, JAL typescript, 1052–1053.

22. Merrill, *The First One Hundred Years of American Geology*, 127 and 209. See also Bean, "State Geological Surveys of Wisconsin," 204.

23. Lyman C. Draper to IAL, December 6, 1854, JAL typescript, 993.

24. Clarke, *James Hall of Albany*, 287.

25. Ibid.

26. Bean, "State Geological Surveys of Wisconsin," 206. See also Hawks, "Increase A. Lapham: Wisconsin's First Scientist," 229.

27. Percival, *Annual Report of the Geological Survey*, 7–8.

28. IAL to E. Desor, January 5, 1857, JAL typescript, 1071.

29. Bean, "State Geological Surveys of Wisconsin," 207.

30. Daniels, *Annual Report of the Geological Survey*, 62.

31. Bean, "State Geological Surveys of Wisconsin," 209.

32. IAL to Moses M. Davis, January 15, 1859, Lapham MSS.

33. Hall, *Report of the Superintendent of the Geological Survey*, 36–37.

34. I. A. Lapham, "The Penokee Iron Range," 392–393.

35. *Milwaukee Free Democrat*, November 11, 1859, as referenced by JAL typescript, 1156. JAL also states on same page, "This discovery was also announced in Silliman's Journal vol. 29, 2d series p. 145."

CHAPTER 18: *A Quaker in Wartime*

1. Some 30 opponents of the Kansas–Nebraska Act met in a schoolhouse in Ripon, Wisconsin, in February 1856 and called themselves Republicans. Ripon advertises itself today as the birthplace of the Republican Party.

2. Kleppner, *The Third Electoral System*, 182.

3. "List of Premiums and Regulations for the Ninth Annual Fair," 174.

4. Nicolay, *A Short Life of Abraham Lincoln*, 144.

5. Lincoln, "Address before the State Agricultural Society, Sept. 30, 1859," 389–394.

6. IAL, journal entry, May 18, 1860, JAL typescript, 1166.

7. Ibid., July 6, 1860, 1167.

8. James R. Albach, *Annals of the West, Embracing a Concise Account of Principal Events Which Have Occurred in the Western States and Territories* (Pittsburgh: W. S. Haven, 1856).

9. IAL to DL, September 7, 1847, JAL typescript, 634–636.

10. IAL to "The Honorable, The Senate and Assembly of the State of Wisconsin," January 1850, JAL typescript, 1356, and original in Lapham MSS.

11. IAL, journal entry, April 8, 1858, JAL typescript, 1113.

12. George Meade to IAL, July 3, 1860, JAL typescript, 1167–1168.

13. Samuel Stone to IAL, August 26, 1860, JAL typescript, 1180.

14. Samuel Stone to IAL, December 17, 1860, JAL typescript, 1161.

15. George G. Meade to IAL, March 4, 1861, JAL typescript, 1184.

16. IAL, journal entry, January 1, 1860, JAL typescript, 1164.

17. Miller, "American Pioneers in Meteorology," 189.

18. Frank Rives Milliken, *Joseph Henry, Father of the Weather Service*, Joseph Henry Papers Project, 1997–2007, available online at the Smithsonian Institution Archives, Washington, DC, at siarchives.si.edu/history/jhp/joseph03.htm.

19. Asa Horr to IAL, March 11, 19, and 27, 1861, April 7, 1861, and May 6 and 29, 1861, JAL typescript, 1185–1189.

20. IAL to Asa Horr, May 10, 1861, JAL typescript, 1189.

21. Still, *Milwaukee: The History of a City*, 168–200.

22. Current, *The Civil War Era*, 374–384.

23. IAL to John F. Potter, January 3, 1862, Lapham MSS.

24. Benjamin Silliman and James D. Dana to "Mssrs. Henry & Seneca Lapham," January 22, 1862, Lapham MSS.

25. Increase's contribution, unattributed, was first published in 1863 and covered twelve columns of small type over six pages. See Ripley and Dana, "Wisconsin," 496–502.

26. IAL to George Ripley, May 9, 1862, Lapham MSS.

27. IAL to William H. Watson, June 6, 1862, Lapham MSS.

28. Charles Lesquereux to IAL, July 14, 1862, JAL typescript, 1211–1213.

29. Ibid.

30. Samuel Stone to IAL, February 27, 1863, Lapham MSS.

31. Charles Hubbard to Julia Lapham, April 7, 1863, Lapham MSS.

32. Leo Lesquereux to IAL, March 21, 1863, JAL typescript, 1220.

33. IAL to Byron Kilbourn, April 7, 1863, JAL typescript, 1222.

34. Byron Kilbourn to IAL, April 8, 1863, JAL typescript, 1222–1223.

35. IAL to Alexander Mitchell, May 5, 1865, JAL typescript, 1247–1248.

36. Edmonds, "Increase A. Lapham and the Mapping of Wisconsin," 165.

37. I. A. Lapham, "On the Climate of the Country Bordering Upon the Great North American Lakes," 58–60. This Lapham insight leads directly to the phrase familiar to all Wisconsin citizens who read or listen to weather forecasts: "Cooler near the lake."

38. IAL to Henry Lapham, August 6, 1864, Lapham MSS.

39. Ibid.

40. The *Milwaukee Sentinel* provided extensive coverage of the efforts of the various local units of the Wisconsin Soldiers Aid Society, reporting details of each meeting and listing attendees and donations, usually on page one.

41. IAL to Mrs. A. E. Stone, January 30, 1864, JAL typescript, 1287.

42. "List of the Executive Committee and Subcommittees of Soldiers Home Fair," *Milwaukee Sentinel*, May 11, 1865.

43. "Plans for Department of Geology and Natural History at Fair," *Milwaukee Sentinel*, May 17, 1865.

44. "Sale of Geology and Natural History Specimens," *Milwaukee Sentinel*, August 3, 1865.

CHAPTER 19: *Prophetic Thinking and Forecasts*

1. Williamson and Daum, *The American Petroleum Industry*, 123.

2. I. A. Lapham, "The Forest Trees of Wisconsin," 196–197.

3. Ibid., 197.

4. Lapham, Knapp, and Crocker, *Report of the Disastrous Effects of the Destruction of Forest Trees*, 5.

5. Ibid., 31.

6. IAL, journal entry, June 24, 1868, JAL typescript, 1296.

7. Ibid.

8. C. H. Doerflinger to IAL, December 16, 1858, JAL typescript, 1298.

9. IAL to John Lawrence Smith, December 22, 1868, JAL typescript, 1299.

10. Smith, "A New Meteoric Iron," 271.

11. Cleveland Abbe to IAL, JAL typescript, 1326.

12. Thompson, *Graveyard of the Lakes*, 210.

13. Holton, "Proceedings of the Second Annual Meeting," 146. The National Board of Trade was a precursor to the Chamber of Commerce of the United States.

14. Miller, "New Light on the Beginnings of the Weather Bureau," 67–68.

15. The story of his California trip is told in letters to Mary, Julia, Henry, Charles, and Seneca Lapham, April 14–May 17, 1870, JAL typescript, 1329–1342.

16. IAL to Mary Lapham, April 14, 1870, JAL typescript, 1331.

17. IAL to Julia Lapham, April 15, 1870, JAL typescript, 1332.

18. IAL to Henry Lapham, April 21, 1870, JAL typescript, 1335.

19. IAL to Charles Lapham, April 25, 1870, JAL typescript, 1336.

20. IAL to Julia Lapham, May 1, 1870, JAL typescript, 1339.

21. IAL to Charles Lapham, May 4, 1870, JAL typescript, 1341.

22. IAL to Mary Lapham, May 17, 1870, JAL typescript, 1342.

23. I. A. Lapham, "On the Discovery of Fish Remains in Glacial Blue Clay of the New Chicago Tunnel Under Lake Michigan," *Milwaukee Evening Wisconsin*, July 13, 1870, and JAL typescript, 1348–1349.

24. Winchell, "Increase Allen Lapham," 11.

25. IAL to Julia Lapham, November 27, 1870, JAL typescript, 1394.

26. IAL to Julia Lapham, December 8, 1870, Lapham MSS.

27. Ibid.

28. Garrick Mallery to IAL, December 20, 1870, Lapham MSS.

29. Albert J. Myer to IAL, December 21, 1870, Lapham MSS.

30. IAL to Albert J. Myer, December 28, 1870, JAL typescript, 1400.

31. IAL to Julia Lapham, January 23, 1871, JAL typescript, 1435.

32. IAL to Henry Lapham, January 24, 1871, JAL typescript, 1436.

33. IAL to Seneca Lapham, January 27, 1871, JAL typescript, 1437. Increase Lapham encountered Mrs. Alexander Mitchell, wife of the railroad magnate and banker of Milwaukee, on the train. She was overseeing a lavish Florida winter home on 140 acres south of Jacksonville known as Villa Alexandria. Her brother, Harrison Reed, formerly an editor of the *Milwaukee Sentinel*, was governor of Florida at the time of Lapham's visit.

34. IAL to Mary Lapham, January 30, 1871, JAL typescript, 1439–1440.

35. IAL to Julia Lapham, February 3, 1871, JAL typescript, 1441.

36. I. A. Lapham, "Oconomowoc Lake and Other Small Lakes of Wisconsin," 32.

37. IAL to Henry Lapham, July 28, 1871, Lapham MSS.

38. Samuel Stone to IAL, October 12, 1871, JAL typescript, 1449.

39. Ibid.

40. Samuel Stone to IAL, October 13, 1871, Lapham MSS.

41. IAL to Samuel Stone, October 14, 1871, Lapham MSS.

42. I. A. Lapham, "The Great Fires of 1871 in the Northwest," 678–681.

43. IAL to Cyrus B. Comstock, December 1, 1871, JAL typescript, 1453–1454.

44. IAL to Mary Lapham, January 19, 1872, Lapham MSS.

45. Joseph Henry to IAL, February 7, 1872, Lapham MSS.

46. H. W. Howgate to IAL, May 11, 1872, JAL typescript, 1433.

47. IAL to Albert J. Myer, June 3, 1872, JAL typescript, 1433.

48. IAL to Mary Lapham, June 26, 1872, JAL typescript, 1465–1466.

49. IAL to Mary Lapham, July 1, 1872, JAL typescript, 1468.

50. IAL, journal entry, July 2, 1872, JAL typescript, 1468.

51. Ibid., July 3, 1872.

52. IAL to Mary Lapham, July 1, 1872, and journal entry, June 30, 1872, JAL typescript, 1467–1469.

53. Nelson P. Hulst to Mary Lapham, December 21, 1896, JAL typescript, 1469–1471.

54. IAL to Cadwallader C. Washburn, October 29, 1872, Lapham MSS.

55. IAL to S. F. Baird, October 30, 1872, JAL transcript, 1479.

56. Ibid., 1478–1479.

CHAPTER 20: *A Complete Survey*

1. IAL, journal entry, February 13, 1870, JAL typescript, 1323.

2. Kroncke, "The Wisconsin Academy of Arts, Sciences and Letters," 163.

3. *Transactions of the Wisconsin Academy of Sciences, Arts and Letters* was the academy's annual report and the vehicle for its scientific papers. Excepting the first volume, which covered two years, it was published annually and sent to other academies, universities, libraries, and societies, usually in exchange for their own reports. *Transactions* published many important papers, especially from University of Wisconsin professors, including Edward Birge, Aldo Leopold, Increase Lapham, Thomas Chamberlin, Roland Irving, and Philo Hoy. Although the academy still exists as a nonprofit that publishes a quarterly magazine, *Wisconsin People and Ideas*, and takes on special projects in the arts and sciences, it is no longer financed by the state. *Transactions* ceased publication in 2003. As its first editor, Increase was in charge of only volume 1. See Lapham, *Transactions of the Wisconsin Academy*.

4. I. A. Lapham, *Transactions of the Wisconsin Academy*, 1:36.

5. Thomas C. Chamberlin to IAL, December 5, 1872, Lapham MSS.

6. "An Act to Provide for a Complete Geological Survey of Wisconsin," in *Geology of Wisconsin*, 2:5.

7. Bean, "State Geological Surveys of Wisconsin," 211.

8. IAL, journal entry, April 8, 1873, JAL typescript, 1487.

9. Kroncke, "The Wisconsin Academy of Arts, Sciences and Letters," 163.

10. Roland D. Irving to IAL, October 22, October 26, and November 18, 1871, JAL typescript, 1451–1453.

11. Roland D. Irving, introduction to "Geology of the Eastern Lake Superior District," in *Geology of Wisconsin*, 3:53–54.

12. Ibid., 54.

13. Hawks, "Increase A. Lapham: Wisconsin's First Scientist," 254.

14. I. A. Lapham, "On the Relation of the Wisconsin Geological Survey to Agriculture," 207–210.

15. Nesbit, *Wisconsin: A History*, 383–385.

16. I. A. Lapham, "Report of Progress and Results for the Year 1873," in *Geology of Wisconsin*, 2:11.

17. Ibid.

18. IAL, journal entry, February 6, 1874, JAL typescript, 1514.

19. Ibid., February 16, 1874, 1519.

20. Ibid., February 26, 1874, 1524.

21. IAL to Samuel Stone, April 16, 1874, JAL typescript, 1525.

22. I. A. Lapham, "Report of Progress and Results for the Year 1874," in *Geology of Wisconsin*, 2:45.

23. *History of Milwaukee, Wisconsin*, 576.

24. These words are found in a dozen letters in the geological notes in the JAL typescript, uu-iii. Julia Lapham numbered the geological notes in the typescript alphabetically.

25. Seneca Lapham to Samuel Stone, February 17, 1875, JAL typescript, 1532.

26. Thomas C. Chamberlin to IAL, February 18, 1875, JAL typescript, 1533–1534; W. W. Daniells to IAL, February 22, 1875, JAL typescript, 1535.

27. Merrill, *The First One-Hundred Years of American Geology*, 484–485.

28. Roland D. Irving to IAL, April 20, 1870, JAL typescript, 1543.

29. "Geological Survey of Wisconsin," 398–400.

30. Bean, "State Geological Surveys of Wisconsin," 213.

31. Dott, "Rock Stars," 30–31.

32. The term "Kettle Moraine" was first used by T. C. Chamberlin, "General Geology, Quaternary Age," in *Geology of Wisconsin*, 1:275.

33. Bean, "State Geological Surveys of Wisconsin," 215.

34. Robert C. Nesbit, *Wisconsin: A History*, 427.

35. Thomas C. Chamberlin, *The State Journal*, n.d., quoted in *Report of the Trigintennial Meeting with a Biological and Statistical Record (1897)*, by Yale University Class of 1867 (New York: J.G.C. Binney, 1897): 274. University of California Digital Library, www.archive.org/stream/reportoftriginteooyalerich #page/274.

36. O. W. Wight, "Report of Progress and Results for the Year 1875," in *Geology of Wisconsin*, 2:67.

37. Thomas C. Chamberlin, introduction to *Geology of Wisconsin*, 1:viii.

CHAPTER 21: *So Calm and Peaceful*

1. IAL to Moses M. Strong, February 17, 1875, JAL typescript, 1536.

2. Moses M. Strong to IAL, February 18, 1875, JAL typescript, 1536.

3. I. A. Lapham, "The Law of Embryonic Development," 257–260.

4. E. A. Calkins to IAL, May 7, 1875, JAL transcript, 1565.

5. IAL to Elias A. Calkins, n.d., JAL typescript, 1566.

6. IAL, journal entry, July 27, 1875, JAL typescript, 1567.

7. Ibid., July 19, 1875, JAL typescript, 1568.

8. IAL to W. W. Field, August 2, 1875, JAL typescript, 1569.

9. IAL, journal entry, August 12, 1875, JAL typescript, 1568.

10. IAL to S. F. Baird, August 21, 1875, JAL typescript, 1570.

11. I. A. Lapham, "Mound Builders. Description of the Samples Shown at the State Fair," *Milwaukee Evening Wisconsin*, Sept. 11, 1975, JAL typescript, 1571.

12. IAL to S. F. Baird, August 21, 1875, JAL typescript, 1570; P. R. Hoy to IAL, May 7, 1875, JAL typescript, 1566.

13. IAL, journal entry, September 13, 1875, JAL typescript, 1579.

14. I. A. Lapham, "Oconomowoc Lake, and Other Small Lakes of Wisconsin," 31.

15. Hoy, "Increase A. Lapham, LL.D.," 264–267. Hoy's paper was also read to the Sixth Annual Meeting of the Wisconsin Academy of Sciences, Arts and Literature, February 8, 1876.

16. Sherman, "Increase Allen Lapham, LL.D.," 51.

17. Ibid., 20–21.

18. Sweet, "Catalogue of the Wisconsin State Mineral Exhibit at Philadelphia, 1876," 3–4. The contents of Lapham's cabinet were returned to Wisconsin and housed in the University of Wisconsin's first Science Hall. They were destroyed

in a fire December 1, 1884. The cabinet resides at the Milwaukee County Historical Society.

19. Ibid.
20. Atwood, *The Centennial Exhibition of 1876*, 490–499.
21. Seneca Lapham to Mary Lapham, November 3, 1875, Lapham MSS.

This book would have been much more difficult to write without the determination of Increase Lapham to preserve and order his papers and the determination of his daughters, primarily Julia, to use his papers to keep his memory alive. Increase kept journals most of his life, beginning when he was sixteen. And he kept drafts of many of the letters he wrote as well as most of the letters he received. Most of the documents in his correspondence he apparently folded to a particular size and then labeled with the writer's or receiver's name, the date of the receipt of the letter, and often the subject. With this method he filed letters and could retrieve them by date, sender, or subject. He also saved and dated his published and unpublished writings.

With this voluminous material at hand, Julia Lapham, probably in the early 1870s, began to hand write and then type or have typed a chronology of her father's life. She called it his "Autobiography." It is more than fifteen hundred legal-sized, typed pages housed at the Wisconsin Historical Society. Though she apparently destroyed many original documents, enough remain that we can compare them to her typescript and see that her changes were of several types. She corrected her father's spelling errors and punctuation. She omitted some family information that she considered too private. She included almost nothing about his financial or business dealings. Her aim seems to have been to establish his reputation as a scientist. This she does admirably, but he had done this himself.

Much of our work in writing his biography depended on this typescript of Julia Ann Lapham's. All quotations from it are cited "JAL typescript" with the page number. If a typed page number was crossed out in pencil, we use the original typed number.

We also refer to many documents from the more than twenty-seven boxes of original documents by or about Increase Lapham held in the fabulous Wisconsin Historical Society archives—not only letters and journals but also books of drawings, drafts and fair copies of articles, collections of photographs, autographs, maps, souvenirs, and many lists.

Finally, Increase Lapham published only what he thought to be important, and his judgment was impeccable. Thus, virtually all of his published writings, books, and articles survive as antiquarian treasures for collectors. Many exist in facsimile editions or have been digitized and made available online. The possible exceptions are unsigned newspaper articles.

Selected Writings of Increase Lapham Available as Reprints

The Antiquities of Wisconsin, as Surveyed and Described by I. A. Lapham. Washington, DC: The Smithsonian Institution, 1855. Reprinted with a foreword by Robert A. Birmingham and an introduction by Robert P. Nurre. Madison: University of Wisconsin Press, 2001.

A Documentary History of the Milwaukee and Rock River Canal. Compiled and Published by order of the Board of Directors of the Milwaukee and Rock River Canal Company. Milwaukee: Office of the Advertiser, 1840. Online facsimile excerpt at http://www.wisconsinhistory.org/turningpoints/search.asp?id=108.

A Geographical and Topographical Description of Wisconsin with Brief Sketches of Its History, Geology, Mineralogy, Natural History, Population, Soil, Productions, Government, Antiquities, &c. &c. Milwaukee: P.C. Hale, 1844. Online facsimile available at http://www.wisconsinhistory.org/turningpoints/search.asp?id=2.

"Fauna and Flora of Wisconsin." *Transactions of the Wisconsin State Agricultural Society* 2 (1852): 337–419.

Report on the Disastrous Effects of the Destruction of Forest Trees, Now Going on So Rapidly in the State of Wisconsin. With J. G. Knapp and H. Crocker. Madison: Atwood & Rublee, 1867. Reprinted by Nabu Public Domain Reprints. Menasha, WI: George Banta, 1967.

Thomas, Samuel W., and Eugene H. Conner, eds. *The Journals of Increase Allen Lapham for 1827–1830.* Louisville, KY: G. R. Clark Press, Inc., 1973.

Wisconsin: Its Geography and Topography, History, Geology, and Mineralogy: Together with Brief Sketches of Its Antiquities, Natural History, Soil, Productions, Population, and Government. Milwaukee: I. A. Hopkins, 1846. Reprint, North Stratford, NH: Mid-American Frontier, Ayer Company, 1999.

Books

Aderman, Ralph, ed. *Trading Post to Metropolis: Milwaukee County's First 150 Years.* Milwaukee: Milwaukee County Historical Society, 1987.

Albach, James R. *Annals of the West: Embracing a Concise Account of Principal Events Which Have Occurred in the Western States and Territories.* Pittsburgh: W. S. Haven, 1856.

Aldridge, Bertha Bortle Beal. *Laphams in America: Thirteen Thousand Descendants including Descendents of John from Devonshire, England to Providence, R.I., 1673, Thomas from Kent, England to Scituate, Mass., 1634, and Genealogical Notes of Other Lapham Families.* New York: Victor, 1953.

Andreas, W. T. *History of Milwaukee, Wisconsin*. Chicago: Western Historical Company, 1881.

Atwood, David. *The Centennial Exhibition of 1876: The Legislative Manual for the State of Wisconsin*. Madison: R. M. Bashford and E. B. Bolens, 1877.

Banta, R. E. *The Ohio*. 1949. Lexington: University Press of Kentucky, 1998.

Bascom, Emma, Anna B. Butler, and Katherine F. Kerr, eds. *Centennial Records of the Women of Wisconsin*. Madison: Atwood and Culver, 1876.

Berquist, Goodwin, and Paul C. Bowers. *Byron Kilbourn and the Development of Milwaukee*. Milwaukee: Milwaukee County Historical Society, 2001.

Birmingham, Robert A., and Leslie E. Eisenberg. *Indian Mounds of Wisconsin*. Madison: University of Wisconsin Press, 2000.

Boudreau, Richard, ed. *The Literary Heritage of Wisconsin*. Vol. 1. La Crosse, WI: Juniper Press, 1986.

Buck, James Smith. *Pioneer History of Milwaukee: 1840–1886*. Vols. 1–4. Milwaukee: Symes, Swain & Co., 1876, 1881, 1884, 1886.

Buley, R. Carlyle. *The Old Northwest Pioneer Period, 1815–1840*. Vols. 1 and 2. Bloomington: Indiana University Press, 1950.

Clarke, John M. *James Hall of Albany, Geologist and Paleontologist*. Albany: n.p., 1921.

Cole, Charles C. *A Fragile Capital: Identity and the Early Years of Columbus, Ohio*. Columbus: Ohio State University Press, 2001.

Current, Richard N. *The History of Wisconsin*. Vol. 2, *The Civil War Era, 1848–1873*. Madison: State Historical Society of Wisconsin, 1976.

Daniels, Edward. *Annual Report of the Geological Survey of the State of Wisconsin for the Year Ending December 31, 1857*. Madison: Atwood & Rublee, 1858.

Draper, Lyman Copeland. *Collections of the State Historical Society of Wisconsin*. Vol. 1. Madison: State Historical Society of Wisconsin, 1903.

Eaton, Amos. *A Geological and Agricultural Survey of the District Adjoining the Erie Canal, in the State of New-York*. Albany: Packard & Van Benthuysen, 1824.

———. *Manual of Botany for North America*. 5th ed. Albany: Websters and Skinners, 1829.

Featherstonehaugh, G. W. *Report of a Geological Reconnoissance made in 1835 by the way of Green Bay and the Wisconsin Territory to the Coteau de Prairie*. Washington, DC: Gales and Seaton, 1836.

Frost, J. William. *The Quaker Family in Colonial America*. New York: St. Martins, 1973.

Geology of Wisconsin Survey of 1873–1879. Vols. 1–4. Published Under the Direction of the Chief Geologist by the Commissioners of Public Printing, Madison, WI. Vol. 1, 1883, Vol. 2, 1877, Vol. 3, 1880, Vol. 4, 1882.

Gray, Asa. *Gray's School and Field Book of Botany*. New York: Ivison, Blakeman, Taylor & Co., 1868.

Greene, John C. *American Science in the Age of Jefferson*. Ames: Iowa State University Press, 1984.

Gregory, John G., ed. *Southeastern Wisconsin: A History of Old Milwaukee County*. Vols. 1–4. Chicago: S. J. Clarke Publishing Co., 1932.

Gurda, John. *The Making of Milwaukee.* Milwaukee: Milwaukee County Historical Society, 1999.

Hall, James. *Report of the Superintendent of the Geological Survey Exhibiting the Progress of the Work.* Madison: E. A. Calkins & Co., 1861.

Hagerty, J. E., C. P. McClelland, and C. C. Huntington. *History of the Ohio Canals: Their Construction, Cost, Use and Partial Abandonment.* Columbus: Ohio State Archeological and Historical Society, 1905. Clarified and copyrighted by Arthur W. McGraw, 1992.

Haven, Samuel F. *Archeology of the United States; or, Sketches Historical and Bibliographical, of the Progress of Information and Opinion Respecting Vestiges of Antiquity in the United States.* Smithsonian Contributions to Knowledge. Vol. 8. Washington, DC: Smithsonian, 1856.

Hedeen, Standley. *Big Bone Lick: The Cradle of American Paleontology.* Lexington: University Press of Kentucky, 2008.

History of the Lower Scioto Valley, Ohio. Chicago: Inter-state Publishing Company, 1884.

Holton, Edward D. *Proceedings of the Second Annual Meeting of the National Board of Trade, Richmond, Virginia, 1869.* Boston: National Board of Trade, 1870.

Howe, Daniel Walker. *What Hath God Wrought: The Transformation of America, 1815–1848.* Oxford: Oxford University Press, 2007.

Kinzie, Juliette M. *Wau-Bun: The "Early Day" in the North-West.* 1856. Urbana: University of Illinois Press, 1992.

Kleppner, Paul. *The Third Electoral System, 1853–1892.* Chapel Hill: University of North Carolina Press, 1979.

Koeppel, Gerard. *Bond of Union: Building the Erie Canal and the American Empire.* Philadelphia: Perseus Books, 2009.

Lapham, I. A., ed. *A Documentary History of the Milwaukee and Rock River Canal.* Milwaukee: Milwaukee and Rock River Canal Company, 1840.

———. "The Great Fires of 1871 in the Northwest." *Report of the Chief Signal Officer.* In *Report from the Secretary of War, Executive Documents of the Third Session of the Forty-second U. S. Congress.* Washington, DC: Government Printing Office, 1873.

———, ed. *Transactions of the Wisconsin Academy of Sciences, Arts and Letters.* Vol. 1, *1870–1872.* Madison: Atwood & Culver, Printers and Stereotypers, 1872.

Larkin, Jack. *The Reshaping of Everyday Life, 1790–1840.* New York: Harper, 1988.

Lentz, Ed. *Columbus: The Story of a City.* Charleston, SC: Arcadia, 2003.

Lurie, Nancy Oestreich. *A Special Style: The Milwaukee Public Museum, 1882–1982.* Milwaukee: The Milwaukee Public Museum, 1983.

Mansfield, Edward Deering, and Benjamin Drake. *Cincinnati in 1826.* Cincinnati: Morgan, Lodge, and Fisher, 1827.

Martin, Lawrence. *The Physical Geography of Wisconsin.* Wisconsin Geological and Natural History Survey. 2nd ed. Madison: State of Wisconsin, 1932.

McMurtrie, Henry. *Sketches of Louisville and Its Environs.* Louisville: S. Penn, 1819.

Merrill, George P. *The First One Hundred Years of American Geology*. 1924. New York: Hafner Publishing Co., 1969.

Murrish, John. *Report of the Geological Survey of the Lead Regions*. Madison: State of Wisconsin, 1871.

Nesbit, Robert C. *Wisconsin: A History*. 2nd ed. Madison: University of Wisconsin Press, 1989.

Nicolay, John G. *A Short Life of Abraham Lincoln*. New York: The Century Co., 1911.

Owen, David Dale. *Report of a Geological Survey of Wisconsin, Iowa, and Minnesota*. Philadelphia: Lippincott, Grambo & Co., 1852.

Percival, James C. *Annual Report of the Geological Survey of the State of Wisconsin*. Madison: Calkins & Proudfit, 1856.

Rhodes, Richard. *John James Audubon: The Making of an American*. New York: Vintage Books, 2004.

Riley, Kathleen L. *Lockport: Historic Jewel of the Erie Canal*. Charleston, SC: Arcadia Publishing, 2005.

Sandburg, Carl. *Abraham Lincoln: The Prairie Years*. New York: Harcourt, Brace, 1926.

Schafer, Joseph. *Four Wisconsin Counties: Prairie and Forest*. Madison: State Historical Society of Wisconsin, 1927.

———. *The Wisconsin Lead Mining Region*. Madison: State Historical Society of Wisconsin, 1932.

Schultz, Gwen M. *Wisconsin's Foundations: A Review of the State's Geology and Its Influence on Geography and Human Activity*. 1986. Madison: University of Wisconsin Press, 2004.

Shaw, Ronald E. *Canals for a Nation: The Canal Era in the United States, 1790–1860*. Lexington: University Press of Kentucky, 1990.

———. *Erie Water West*. Lexington: University Press of Kentucky, 1966.

Smith, Alice E. *The History of Wisconsin*. Vol 1, *From Exploration to Statehood*. Madison: State Historical Society of Wisconsin, 1985.

Spafford, Horatio Gates. *A Pocket Guide for the Tourist and Traveller, Along the Line of the Canals, and the Interior Commerce of the State of New York*. New York: T. and J. Swords, 1824.

Squier, Ephraim G., and Edwin H. Davis. *Ancient Monuments of the Mississippi Valley*. 1848. Washington, DC: Smithsonian Institution, 1988.

Still, Bayrd. *Milwaukee: The History of a City*. Madison: State Historical Society of Wisconsin, 1945.

Strong, Moses M. *History of the Territory of Wisconsin*. Madison: Democratic Printing Co., 1885.

Stuart, Charles B. *Lives and Works of Civil and Military Engineers of America*. 1871. New York: D. Van Nostrand, 2009.

Thomas, David. *Travels through the Western Country in the Summer of 1816*. Auburn, NY: David Rumsey, 1819.

Thompson, Mark L. *Graveyard of the Lakes*. Detroit: Wayne State University Press, 2000.

Trevorrow, Frank W. *Ohio's Canals*. Oberlin, OH: Trevorrow, 1973.

Triplett, Boone, and Bill Oeters, eds. *Towpaths: A Collection of Articles from the Quarterly Publication of the Canal Society of Ohio*. Akron: Canal Society of Ohio, 2011.

Trollope, Frances. *Domestic Manners of the Americans*. New York: Knopf, 1949.

Van Hise, Charles R. *The Conservation of Natural Resources in the United States*. New York: The Macmillan Co., 1910.

Way, Peter. *Common Labor: Workers and the Digging of North American Canals, 1780–1860*. Baltimore: Johns Hopkins University Press, 1993.

Wheeler, A. C. *The Chronicles of Milwaukee*. Milwaukee: Jermain & Brightman, 1861.

Williamson, Harold F., and Arnold R. Daum. *The American Petroleum Industry*. Vol. 1. Evanston, IL: Northwestern University Press, 1959.

Woods, Terry K. *Ohio's Grand Canal: A Brief History of the Ohio & Erie Canal*. Kent, OH: Kent State University Press, 2008.

Wright, Louis B. *Culture on the Moving Frontier*. New York: Harper, 1961.

Wyman, Mark. *The Wisconsin Frontier*. Bloomington: Indiana University Press, 1998.

OBITUARIES, APPRECIATIONS, BIOGRAPHIES, ARTICLES

Baker, Charles Minton. "Charles Minton Baker's Journal from Vermont to Wisconsin." *Wisconsin Magazine of History* 5, no. 4 (1922): 398.

Bean, E. F. "State Geological Surveys of Wisconsin." *Transactions of the Wisconsin Academy of Sciences, Arts, and Letters* 30 (1937): 203–220.

Clark, James I. "Increase A. Lapham, Scientist and Scholar." Booklet. Madison: State Historical Society of Wisconsin, 1957.

"Death of Increase Lapham, LL.D." *Wisconsin State Journal*, September 15, 1875.

Dexter, Ralph W. "The Early Career of John L. Riddell as Science Lecturer in the 19th Century." *Ohio Journal of Science* 88, no. 5 (1988): 184–185.

Dott, Robert H., Jr. "Rock Stars." *GSA Today* (October 2006): 30–31.

Edmonds, Michael. "Increase Lapham and the Mapping of Wisconsin." *Wisconsin Magazine of History* 68, no. 3 (1985): 163–187.

Forman, Jonathon. "The First Cholera Epidemic in Columbus, Ohio (1833)." *Annals of Medical History* 6, no. 5 (1934): 410–426.

"Geological Survey of Wisconsin." *American Journal of Science and Arts* 9, no. 53 (May 1875): 398–400.

Gregory, John G. "Museum Origins in Milwaukee." *The Wisconsin Archeologist* 12, no. 1 (1933): 54–62.

Hawks, Graham P. "Increase A. Lapham: Wisconsin's First Scientist." PhD diss., University of Wisconsin, 1960.

Hayes, Paul G. "Increase A. Lapham: A Useful and Honored Life." *Wisconsin Academy Review* 41, no. 2 (1995): 10–15.

———. "Increase Allen Lapham and the Milwaukee Public Museum." *Lore* 47, no. 2 (1997): 36–41.

———. "Increase Allen Lapham: Wisconsin's First Geologist." *Geoscience Wisconsin* 18 (2001): 1–6.

Hoy, P. R. "Increase A. Lapham, LL.D." *Transactions of the Wisconsin Academy of Sciences, Arts and Letters* 3 (1875–1876): 264–267.

Hoyt, John Wesley. "Some Personal Recollections of Abraham Lincoln." *Transactions of the Wisconsin Academy of Sciences, Arts and Letters* 16, part 2, no. 6 (1910): 1305–1307.

Iltis, Hugh H., and Theodore S. Cochrane. "'A Cabinet of Natural History': The University of Wisconsin–Madison Herbarium's Sesquicentennial, 1849–1999." *Wisconsin Academy Review* 45, no. 2 (1999): 30–36.

Janik, Erika. "Citizen Scientist." *Wisconsin Natural Resources Magazine.* February 2007. http://dnr.wi.gov/wnrmag/html/stories/2007/feb07/lapham.htm.

Kellogg, Amherst Willoughby. "Documents: Reflections of Life in Early Wisconsin." *Wisconsin Magazine of History* 8, no. 1 (1924–1925): 88–110.

Kellogg, Louis Phelps. "The Origins of Milwaukee College." *Wisconsin Magazine of History* 9, no. 4 (1926): 386–408.

Kieckhefer, Grace Norton. "Milwaukee-Downer College Rediscovers Its Past." *Wisconsin Magazine of History* 34, no. 4 (1950–1951): 210–216.

Kroncke, Mary Frost. "The Wisconsin Academy of Arts, Sciences and Letters: A Centennial History." *Wisconsin Magazine of History* 53, no. 3 (1970): 163-174.

Lapham, Darius, and I. A. Lapham. "Observations on the Primitive and Other Boulders of Ohio." *American Journal of Science and Arts* 22, no. 2 (1832): 300–303.

Lapham, I. A. "Agriculture in Ohio," *The Genesee Farmer* 1, no. 1 (1834).

———. "The Forest Trees of Wisconsin." *Transactions of the Wisconsin State Agricultural Society* 4 (1854–1857): 195–251.

———. "The Grasses of Wisconsin." *Transactions of the Wisconsin State Agricultural Society* 3 (1853–1854): 397–488.

———. "The Law of Embryonic Development the Same in Plants as in Animals." *The American Naturalist* 9, no. 4 (1875): 257–260.

———. "Oconomowoc Lake, and Other Small Lakes of Wisconsin, Considered With Reference to Their Capacity for Fish-Production." *Transactions of the Wisconsin Academy of Sciences, Arts and Letters* 3 (1875–1876): 31.

———. "On the Discovery of Fish Remains in Glacial Blue Clay of the New Chicago Tunnel Under Lake Michigan." *Milwaukee Evening Wisconsin,* July 13, 1870.

———. "On the Relation of the Wisconsin Geological Survey to Agriculture." *Transactions of the Wisconsin Agricultural Society* 11 (1873–1874): 207–210.

———. "The Penokee Iron Range." *Transactions of the Wisconsin State Agricultural Society* 5 (1858–1860): 392–393.

Lapham, Mary J. "Increase A. Lapham." *Wisconsin Archaeologist* 1 (1902): 32–35.

Lincoln, Abraham. "Address before the State Agricultural Society, Sept. 30, 1859." *Wisconsin Farmer and North-Western Cultivator* 11, no. 11 (November 1859): 389–394.

"List of Premiums and Regulations for the Ninth Annual Fair." *Wisconsin Farmer and North-Western Cultivator* 11 no. 5 (May 1859): 174.

Mann, Charles. "A Memorial: Increase Allen Lapham." Pamphlet. Read before the

Wisconsin Natural History Society, State Historical Society of Wisconsin, 1876. Wisconsin Historical Society Pamphlets and Archives.

Miller, Eric R. "American Pioneers in Meteorology." *Monthly Weather Review* 61, no. 7 (1933): 189–193.

———. "New Light on the Beginnings of the Weather Bureau from the Papers of Increase A. Lapham." *Monthly Weather Review* 59 (February 1931): 67–68.

P. W. "Dr. Lapham's Third Street Garden." *Milwaukee Sentinel*, November 8, 1896.

Quaife, Milo M. "Increase Allen Lapham, First Scholar of Wisconsin." *Wisconsin Magazine of History* 1 (1917): 3–15.

Ripley, George, and Charles A. Dana, eds. "Wisconsin." In *The Cyclopaedia*. Vol. 16. New York: D. Appleton, 1863.

Scott, Walter E. "An Appreciation of Increase Allen Lapham." *Wisconsin Academy Review* 22, no. 1 (1975): 20–28.

Seville, James. "Milwaukee's First Railway." Paper presented to the Old Settlers' Club of Milwaukee, n.d. In *Early Milwaukee*, 89–90. Madison: Robert H. Hunt, 1977. Facsimile of *Early Milwaukee*. Madison: Gregory et al., 1916.

Sherman, S. S. "Increase Allen Lapham, LL.D.: A Biographical Sketch Read Before the Old Settlers' Club, Milwaukee, Wis. Dec. 11, 1875." Milwaukee: Milwaukee News Company, Printers, 1876.

Smith, John Lawrence. "A New Meteoric Iron—the Wisconsin Meteorites." *American Journal of Science*, 2nd ser., 47 (March 18, 1869): 271.

Sweet, E. T. "Catalogue of the Wisconsin State Mineral Exhibit at Philadelphia, 1876." Madison: Atwood & Culver, 1876.

Taylor, Richard C. "Notes Respecting Certain Indian Mounds and Earthworks, in the Form of Animal Effigies, Chiefly in the Wisconsin Territory." *American Journal of Science and Arts* 34 (1838): 88–101.

Trescott, Paul B. "The Louisville and Portland Canal Company, 1825–1874." *The Mississippi Valley Historical Review* 44, no. 4 (1958): 692.

Trevorrow, F. W. "The Laphams—A Canal Family." *Towpaths* 27 (1989): 31.

Wallin, Helen. "Douglass Houghton: Michigan's First State Geologist, 1837–1845." Pamphlet 1. Lansing: State of Michigan, Geographical and Land Management Division, 2004. http://www.michigan.gov/documents/deq/GIMDL-PA01 _216188_7.pdf.

Winchell, Newton H. "Increase Allen Lapham." *The American Geologist* 13, no. 1 (1894): 1–38.

Acknowledgments

We want to thank those who read early drafts and encouraged us to keep on: Betty Adelman, Doug Armstrong, Brita Bergland, Jean Eisman, MaryLee Knowlton, Barbara Miner, Kathy Mooney, Mike Mooney, Susan Moran, Bob Peterson, Paul Salsini, Carol Sklenicka, Terri Sutton. And to the many more who listened to Increase Lapham stories at dinner.

Thanks to Anne Kingsbury and Karl Gartung for keeping Increase's memory alive with a presentation and birthday party for him at Woodland Pattern Book Center in Milwaukee.

Thanks to Harry Anderson, who sent us off to good sources on the history of Milwaukee.

Thanks to John Gurda, who sent us to Wisconsin Historical Society Press and later reviewed the manuscript. Thanks also to Margot Peters for her review.

Thanks to Kurt W. Bauer, surveyor and engineer, who made sure we noted that Lapham was a surveyor and engineer.

Special thanks to the archivists who found wonderful things: Lisa Long at the Ohio Historical Society found the previously unknown photograph of Darius Lapham and more treasures; Carolyn Cottrell at the Portsmouth, Ohio, Public Library suggested we look at Cleveland's map of Portsmouth; Pat Burke, Neil Luebke, and Susan Ott at the Milwaukee Public Museum showed us wonderful artifacts and books, and Claudia Jacobson helped us with photographs; Pat DeFrain at the Milwaukee Public Library kept us in touch with *Silliman's Journal* and more; Ellen Engseth, archivist at the University of Wisconsin–Milwaukee Library, showed us Increase Lapham's shaving kit and helped with the Area Research Center program; and Mark Allen Wetter at the University of Wisconsin Herbarium showed us Increase Lapham's botanical specimens and a surprise book of his mosses. Stephanie Irving helped us dig out Laphamia from the Harry H. Anderson Research Library of the Milwaukee County Historical Society.

But very special thanks go to Harry Miller and Simone Munson at the Wisconsin Historical Society Archives. This book rests heavily on Simone's many kinds of help.

And to all at Wisconsin Historical Society Press who worked so hard and well and patiently to make this book happen: Kathy Borkowski, editorial director; Kate Thompson, acquisitions editor; Elizabeth Boone, production editor; and Ben Genzer and Jordin Barber, editorial interns. Katherine Pickett, our copy editor, raised good questions.

Very special thanks to our editor, Sara Phillips, whose quiet, intelligent guidance and amazing eye for detail made this a much better book than the one we sent her.

INDEX

┤ I ├